Springer Hydrogeology

More information about this series at http://www.springer.com/series/10174

Dipankar Saha · Sanjay Marwaha
Arunangshu Mukherjee
Editors

Clean and Sustainable Groundwater in India

Springer

Editors
Dipankar Saha
Ministry of Water Resources, River Development and Ganga Rejuvenation
Central Ground Water Board
Faridabad, Haryana
India

Sanjay Marwaha
Ministry of Water Resources, River Development and Ganga Rejuvenation
Central Ground Water Board
Faridabad, Haryana
India

Arunangshu Mukherjee
Ministry of Water Resources, River Development and Ganga Rejuvenation
Central Ground Water Board
Faridabad, Haryana
India

and

CAWTM
Manav Rachna International University
Faridabad, Haryana
India

ISSN 2364-6454 ISSN 2364-6462 (electronic)
Springer Hydrogeology
ISBN 978-981-10-4551-6 ISBN 978-981-10-4552-3 (eBook)
https://doi.org/10.1007/978-981-10-4552-3

Library of Congress Control Number: 2017937522

Printed on acid-free paper

This Springer imprint is published by Springer Nature
The registered company is Springer Nature Singapore Pte Ltd.
The registered company address is: 152 Beach Road, #21-01/04 Gateway East, Singapore 189721, Singapore

The chapters included in the book were presented in ***Bhujal Manthan–2015***.

Foreword by Dr. Amarjit Singh

Ensuring clean and sustainable groundwater development in a sustainable manner would be one of the most challenging tasks for the professionals, policy makers, researchers and other stake holders. The beginning of Green Revolution during the 1970s brought in a significant increase in groundwater extraction, which has continued and increased rapidly ever since, with consequent manifestations of adverse environmental impact in the form of declining water levels, dwindling well yields and reduction in sustainability of irrigation wells. Besides, groundwater in several pockets of the country has become unfit for drinking due to contamination, attributed to both natural and anthropogenic causes.

In order to evolve strategies for better groundwater governance and management through wider consultations among various stakeholders, Ministry of Water Resources, River Development and Ganga Rejuvenation has initiated a National Dialogue ***"Bhujal Manthan"***. In the first year in 2015, this was organized at Kurukshetra University, Haryana, and was inaugurated by Honourable Minister Sushree Uma Bharati. It was aimed primarily at emphasizing the need for a collective interaction among various stakeholders engaged in groundwater resource development and management to attain synergy and harmony with the ecology for ensuring the long-term sustainability of this precious resource. The themes deliberated during the Bhujal Manthan were geogenic groundwater pollution—special reference to arsenic and fluoride; anthropogenic groundwater pollution—mitigation measures; groundwater stressed areas—management interventions for sustainable use; groundwater mapping and application of recent techniques; water conservation and conjunctive use of surface and groundwater in an efficient manner; groundwater system response to climate change and strategies; efficient use of groundwater and peoples' participation for its sustainable management. A total of 20 papers bringing out the finer nuances of the key issues around groundwater were presented during the technical sessions.

The present compilation is an outcome of selected papers from wide spectrum of disciplines. The editors have undertaken painstaking efforts and have carried out a wonderful task to edit and compile the papers. The fragrance of the interdisciplinary fusion can be felt as one goes through this enriching volume. I am sure that this book will help in understanding and tackling emerging issues related to governance of groundwater so as to facilitate effective management of the competing demands and efficient use of this invaluable resource.

Dr. Amarjit Singh
Secretary, Ministry of Water Resources
River Development and Ganga Rejuvenation
Government of India

Foreword by Dr. Mihir Shah

Groundwater is arguably India's single most important natural resource. It is the foundation of the livelihood security of millions of Indian farmers and the main source of drinking water for a vast majority of Indians in rural and urban India. The prospects of continued high rates of growth of the Indian economy will depend critically on how judiciously we are able to manage groundwater in the years to come.

Over the past three decades, India has emerged as by far the single largest consumer of groundwater in the world. Even as groundwater has made the country self-sufficient in food, we are now faced with a crisis of depleting water tables and water quality. The deep drilling by tube wells that was once part of the solution to the problem of water shortage now threatens to become a part of the problem itself. We, therefore, need to pay urgent attention to the sustainable and equitable management of groundwater.

In this context, it is heartening to note the renewed emphasis being placed on groundwater management by the Government of India. The National Aquifer Management Programme (NAQUIM) is the most ambitious such programme ever launched in human history. The commitment by the Government towards a participatory approach to groundwater management is most welcome. **Bhujal Manthan** is a landmark initiative in the direction of deepening dialogue across a wide spectrum of stakeholders who urgently need to come together to make groundwater management a success in India. I was delighted to browse through the papers put together in this volume, which will carry much further our understanding of the issues involved in and the approaches necessary for making groundwater management a people's movement in our country.

I congratulate the editors of the book, Ministry of Water Resources, River Development and Ganga Rejuvenation and the Central Ground Water Board for bringing out this volume.

Dr. Mihir Shah
Distinguished Visiting Professor, Shiv Nadar University

Contents

About the Editors

Dipankar Saha is presently working as a member in Central Ground Water Board, Ministry of Water Resources River Development and Ganga Rejuvenation, Government of India. He is coordinating the ambitious National Aquifer Mapping and Management Programme taken up by Government of India to cover the entire country. He has more than 30 years experience in the domain of groundwater investigation, monitoring and assessment. He has obtained Ph.D. on *Groundwater Management* from ISM IIT, Dhanbad. He has authored number of state and national reports and over 35 publications in peer-reviewed international journals. He is professionally trained on *Applied Quaternary Geology* by Asian Institute of Technology, Bangkok and on *Integrated Water Resources Management*, by JICA, Tokyo. He has presented many keynote addresses at national and international level conferences including World Water Week—Stockholm. He is recipient of *National Geoscience Award—2011* on *Groundwater Exploration* and Groundwater Excellence Award in 2014, conferred by Government of India and International Association of Hydrogeologist—India Chapter, respectively.

Sanjay Marwaha is working as Regional Director with Central Ground Water Board, Ministry of Water Resources, River Development and Ganga Rejuvenation, Government of India. He has done his post-graduation from Panjab University, Chandigarh, and was awarded university medals for standing first class first both in under graduation and post-graduation. He had taken courses on groundwater management, geographic information system and remote sensing at ITC, Netherlands, and was adjudged best trainee. During past 30 years, he is credited with number of studies related to various facets of groundwater, especially managed aquifer recharge in parts of semiarid areas of country. He has been associated with compilation of five volumes issued by Government of India on various topics such as groundwater recharge—techniques, perspective plan of groundwater resources of India, master plan for artificial recharge in India, Punjab water Quality—Issues and

Challenges. He has also authored more than 40 publications in national and international forums on various aspects of hydrogeology. Currently, he is associated with one of the most ambitious projects ever taken for mapping of aquifers of India.

Arunangshu Mukherjee has obtained his M.Sc. and Ph.D. degrees from Pt. Ravishankar Shukla University, Raipur in geology. A hydrogeologist by profession, he has worked for Central Ground Water Board for 24 years. At present he is working as Professor in Civil Engineering Dept., Manav Rachna International University (MRIU), Faridabad, India. He is actively engaged in research and teaching of applied groundwater aspects. He gained experience in management and implementation aspects of groundwater including application of computer models in scientific evaluation of hydrogeologic systems, groundwater resources development and protection. He has published over 70 research papers in international and national journals, and conference proceedings, authored many books and written more than 35 project reports. He is currently associated with arsenic research, spatial distribution of As pollution in groundwater of hard rock areas. He is the recipient of "*M P Young Scientist Award*" 1990 and INC-IAH Award of Excellence in "*Ground Water Implementation and Management*" 2015. He is member of International Association of Hydrogeologist and National Cave Research and Protection Organization.

Groundwater Resources and Sustainable Management Issues in India

Dipankar Saha, Sanjay Marwaha and Arunangshu Mukherjee

Abstract Groundwater is a critical component for socioeconomic development in India. The country exhibits wide spectrum on geology, climatic condition, and terrain, which is reflected in considerable variation in groundwater occurrence and movement. In addition, excessive withdrawal, in comparison with its annual replenishment, has created overexploitation of this precious natural resource, obliterating natural groundwater regime in large areas of the country. Besides, in many parts, the groundwater is contaminated both geogenically and anthropogenically. India as a country has established himself as the largest groundwater extractor in the world. The looming issues of overexploitation and deteriorating quality call for sustainable management of groundwater resource in long-term perspective. With the preamble on the review of the hydrogeology of the country, this paper summarizes 20 contributions that have been included in this volume. The papers are rich in their content and present a wide array of groundwater issues of the country, ranging from quality, conjunctive use of surface and groundwater, artificial recharge, community participation in its management, urban hydrogeology, coastal aquifer dynamics, mining hydrogeology, and application of state-of-the-art investigation techniques in groundwater survey and management. The aroma of the volume will enrich which are directly or indirectly linked with groundwater resource management of the country.

Keywords Groundwater · Sustainable management · India · Aquifer · Water level · Quality

D. Saha (✉) · S. Marwaha · A. Mukherjee
Central Ground Water Board, Bhujal Bhawan, NH-IV, Faridabad 121001, India
e-mail: dsaha002@yahoo.com

S. Marwaha
e-mail: smarwaha-cgwb@nic.in

A. Mukherjee
e-mail: arunangshu.fet@mriu.edu.in

Present address
A. Mukherjee
Manav Rachna International University, Faridabad, India

D. Saha et al. (eds.), *Clean and Sustainable Groundwater in India*,
Springer Hydrogeology, https://doi.org/10.1007/978-981-10-4552-3_1

1 Introduction

Groundwater is extremely important for drinking water and food security, besides rendering a significant service toward human health and ecosystem (Gleeson et al. 2016). At many places, aquifers are the only source of available water for drinking and irrigation, rendering groundwater as the most abstracted raw material on the Earth (NGWA 2016). In the changing climate scenario, the role of groundwater is significant, due to its relative stability in terms of both quality and quantity. India, as a country, is tilted toward groundwater for its societal needs. With 243 km^3 extraction annually, India is the largest consumer of groundwater in the world, consuming more than the cumulative draft of second and third highest consumers, i.e., Republic of China and USA, respectively (NGWA 2016). Nearly 85% of rural drinking need, 62% of irrigation need, and more than 50% of urban water demand of the country are met up from aquifers.

India's groundwater story is unique and interwoven in its journey toward achieving food security. The country has witnessed atomistic groundwater development, through private participation, where the million of landowners drilled their own wells. At present, the number of wells used for irrigation in the country is pegged to be more than 25 million. Besides providing irrigation, the wells boast rural economy and employments.

Growing population, urbanization, industrialization, and overall development of infrastructural facilities with improved lifestyle have put water in a competitive sectoral demand more than other parts of the world. There is a radical reduction of per capita annual availability from 5177 to 1545 m^3 since India's independence. All global prediction say large part of India is going to be water scarce by the year 2030. Aquifers, being comparatively safer source of potable water and being distributed in all types of terrain and in agroclimatic zones, the pressure from increasing water demand will be mainly focus on groundwater development and management. Central Ground Water Board being the apex organization under Govt of India in the domain of groundwater exploration, monitoring, and management has a stupendous task in fostering sustainable development of this precious natural resource involving the stakeholders, planners, researchers, and implementing agencies in state level.

2 Groundwater Occurrence

India is a vast country having geographical area over 32.87 lakh km^2 and can be divided into three broad geomorphic divisions, the Peninsular India, the Indo-Ganga-Brahmaputra Plain, and the Extra Peninsular India. The Peninsular India and Extra Peninsular India predominantly represent hilly and undulating topography, interspersed with alluvial deposits of rivers. These two units are predominantly characterized by hardwork aquifers, where groundwater occurs within the top weathered zones (generally <30 m thick) and underlying fracture/joints generally confined within 200 m below ground. The fractured aquifers are highly heterogeneous and largely local to subregional in extent. In these areas, the recharge is mainly through point to local recharge. Groundwater remains in phreatic condition in the weathered zone, whereas groundwater remains semi-confined to confined in the fractures below. The groundwater in this hard rock terrain is comparatively younger in age. Indicating recharge in annual cycle (Saha et al. 2013). Whereas the Indo-Ganga-Brahmaputra Plain is underlain by thick unconsolidated deposits of Quaternary age holding some of the most potential aquifer systems in the world (Mukherjee et al. 2015b).

Beside the Indo-Ganga-Brahmaputra Plains, the wide costal tract in Bangal deltaic regions and along the East Coast is largely occupied by pile of thick unconsolidated sediments mainly of Quaternary age. In total, the unconsolidated deposits roughly occupy one-third of geographical area of the country. They are characterized by multilayered aquifers of regional to subregional extent which are reported down to several hundreds of meter below ground. These aquifers dominantly pose primary granular porosity, rendering substantial yield potential to the wells. The recharge is both through local or regional flows. Groundwater remains under varied condition, ranging from phreatic to semi-confined in shallow and middle depth (up to 100–120 m below ground in general), while under confined in deeper levels. The shallow and middle depth aquifers get modern recharge from rainfall, while deeper aquifers may contain relatively older water even up to several thousand years old (Saha et al. 2011).

2.1 Groundwater Regime

Groundwater levels in India vary considerably in response to various geologic, terrain, and climatic conditions and anthropogenic interventions. A simplified groundwater level map of premonsoon phreatic groundwater level data (CGWB 2016) shows four broad zones running across SE to NW of India (Fig. 1a). The trans-Himalayan foothill zone and the East Coast have nearly continuous band of <5 m depth to water level, apart from few isolated patches largely concentrated in the eastern half of India and in parts of Maharashtra coast. 5–10 m below ground level (bgl) occurs in large part of India, except in northwest states of Rajasthan and Punjab. This is followed by deeper levels of 10–20 mbgl, largely spotted by small pockets concentrated in the western part of the country. The deeper levels are recorded in arid parts of Rajasthan and central part of Punjab where levels are varying from 20 to 40 mbgl and even deeper. Apart from this, patches of deeper levels are also observed in parts of Gujarat, Haryana, Telangana, and Bundelkhand region of Uttar Pradesh and Madhya Pradesh. The perennial shallow groundwater levels (2–5 mbgl) are largely the regional discharge zones, either along the foothills of Himalaya or along the East Coast. Similarly, the deepest groundwater level zones are areas of negligible groundwater recharge in the arid parts of India.

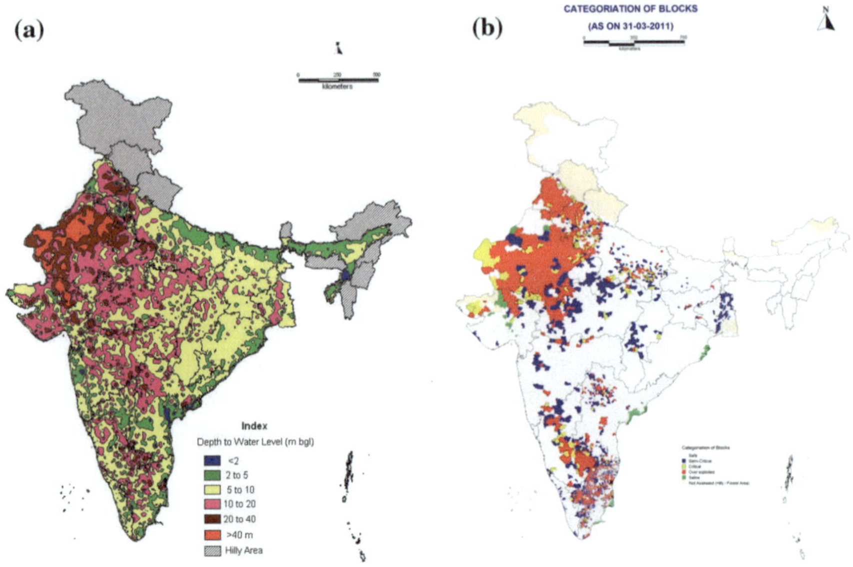

Fig. 1 **a** Depth to water level map of premonsoon 2016. **b** Categorization of assessment units based on SOD. Colors represent as (*Red* = OE, *Yellow* = Critical, *Blue* = Semi-critical, *Green* = Saline and *White* = Safe). *Source* CGWB

2.2 *Level of Groundwater Extraction*

The rainfall is the main source of recharge to aquifers in India. The ratio of groundwater recharge to draft in an annual scale provides the stage of groundwater development (S0D). SOD is considered in India as an earmark to assess the stress on the dynamic groundwater resource. The latest estimation of dynamic groundwater resources (CGWB 2016) has indicated the overall SOD of the country as 62%. The assessment unit, i.e., blocks (talukas, firkas in some states) are categorized based on their SOD as- "safe" (S0D < 70%), "semi-critical" (SOD 70–90%), "critical" (SOD 90–100%), and "overexploited" (SOD > 100%). The SOD is not uniform throughout the country due to obvious reasons that govern the aquifer potential, recharge capability, and extent of groundwater exploitation. Groundwater overexploitation is confined in 1071 units of the country and in another 914 significantly where level of exploitation has exceeded 70%. All these units are largely concentrated in the NW parts, as well as deep in Penninsular India. These areas are also characterized by deeper groundwater levels (Fig. 1b).

Groundwater overextraction has resulted in decline in water levels, dwindling well yields and drying up of dug wells. Besides, there is an overall increase of background salinity in groundwater. This is apart from the known ambient higher values of EC in arid regions of India, as well as in coastal tract. This has prompted to conclude that groundwater overextraction is deteriorating the ambient quality of groundwater also (Mukherjee et al. 2015b). Shift to drinking water abstraction from deeper zones might have triggered certain pollutants to enter into aquifer-based drinking water cycle (Saha et al. 2009; Mukherjee et al. 2015a).

The India's groundwater resource management is at crossroad, because of accelerated pace and skewed extraction leading to significant decline in groundwater level on the one hand and expanding areas under waterlogging and deteriorating water quality resulting in considerable reduction in availability of fresh and potable water. Strategies for better groundwater governance and management need to be developed through wide consultation among various stakeholders, academia, researchers, and state departments dealing with groundwater. To achieve this Central Ground Water Board, Ministry of Water Resources, River Development and Ganga Rejuvenation has organized a one-day thread-bearing discussions 'Bhujal Manthan' on August 21, 2015, at Kurukshetra University, Kurukshetra, Haryana. This volume derives some of the best papers presenting and deliberating in the workshop. The volume contains 22 papers which are grouped into the following four subthemes.

3 Layout of the Book

The articles selected from the Bhujal Manthan-2015 are broadly grouped into following four themes (i) groundwater quality, (ii) conjunctive use of surface and groundwater, (iii) management intervention for sustainable water, and (iv) groundwater problem and application of techniques.

3.1 Groundwater Quality

Qualitatively, groundwater is considered as a safer source in comparison with surface water. Mineralization of groundwater up to certain extent is considered to be good for human health and crops. By and large, the ambient quality of groundwater in India is still good and requires no external treatment before drinking, irrigation, and industrial uses. The quality of groundwater in the operational zone of open dug wells was found potable since ages. But due to various anthropogenic reasons, slowly this zone became contaminated with fecal collies, forced shifting of abstraction from deeper zones through mechanized wells, and hand pumps for drinking needs. Deeper zones hold groundwater with relatively more mineralization that under reducing condition may produce toxicity. Apart from geogenic contamination such as arsenic and fluoride, seawater intrusion in the costal tract, agricultural practice, urban wastewater and land fields, and liquid and solid disposed from industries and mines are the main source of contamination of groundwater (Banerjee et al. 2012).

During the recent years, much of the emphasis in groundwater for drinking use has shifted from problem of availability to issues of quality. The quality problems are broadly grouped into two: 'geogenic' and 'anthropogenic.' Geogenic pollution refers to naturally occurring elevated concentration of certain constituents in groundwater having adverse health effects. In India, contamination of fluoride and arsenic is geogenic in nature and affecting several parts of the country and has became major health concern and posses challenge for safe water supply (Bajerjee et al. 2012). Besides these, elevated levels of salinity, iron, manganese, uranium, radon, chromium selenium, and other traced elements reported at places may also be of geogenic origin. In India, approximately 40 million people are residing within the risk zone of arsenic contamination (Saha and Sahu 2015). The states from which arsenic contamination has been reported are West Bengal, Bihar, Uttar Pradesh, Jharkhand, Assam, Manipur, Punjab, Haryana, Chhattisgarh, and Karnataka. However, the elevated concentration is mostly confined in the Indo-Gangetic Plains (Saha 2009) CGWB has been advocating exploitation of arsenic safe deeper aquifers in the Indo-Gangetic Plains under regulated pumping. The other remedial measures include removing arsenic from groundwater after it is extracted and reducing the level within the aquifer itself through in situ treatment and through dilution of the contaminants by artificial recharge. Fluoride concentration has pan-India presence. Its concentration above maximum permissible limit of 1.5 mg/L has been reported in groundwater from parts of many states of India, namely Andhra Pradesh, Telangana, Assam, Bihar, Chhattisgarh, Delhi, Gujarat, Haryana, Jammu and Kashmir, Jharkhand, Karnataka, Kerala, Madhya Pradesh, Maharashtra, Odisha, Punjab, Rajasthan, Tamil Nadu, Uttar Pradesh, and West Bengal. Fluoride-contaminated areas are mostly characterized by the presence of crystalline basement, and volcanic and sedimentary rock areas particularly characterized by arid and semiarid climatic conditions. Dissolution and weathering of F− bearing minerals, pulled by calcite precipitation, are considered to be the dominant mechanism responsible for groundwater fluoride (F−) contamination. Long-term solution involves providing alternate source of potable water by construction of wells tapping fluoride-free aquifers, installation of community-based de-fluoridation plants, and surface water-based supply, wherever available.

Six papers have been incorporated under this theme and amply signify the importance of this issue. Out of that four papers discussed, the issues are related to elevated levels of fluoride and its mitigation and one deals with uranium contamination. The remaining paper discusses on deteriorating groundwater quality because of saline water ingress in East Coast area. Prof. **L. Elango and Jagadeshan** in their paper are of opinion that managed aquifer recharge can be an effective mean for dilution of fluoride concentration in groundwater. Through their experiment by dug well recharge, they have able to obtain desired results in a contaminated phreatic aquifer in hard rock areas of Dharmapuri district in Tamil Nadu state. **Dr. Santhil Kumar** and their co-authors in their study had shown the example of seawater ingress and deterioration of groundwater quality due to excessive pumping driven by urbanization in coastal Chennai area of Tamil Nadu state. Bromide has been used as an indicator to delineate the seawater–freshwater interface. **Sh. Supriya Bramha** has discussed the level of fluoride concentration in different aquifers and related issues in varied hydrogeological setup represented by hard rock and older alluvium in West Bengal. **Prof. H.K. Pandey** and co-authors discussed the petro-mineralogical and geochemical significance of fluoride contamination and its release mechanism in groundwater of hard rock aquifer from Sunbhadra district of Uttar Pradesh. In another study made by **Dr. P. Madhnure** and co-authors, the evolution of fluoride-rich groundwater has been dealt in detail in Cherlapalli water shed in Nalgonda district of Telangana state represented by hard rock aquifer. They are of opinion that fluoride enrichment is related to Ca^{+2} removal during rock–water interaction.

Uranium contamination of groundwater in southwest part of Punjab has been discussed by **Prof. K.P. Singh** along with his co-authors. They have opined that the genesis of uranium is not related to basement granite or the phosphate fertilizer. They have found uranium concentration in sediments of the area which may be one of the sources for its elevated concentration at places in groundwater.

3.2 Conjunctive Use of Surface and Groundwater

To improve the overall efficiency of water resource development and management, 'conjunctive use' has been considered as an affective tool. It is defined by Foster et al. (2010) as a situation where both groundwater and surface water are developed (or coexist and can be developed) to supply a given irrigation canal command—although not necessarily using both sources continuously over time nor providing each individual water user from both sources. Alternatively, FAO (1995) described conjunctive use as 'use of surface water and groundwater consists of harmoniously combining the use of both sources of water in order to minimize the undesirable physical, environmental and economical effects of each solution and to optimize the water demand/supply balance.' Because different stakeholders are involved in the conjunctive use of surface and groundwater resources, a conflict-resolution technique should be employed to resolve cross-interests. A major issue related to conjunctive use of surface and groundwater in efficient manner is to have some form of institutional arrangement and a supportive regulatory framework. The number of studies has been undertaken by institutions and government departments for establishing feasibility of conjunctive use in India. The studies have established that the isolated use of surface water ignoring groundwater in irrigation command area has resulted in various environmental problems such as waterlogging and salinity. Further, it is also felt that there is a need to adopt groundwater management which incorporates a groundwater simulation model as constraint in the management model which can be efficiently used in planning the conjunctive use of water. Such planning will need to be supported by a national policy and to occur within a framework that ensures sustainable use of the resources. This will require significant technical inputs, especially within the context of the need to assess the available consumptive pool.

Dr. D.K. Chadha, former Chairman, CGWB, in his contribution pointed toward decimal efforts being made toward the conjunctive use of surface and groundwater in the country and suggested means and ways of efficient conjunctive use. He strongly advocated implementation of its use in the areas where such studies have already been completed. **Prof. Shasank Shekhar** and co-authors advocated that the conjunctive use practice can prevent the rivers from getting dried up because of depleted contribution from aquifers. They have advocated that conjunctive use can provide the requisite environmental sustainability to the river ecosystem.

3.3 Management Interventions for Sustainable Use

Globally, irrigated agriculture is the largest consumer of groundwater resources. However, in many arid and semiarid areas, recharge from rainfall is limited and unconstrained use is causing serious aquifer depletion and environmental degradation. The interactions between irrigation, surface water, and groundwater resources are often very close, such that active cross-sector dialogue and integrated vision are needed to promote sustainable land and water management. Clear policy guidance and focused local action involving communities are required to make efficient use of groundwater in view of drought mitigation and climate change adaptation. Policies must be tailored to local hydrogeological settings and agroeconomic realities, and their implementation will require active involvement of the farming community.

Dr. Himanshu Kulkarni and co-authors in their paper discussed alternative way of calculating efficiency of groundwater in agriculture system. They presented an interesting case of groundwater efficiency assessment involving the inherent aquifer characteristics like specific yield which are often ignored. The paper illustrates two examples from the Deccan basalt aquifer system in western Maharashtra state, where use of specific yield parameter has helped in a robust understanding of groundwater use efficiency for its better management.,

Prof. Tushar Shah, Chairman of Task Force of Managed Aquifer Recharge Project of Gujarat State, from his experience and expertise has sketched the impact of groundwater overexploitation and possibilities of artificial recharge for water-stressed areas of Gujarat. He has opined that apart from hydrogeologists, water experts, economist, and social scientist can also made important contribution to water strategy formulation. **Dr. P.K. Gangopadhyay** and co-authors in their interesting article presented a case study of co-solving groundwater level depletion as well as flooding problem using managed aquifer recharge in the Ram Ganga subbasin of the Gangetic Plain through participatory mode. They use the terminology UTFI (underground taming of floodwater through irrigation) where floodwater is diverted into the de-saturated part of the aquifer below. This has double benefit of reducing the flood risk and enhancing the groundwater resource of the area.

The impact of artificial recharge on groundwater is presented by **Sh. Subburaj** and co-authors in the hard rock aquifer of Salem district of Tamil Nadu. They have shown the increase of sustainability of groundwater due to implementation of artificial recharge projects. The pumping hours by the farmers have increased, and also, there is rise in water level in the area. **Dr. S.S. Vittala** and co-authors presented the result of soil infiltration studies carried out in Ankasandra water shed in Tumkur district of Karnataka state. They found that lowest infiltration rate is noted at tank belts. Infiltration rate is an extremely important component for estimating the recharge. **Dr. B. Ghosh** and co-authors presented interesting case studies of participatory groundwater management and self-regulation in varied hydrogeological typologies in India. They argued that decentralized groundwater management can offer better solution for its long-term sustainability and social equity. **Dr. Bidyut Kumar Bhadra** and co-authors have presented interesting case study of groundwater recharge in arid region of Rajasthan. The study was carried out at three places of Rajasthan underlain by different aquifer systems. All the areas exhibits decline in groundwater level at various rates. They have opined for adopting preventive measures for increasing the recharge and efficient use of groundwater. **Prof. P.K. Singh** and others presented a case study of groundwater regime in coal mining area. The mining activity is impacting the water level and flow regime in the surrounding areas. In case studies around Jharia, East Bocaro and West Bocaro cold field of Jharkhank State detailed behavior of water level are discussed.

3.4 Groundwater Mapping and Application of Recent Techniques

Groundwater mapping involves multidisciplinary techniques to understand, map and assess various issues like, delineating the aquifers, hydraulic parameters of aquifers, well sustainability, assessing the resource availability, quality variation—both specially and vertically, groundwater pollution etc. As the groundwater resources are becoming scarce and its social and economic importance is increasing rapidly, there is a spurt in application of new technique. In the domain of groundwater lot of researches are going on to understand science and developing new technologies on accurately delineating the aquifers and characterizing them through various geophysical techniques and applications of isotope and tracers. Challenges are particularly to delineate deeper fractures which are water bearing and have good yield potential in hard rock terrain. Significant innovations are taking place for automatic water level measurement and quality monitoring through telemetry. Though groundwater modeling is being applied to understand the flow regime, wider applications are needed for predicting the future scenario under various stress conditions and pollution studies through solute transport modeling. Researches are also required for comprehensive understanding on impact of various climate change scenario and extreme rainfall event on groundwater regime. Besides, lot of technological development is taking place for analysis of harmful consequent in water even in nanogram level. Application of remotely sensed data for groundwater mapping is well established, but the new domains are emerging with an advent of gravity satellite which is being applied in assessing groundwater resource and its depletion.

Prof. P.K. Sikader and P. Sahu highlighted the stress on urban aquifer system of Kolkata and its impacts on natural recharge pattern using groundwater modeling. They provoked for using groundwater modeling as decision-making tool for better management of the confined aquifer surrounding Kolkata urban area. The recommendation can be replicated in other urban areas with similar hydrogeology setting. **Prof. Abhijit Mukherjee** discussed the hydrodynamics of groundwater flow using arsenic as tool for modeling. A regional hydrostratigraphic model has been developed in the Gangetic Plain of West Bengal to indicate that groundwater flow is dependent on the amount and timing of precipitation and is controlled by the topography and alignments of major streams, which can be heavily distorted in case the region experiences over pumping. **Dr. Subhash Singh** and co-authors have used gama logging in a very effective manner to detect the radio activity in fluoride-contaminated aquifer of Shivepuri district of Madhya Pradesh state to understand the extent of contamination.

4 Conclusion

Groundwater is rendering a yeomen service in India by providing significant contribution to water supply for drinking, irrigation, and industrial sector. The dependence on this resource is increasing day by day because of its three unique characteristics: (i) special distribution and use as a common pool resource, (ii) its availability in all agroclimatic region and terrain condition, and (iii) drought—proofing character of groundwater and its relative immunity to extreme events under climate change scenario.

In view of its overexploitation in large parts of the country, quality deterioration even in areas where they are available in plenty and rapidly increasing dependence on this resource for food and drinking security has thrown open serious challenges to all its stakeholders—the management of this resource in a sustainable manner. Sustainable management of this hidden resource is only possible when we have a comprehensive understanding of the system, both on a sound scientific understanding and social-economic prospective. The selected 21 papers from the Bhujal Manthan-2015, the largest groundwater discussion platform in India, organized by CGWB, embodies the groundwater issues of India. The papers extend valuable thoughts which would be useful for laying the road map for sustainable management of this resource. This volume will immensely benefit the researcher, policy makers, academia, and the stakeholders in understanding the various facts of this resource.

References

Banerjee DM, Mukherjee A, Acharya SK, Chatterjee D, Mahanta C, Saha D, Kumar S, Singh M, Sarkar A, Dubey CS, Shukla D, Raju NJ (2012) Contemporary groundwater pollution studies in India: a review, Proc Ind Nat Sci Acad 78(3)

CGWB (2016) Dynamic groundwater resources of India. www.cgwb.gov.in

Foster S, van Steenbergen F, Zuleta J, Garduno H (2010) Conjunctive use of ground water and surface water: from spontaneous coping strategy to adaptive resource management. GW-MATE strategic overview series 2, World Bank, Washington

Gleeson T, Befus KM, Jasechko S, Luijendijk E, Bayani Cardenas M (2016) The global volume and distribution of modern groundwater. Nat Geosci 9:161–167. doi:10.1038/NGEO2590

Mukherjee A, Gupta A, Ray RK, Tewari D (2015a) Aquifer response to recharge—discharge phenomenon: inference from well hydrographs for genetic classification. J Appl Water Sci. doi:10.1007/s13201-015-0293-z

Mukherjee A, Saha D, Harvey CF, Taylor RG, Ahmed KM, Bhanja SN (2015b) Groundwater systems of the Indian Sub-Continent. J Hydrol Reg Stud. http://dx.doi.org/10.1016/j.ejrh.2015.03.005

NGWA (2016) Facts about global groundwater usage. Compiled by the Ground Water Association. http://www.ngwa.org/Fundamentals/Documents/global-groundwater-use-fact-sheet.pdf

Saha D (2009) Arsenic groundwater contamination in parts of Middle Ganga Plain, Bihar. Curr Sci 97(6)

Saha D, Sahu S (2015) A decade of investigations on groundwater arsenic contamination in Middle Ganga Plain, India. Environ Geochem Health. doi:10.1007/s10653-015-9730-z

Saha D, Sarangam H, Diwedi SN, Bhartariya KJ (2009) Evaluation of hydrochemical processes in the arsenic contaminated alluvial aquifers in parts of Mid-Ganga basin, Bihar, India. Environ Earth Sci. doi:10.1007/s12665-009-0392-y

Saha D, Sinha UK, Dwivedi SN (2011) Characterization of recharge processes in shallow and deeper aquifers using isotopic signatures and geochemical behavior of groundwater in an arsenic-enriched part of the Ganga Plain. Appl Geochem. doi:10.1016/j.apgeochem.2011.01.003

Saha D, Dwivedi SN, Roy GK, Reddy DV (2013) Isotope-based investigation on the groundwater flow and recharge mechanism in a hard-rock aquifer system: the case of Ranchi urban area, India. Hydrogeol J. doi:10.1007/s10040-013-0974-3

Fluoride Contamination in Groundwater: A Pilot Study on Dug Well Recharge System for In situ Mitigation

L. Elango and G. Jagadeshan

Abstract Dissolved ions concentration in groundwater beyond the recommended limits is a major problem as they make the water unsuitable for drinking purpose. Fluorine commonly found in certain rocks is released into groundwater due to the processes of rock–water interaction. This leads to increase in the concentration of fluoride in groundwater which is a major problem in several parts of the world including India. Presence of fluoride beyond the prescribed limits causes health problems to humans due to prolonged consumption of water, which is common in many parts of India. Dental and skeletal fluorosis is observed due to prolonged drinking of water with fluoride concentration above 1.5 mg/l. The objective of the study is to know how fluoride get released from the host rock and spot out suitable location for installing a dug well recharge system to decrease the fluoride concentration in groundwater. Several methodologies exist for in situ or exsitu removal of fluoride from groundwater. Exsitu methods can be enforced at community level or even at household level for the reduction of fluoride before its consumption, through ion exchange, reverse osmosis, adsorption, electrodialysis, coagulation, Nalgonda technique, electrodialysis, coagulation, precipitation, etc. Even artificial recharging structures can also be built in suitable location for diluting fluorite concentration in groundwater. Rainwater harvesting is also found effective to reduce the fluoride concentration of groundwater in existing wells. A pilot study was carried out by construction of a dug well recharge system in Dharmapuri district, Tamil Nadu, India. The study successfully demonstrated the applicability of dug well recharge system at a carefully selected site based on the systematic long-term hydrogeochemical studies to solve the problem of fluoride contamination affecting millions of rural people.

Keywords Groundwater · Fluoride contamination · Rainwater harvesting
Hard rock aquifers · Geogenic contamination

L. Elango (✉) · G. Jagadeshan
Department of Geology, Anna University, Chennai 600025, India
e-mail: elango34@hotmail.com

D. Saha et al. (eds.), *Clean and Sustainable Groundwater in India*,
Springer Hydrogeology, https://doi.org/10.1007/978-981-10-4552-3_2

1 Introduction

Geogenic and anthropogenic contamination may make groundwater unsuitable for various uses including its use as drinking water. Consumption of such water may cause many health problems. Presence of fluoride, arsenic, and iron above recommended limits is found in groundwater due to geogenic contamination in many parts of India and world (Mukherjee et al. 2015). Release of arsenic from arseno-pyrite minerals is responsible for higher concentration of arsenic in groundwater of Bengal basin (Farooq et al. 2011; Saha 2009). As about 80% of rural population of India depends on the use of groundwater for domestic use, the water needs to be treated before it is used for drinking purpose. The initial concentration of fluorite, occurrence and removal of co-contaminants, and disposal of sludge, etc., decide treatment methods of water to be treated (Brindha et al. 2016). Bhagavan and Raghu (2005) studied the effect of increase in recharge by check dams, and Pettanati et al. (2014) highlighted the use of percolation ponds in regions with fluoride problem in groundwater. Some studies have also indicated an increase in fluoride concentration in groundwater during the process of recharge when check dams are constructed (Bhagavan and Raghu 2005). Thus, contradicting findings are also observed from recharge for in situ mitigation of fluoride concentration (Brindha et al. 2016). Hence, it is essential to reveal the release mechanism of fluoride during the process of recharge based on a long-term temporal variation in fluorite concentration and in the level of groundwater in the area. The objective of this study is to understand the release processes of fluoride and identify a suitable location for construction of dug well recharge system in the area for reducing the concentration of fluoride in groundwater to an acceptable level. This pilot study was carried out in Dharmapuri district, Tamil Nadu (Fig. 1), where about 72,000 out of about 190,000 school students were found having dental fluorosis (UNICEF 2009).

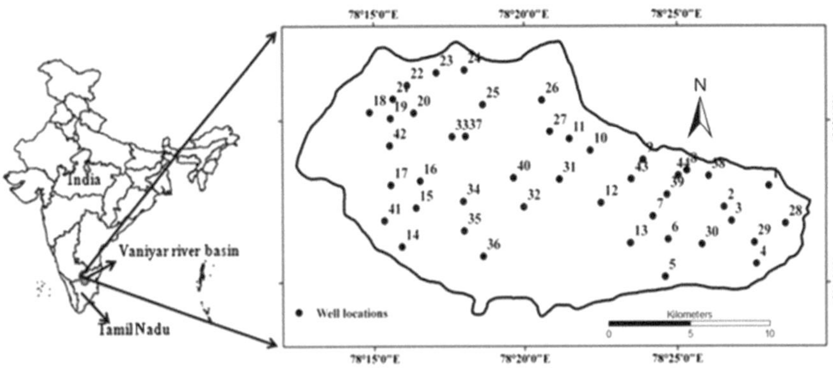

Fig. 1 Location map of study area and sampling wells

2 Methodology

Systematic investigation during June 2011 helped in identifying 44 key monitoring well to represent the area (Fig. 1). These key wells were used to collect bimonthly groundwater level and water sample between June 2011 and August 2014. Largely, the key wells are >30 m in depth and are utilized for obtaining drinking and agricultural water. Water level indicators (Solinist) were used to collect water level data. Pre-washed and treated 1000-ml capacity clean polyethylene bottles were used to collect groundwater samples after rinsed with sample. Merck 0.22-μm filter paper (Merck 755004722) were used to filter the samples before collection. To know the in situ temperature, EC and pH of the groundwater in field portable digital meter (YSI 556MPS) were used. Total dissolved solids (TDS) were obtained using EC = TDS/0.64 (Carroll 1962). Aquamerck (1.11109.0001) test kit was used to measure CO_3^{2-} and HCO^{3-} in the field itself. Ion chromatograph (Metrohm 861) was used for measuring the concentrations of Na^+, K^+, Ca^{2+}, Mg^{2+}, Cl^-, F^-, NO_3^-, and SO_4^{2-} of samples in the laboratory. The concentration of silica was determined by using spectrophotometer (Systronics 201) for the groundwater samples collects in August 2014. Batch test experiment's results were obtained to know the fluoride release processes from the rocks/soil, at laboratory.

3 Sources of Fluoride in Groundwater

Fluoride-bearing minerals are common in nature; therefore, contamination of fluorite is vast. Sometime it is intensive and became alarming in many places. The permissible range for fluoride in drinking water as per the Bureau of Indian Standards (BIS 2012) is from 0.6 to 1.5 mg/l. A desired concentration of fluoride in drinking water is favorable to determine the physiological activities in human bodies, whereas the consumption of fluoride above or below the permissible limits may cause health problems such as fluorosis of dental and skeletal structure. In India, northwestern and southern states are severely affected with fluorosis (Agarwal et al. 1997; Ali et al. 2016). Weathering of rocks rich in fluoride is the common natural cause for the fluoride in groundwater. The marine sediments and sediments of foothill areas (WHO 2002; Fawell et al. 2006), as well as marginal alluvial terrain of the Gangetic Plains (Saha et al. 2008; Saha and Alam 2014), are more vulnerable for water with high concentration of fluoride. Fluorine occurs naturally in igneous and sedimentary rocks. Common fluoride minerals are sellaite, fluorite or fluorspar, cryolite, fluorapatite, apatite, topaz, phlogopite, biotite, epidote, tremolite and hornblende, mica, clays, villuanite, and phosphorite (Brindha and Elango 2011; Haidouti 1991; Gaumat et al. 1992; Gaciri and Davies 1993; Kundu et al. 2001). Nalgonda district, Andhra Pradesh, known for its fluorite contamination is due to the inherent fluoride-rich granitic rocks. The fluoride content in granitic rocks in Nalgonda district varies from 325 to 3200 mg/kg with a mean of 1440 mg/kg. Thus, the Nalgonda granites having higher fluoride content than world granite, where it occurs average 810 mg/kg (Brindha and Elango 2011), whereas in Hyderabad granites, it is estimated 910 mg/kg (Brindha and Elango 2011; Ramamohana Rao et al. 1993). However, groundwater in northeastern districts of Tamil Nadu has high fluoride because of their fluoride-rich mineral such as epidote, hornblende, biotite, apatite, carbonatite, mica, and fluorapatite (Jagadeshan 2015; Jagadeshan et al. 2015a, b; Brindha et al. 2016). Jagadeshan et al. (2015b) carried out a study of rocks of Vaniyar river basin in Dharmapuri district, Tamil Nadu, to understand the mineralogical composition. He made thin sections of the charnockite and epidote hornblende biotite gneiss rocks (Fig. 2), where plagioclase, hypersthene, biotite, quartz, hornblende, orthoclase, and muscovite opaque (probably iron oxide) were the major minerals. Biotite which consists of fluorine was also identified in charnockite rock.

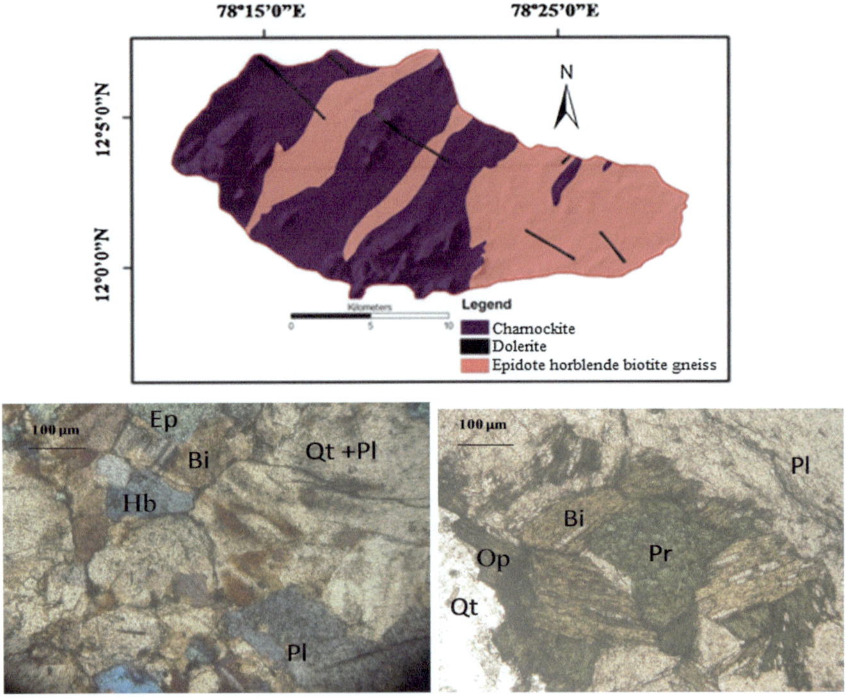

Fig. 2 Geological map of the study area and thin section photographs (X5 magnification, field of view 2500 μm wide, plan polarized illumination) showing quartz (*Qt*), plagioclase (*Pl*), epidote (*Ep*), biotite (*Bi*), hypersthene (*Op*), hornblende (*Hb*), pyroxene, and other opaque minerals (*Op*) in epidote hornblende gneiss (*left*) and charnockite (*right*)

4 Geochemical Analysis of Groundwater

Largely, the elevated amount of fluorite in groundwater is due to its release from host rock. The climatic conditions, pH, host rock mineralogy, and hydrogeological condition of an area are generally governing the release of fluoride into groundwater (Shan et al. 2013; Raju et al. 2009). The phosphatic fertilizers use may also lead to rise in fluoride concentration in groundwater anthropogenically.

Utilization of fertilizers and industrial activities such as brick kilning are some of the other causes for elevated fluoride concentration in groundwater (Selvam 2015). The weathering of these rocks results in increased fluoride content in groundwater. Longer residence time in aquifers with fractured fluoride-rich rocks enhances fluoride levels in the groundwater. Granite and granitic gneisses in Nalgonda, India, contain fluoride-rich minerals such as fluorite (0–3.3%), biotite (0.1–1.7%), and

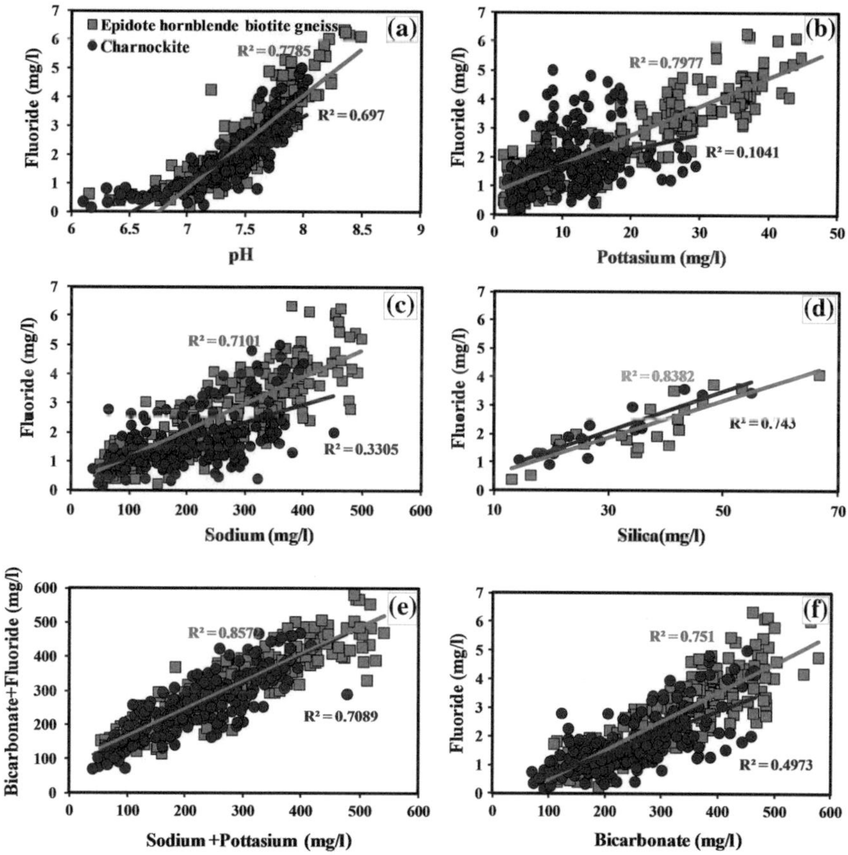

Fig. 3 Bivariate diagram of fluoride versus pH (**a**), potassium (**b**), sodium (**c**), silica (**d**), bicarbonate + fluoride versus sodium + potassium (**e**), and bicarbonate versus fluoride (**f**)

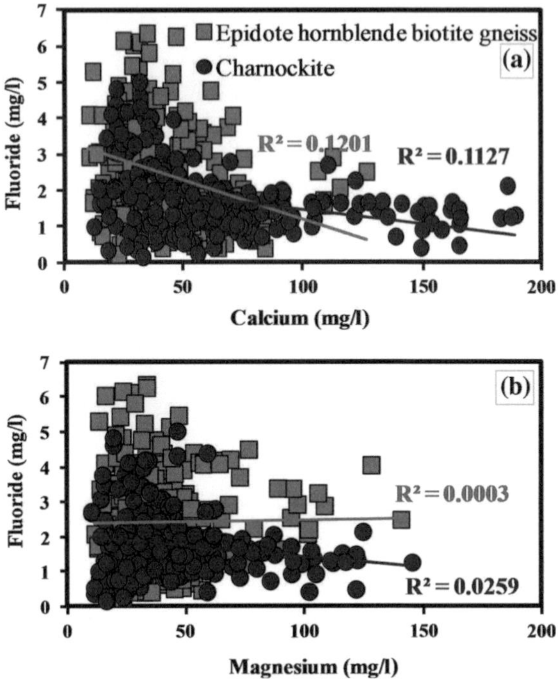

Fig. 4 Bivariate diagram of fluoride versus **a** calcium and **b** magnesium

hornblende (0.1–1.1%) (Brindha and Elango 2011; Brindha et al. 2016; Ramamohana Rao et al. 1993). A strong positive correlation is found among sodium, potassium, silica, bicarbonate + fluoride versus sodium + potassium and bicarbonate versus fluoride (Fig. 3).

However, negative correlation was obtained for fluoride versus calcium and magnesium (Fig. 4) (Jagadeshan et al. 2015b). These diagrams support the leaching of fluoride from minerals.

During laboratory experiments Jagadeshan et al. (2015b) found leaching of fluoride from charnockite and epidote hornblende biotite gneiss rocks resulted in enhanced fluoride in water from 1.41 to 3.54 mg/l, respectively, after 2000 hours (Fig. 5).

For reducing or removal of fluoride concentration from drinking water, there exist various methods at domestic or community level. The various techniques of fluoride removal are ion exchange method, precipitation method, adsorption method, Nalgonda method, and reverse osmosis. Thus, the existing treatment methods of water to remove fluoride for domestic use have several limitations. Further the people's mind-set is also not to use treatment methods as it is time consuming, expensive, and also requires trained personal. To overcome the problems and limitation of fluoride removal methods from the water, induced recharge may be adopted. David et al. (2014) reported in tropical countries such as India, rainwater harvesting could be useful for dilution of high-fluoride groundwater.

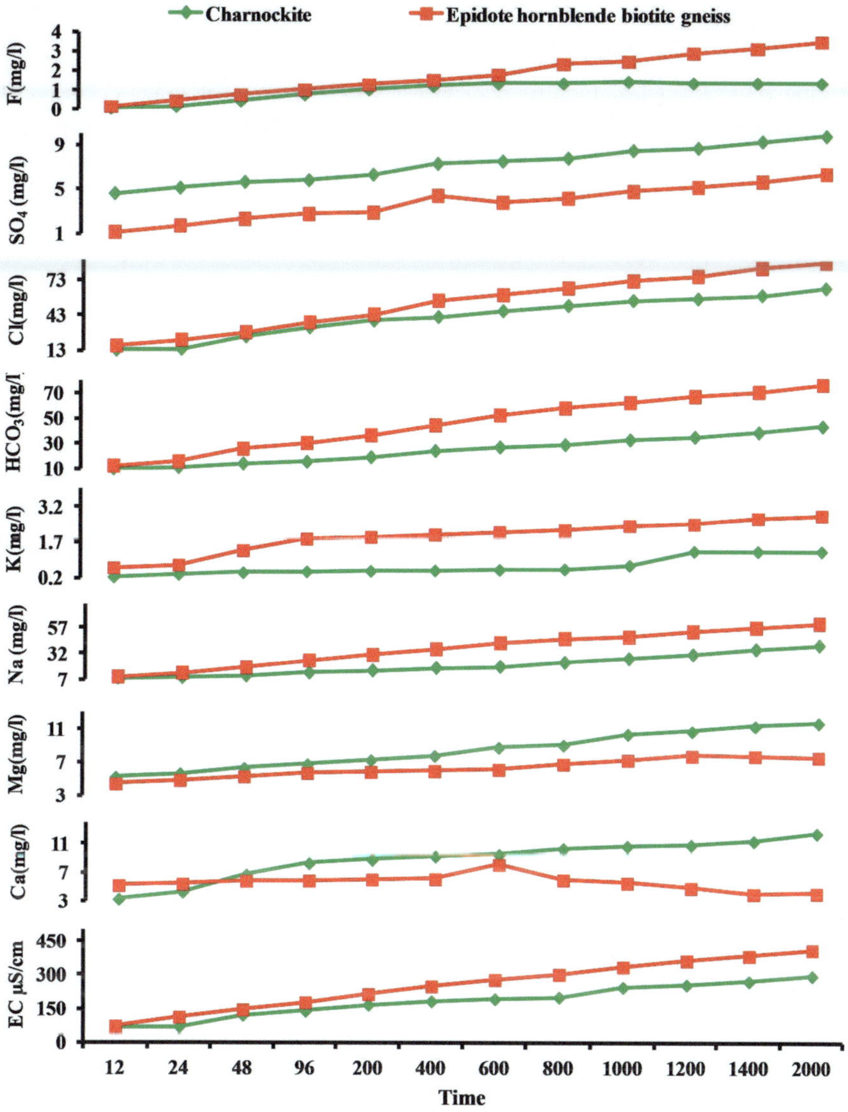

Fig. 5 Concentrations of ions released during laboratory experiment with respect to time from powder of (45 µ size) charnockite and epidote hornblende biotite gneiss

The recharge from the storage tank (Gupta and Deshpande 1998), subsurface barrier (Rolland et al. 2011), and check dams (Bhagavan and Raghu 2005) has reduced the concentration of fluoride in groundwater by dilution.

5 Pilot Study

A pilot study was carried out by the construction of a dug well recharge system in Dharmapuri district, Tamil Nadu. A few photographs taken during the construction of this structure are shown in Fig. 6. The schematic diagram of the constructed recharge structure is shown in Fig. 7. The site was chosen on the basis of three-year temporal variation in the fluoride concentration and fluctuation in groundwater level (Jagadeshan et al. 2015b). The groundwater level in induced recharge well (Well no. 44) raised to 9.1 m from 14.5 m (bgl), and the electrical conductivity has decreased from 1342 to 945 μS/cm (Jagadeshan et al. 2015b). Geochemical results indicate that fluoride concentration has decreased from 3.1 to 1.41 mg/l in well no. 44 due to dilution by the rainwater passing through induced recharge structure into dug well. This clearly indicates that the raise in groundwater level reduces the electrical conductivity and fluoride concentration of groundwater. The induced recharge from the structure constructed benefited an area of about 1 km^2 (Fig. 8). This study promises that the concentration of fluoride in groundwater can be decreased in other fluoride-affected areas if such low-cost induced recharge structures are constructed.

Fig. 6 Photographs taken during the construction

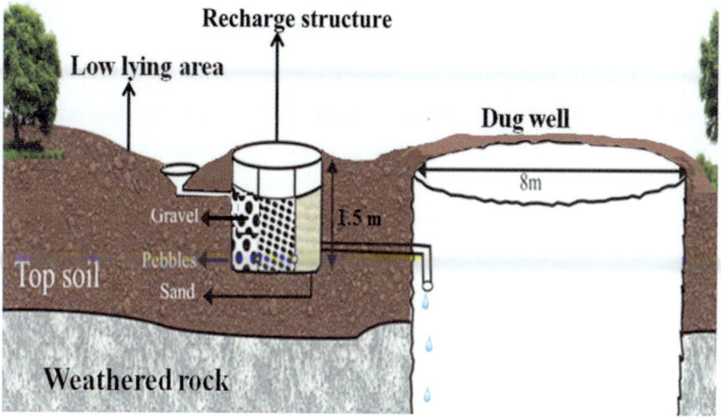

Fig. 7 Conceptual diagram of induced recharge structure

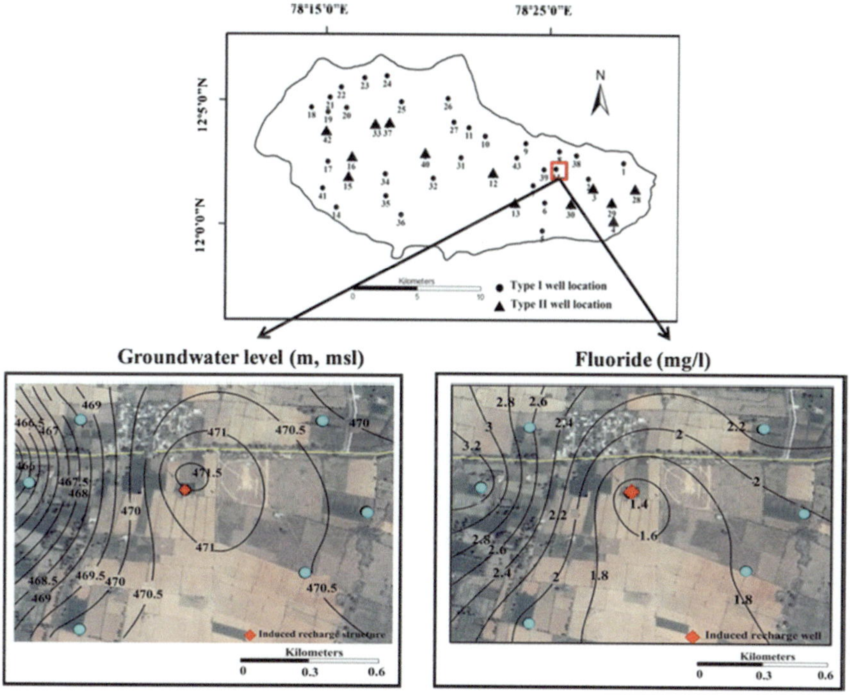

Fig. 8 Spatial variation diagram for groundwater level (m msl) and fluoride concentration (mg/l) around the induced l recharge structure

6 Conclusion

Groundwater with high fluoride is a major problem as evident from studies carried out by numerous researchers in different parts of the world. The host rock rich in fluorine is the major source of fluoride in groundwater. Concentration of fluoride in groundwater increases due to host rocks weathering and prolonged residence time of water. Treatment of the high-fluoride groundwater is possible by several methods, but they have limitations and cost intensive. Technique such as rainwater harvesting, constructing check dams and percolation ponds, and facilitating artificial recharge of rainwater through existing wells are suitable on-site treatment available. A pilot study carried out by a dug well recharge system in Dharmapuri district, Tamil Nadu, demonstrated the potential of this scheme in reducing the concentration of fluoride in groundwater over an area of about 1 km^2. The recharge from the structure has reduced concentration of fluoride in groundwater from 3.1 to 1.44 mg/l due to dilution. The study successfully demonstrated the applicability of dug well recharge system at a carefully selected site based on the systematic long-term hydrogeochemical studies to solve the problem of fluoride contamination affecting millions of rural people.

Acknowledgements The authors would like to thank the University of Grant Commission, New Delhi, for the financial support under Centre with Potential for Excellence in Environmental Sciences scheme (CPEPA Grant no. F. No. 1-9/2002 NS/PE). Authors also thank the Jai-Kranti Abhiyan scheme of the Ministry of Water Resources, Government of India and Dipankar Saha, Member, CGWB, for motivating them to prepare this chapter.

References

Agarwal V, Vaish AK, Vaish P (1997) Ground water quality: focus on fluoride and fluorosis in Rajasthan. Curr Sci 73(9):743–746

Ali S, Thakur SK, Sarkar S, Shekhar S (2016) Worldwide contamination of water by fluoride. Environ Chem Lett. doi:10.1007/s10311-016-0563-5

Bhagavan SV, Raghu V (2005) Utility of check dams in dilution of fluoride concentration in ground water and the resultant analysis of blood serum and urine of villagers, Anantapur District, Andhra Pradesh, India. Environ Geochem Health 27:97–108

BIS (2012) Bureau of Indian standards drinking water-specification. IS: 10500, New Delhi

Brindha K, Elango L (2011) Fluoride in groundwater: causes, implications and mitigation measures fluoride properties. Appl Environ Manag 111–136

Brindha K, Jagadeshan G, Kalpana L, Elango L (2016) Fluoride in weathered rock aquifers of southern India: managed aquifer recharge for mitigation. Environ Sci Pollut Res. doi:10.1007/s11356-016-6069-7

Carroll D (1962) Rainwater as a chemical agent of geological processes—a review. U.S. Geological Survey Water-Supply Paper 1535-G, p 18

David E, Reisner T, Pradeep (2014) Aquananotechnology. Global Prospects CRC Press, Science, p 887

Farooq SH, Chandrasekharam D, Norra S, Berner Z, Eiche E, Thambidurai P, Stüben D (2011) Temporal variations in arsenic concentration in the groundwater of Murshidabad District, West Bengal, India. Water Res 44:5575–5578

Fawell J, Bailey K, Chilton J, Dahi E, Fewtrell L, Magara Y (2006) Fluoride in drinking water. WHO, IWA Publishing, pp 1–144

Gaciri SJ, Davies TC (1993) The occurrence and geochemistry of fluoride in some natural waters of Kenya. J Hydrol 143:395–412

Gaumat MM, Rastogi R, Misra MM (1992) Fluoride level in shallow groundwater in central part of Uttar Pradesh. Bhu-Jal News 7(2 & 3):17–19

Gupta SK, Deshpande RD (1998) Depleting Groundwater levels and increasing fluoride concentration in villages of Mehsana District, Gujarat, India. Cost to Economy and Health, Water Resource and Research Foundation, Ahmadabad, p 74

Haidouti C (1991) Fluoride distribution in soils in the vicinity of a point emission source in Greece. Geoderma 49:129–138

Jagadeshan G (2015) Geochemical reaction responsible for fluoride rich groundwater and remediation by induced recharge in Vaniyar River Basin, Tamil Nadu. Ph.D. thesis, Anna University, Chennai, India

Jagadeshan G, Kalpana L, Elango L (2015a) Major ion signatures for identification of geochemical reactions responsible for release of fluoride from geogenic sources to groundwater and associated risk in Vaniyar River basin, Dharmapuri district, Tamil Nadu, India. Environ Earth Sci 73(7):67–80

Jagadeshan G, Kalpana L, Elango L (2015b) Hydrogeochemistry of high fluoride groundwater in hard rock aquifer in a part of Vaniyar River basin, Tamil Nadu, India. Geochem Int 53(6): 554–564

Kundu N, Panigrahi MK, Tripathy S, Munshi S, Powell MA, Hart BR (2001) Geochemical appraisal of fluoride contamination of groundwater in the Nayagarh District of Orissa, India. Environ Geol 41:451–460

Mukherjee A, Saha D, Harvey CF, Taylor RG, Ahmed KM, Bhanja SN (2015) Groundwater systems of the Indian Sub-Continent. J Hydrol Reg Stud (New online journal of Journal of Hydrology, whose Impact Factor is 3.053). http://dx.doi.org/10.1016/j.ejrh.2015.03.005

Pettenati M, Picot CG, Thiery D, Boisson A, Alazard M, Perrin J, Dewandel B, Maréchal JC, Ahmed S, Kloppmann W (2014) Water quality evolution during managed aquifer recharge (MAR) in Indian crystalline basement aquifers: reactive transport modeling in the critical zone. Procedia Earth Planet Sci 10:82–87

Raju NJ, Dey S, Das K (2009) Fluoride contamination in ground waters of Sonbhadra district, Uttar Pradesh, India. Curr Sci 96:979–985

Ramamohana Rao NV, Suryaprakasa Rao K, Schuiling RD (1993) Fluorine distribution in waters of Nalgonda District, Andhra Pradesh, India. Environ Geol 21:84–89

Rolland A, Muralidharan D, Rangarajan R, Sathyanarayana U, Deshmukh SD (2011) Mitigation of fluoride problem through artificial recharge strategies, case study from Nalgonda District, Andhra Pradesh. J Geol Soc India 2:45–53

Saha D (2009) Arsenic groundwater contamination in parts of middle Ganga plain, Bihar. Curr Sci 97(6)

Saha D, Alam F (2014) Groundwater vulnerability assessment using DRASTIC and pesticide DRASTIC models in intense agriculture area of the gangetic plains. Environ Monit Assess, India. doi:10.1007/s10661-014-4041-x

Saha D, Dhar YR, Sikdar PK (2008) Geochemical evolution of groundwater in the pleistocene aquifers of South Ganga Plain, Bihar. J Geol Soc India 71:473–482

Selvam S (2015) A preliminary investigation of lithogenic and anthropogenic influence over fluoride ion chemistry in the groundwater of the southern coastal city, Tamil Nadu, India

Shan H, Ting L, Chuanyong J (2013) Principal component analysis of fluoride geochemistry of groundwater in Shanxi and Inner Mongolia, China. J Geochem Explor 135:124–129

UNICEF (2009) Position on water fluoridation. Fluoride in water: an overview. http://www.nofluoride.com/Unicef_fluor.cfm

WHO (2002) Fluorides, Geneva, World Health Organization. Environmental Health Criteria, pp 227, 268

Deciphering Freshwater/Saline Water Interface in and Around Northern Chennai Region, Southern India

M. Senthilkumar, D. Gnanasundar and E. Sampath Kumar

Abstract Chennai, a city of almost seven million people, is one of the most water-stressed cities of India. While the population keeps skyrocketing, the volume of water available for them is dwindling. Scarcity intensifies conflict between Chennai and its peri-urban areas, which warrants urgent attention. The city receives about 985 million litres a day (mld) against the demand of 1200 mld. The present demand of 1200 mld includes the drinking water requirement for the city and its urban agglomerations that is 950 mld and the industrial demand of 250 mld. This demand is projected to increase to 2100 mld by 2031. Presently, the demand is met from the five freshwater reservoirs besides groundwater is pumped from 6 well fields in the inter-fluve area of the Araniyar and Korattalaiyar rivers. In the last decade, 3 well fields had to be abandoned due to seawater intrusion. In the dry months, seawater intrusion is observed 16 km inland, and after the monsoon, the seawater intrusion moves seaward side and is observed 14 km inland. There is an urgent need to monitor and regulate the fresh groundwater resource of the region. This paper describes the exact hydrological frame walk of the area. Rainwater harvesting and artificial recharge at apt locations in the Korattalaiyar River Plain so that the saline and fresh water interface is pushed further towards the seaward side.

Keywords Seawater intrusion · Minjur · Coastal aquifer · Chennai city
Groundwater salinity

M. Senthilkumar (✉) · D. Gnanasundar
Central Ground Water Board, SECR, MoWR, RD & GR, Chennai 600090, India
e-mail: msent50@yahoo.com

E. Sampath Kumar
Central Ground Water Board, NH IV, Faridabad, India

D. Saha et al. (eds.), *Clean and Sustainable Groundwater in India*,
Springer Hydrogeology, https://doi.org/10.1007/978-981-10-4552-3_3

1 Introduction

Groundwater is a major source of freshwater that is widely used for domestic, industrial and agricultural purposes in most parts of the world. Increasing demand and quest for groundwater due to ever-increasing population has thrown light on the unscientific techniques to quench these precious resources leading to declining water levels and deterioration in groundwater quality (Saha 2009; Mukherjee et al. 2015). In coastal regions, when the groundwater level lowers seawater intrudes immediately into the freshwater regions to maintain the hydrodynamic balance of the region. It is necessary effective management of available groundwater resources in the coastal regions. One such area is the Minjur/eastern part of Tiruvallur district that holds six well fields that supply drinking water to the ever-expanding Chennai city. The well fields are located in the thick alluvium lying in between the river. The excessive withdrawal of groundwater from this alluvium has led to seawater intrusion. Studies by various government department/academic institutes have been carried out from time to time. United Nation took up project studies in Araniyar and Koratalaiyar basin during 1967–1970. Studies included construction of tubewells in various parts of the basin to understand the subsurface geology and aquifer parameter and identify the possibilities of seawater intrusion in the Minjur area. Central Ground Water Board (CGWB 1981) also carried out the exploration in Defense establishments at Puduvayal down to a depth of 764 mbgl to assess the groundwater resources of the area. The exploratory drilling at Puduvayal revealed the absence of crystalline basement down to a depth of 764.20 m and the sandstone aquifer with intercalation of shale and compact in nature with poor permeability. Even though several research works have been carried out from time to time to demarcate the freshwater/saline water interface and identify seawater intrusion (Gnanasundar and Elango 1999; Kim et al. 2003; Davis et al. 1996; Sathish et al. 2011), this study was carried out with the objectives (i) to determine existing scenario of groundwater regime and (ii) decipher freshwater/saline water interface in Minjur area using bromide and to provide suitable management strategies for the region.

2 Study Area

The area lies in the northern part of the Tamil Nadu state. The area covering 1450 km is bounded by north latitudes of 13° 05′ 35″ and 13° 33′ 20″ and east longitude of 79° 52′ 15″ and 80° 18′ 30″ forming parts of the survey of India Toposheets of 66 C/3, 4, 7 and 8. The area is bounded in the north by the Andhra Pradesh state, Pulicat Lake; in the west by Uttukottai, Tiruvallur district; and in the south by Chennai district.

The study area includes 04 blocks, namely Gummidipondi, Puzhal, Sholavaram, Minjur, and few areas of greater Chennai. The region enjoys subtropical monsoon climate with January and February as the dry periods and March to May as summer period, followed by the monsoon period. The maximum temperature in this area is about 42 °C during the months of May and June. The minimum temperature is about 21 °C recorded during the months of December and January. The southwest monsoon (June to September), the north-east monsoon (October to December) and

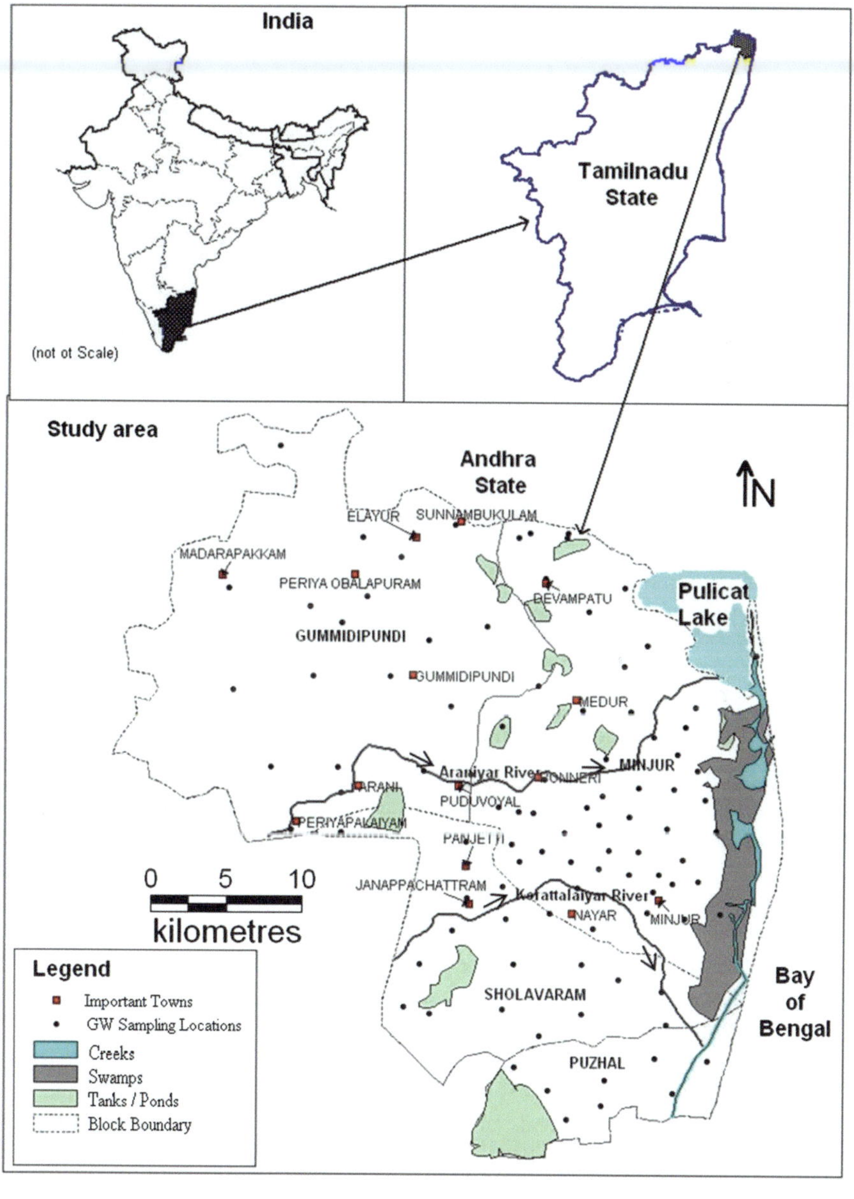

Fig. 1 Administrative map with representative key wells

the transition period contribute 40, 51 and 9%, respectively, of the total average annual rainfall (1207 mm/year). Araniyar and Korattalaiyar are the two important rivers into the study area (Fig. 1).

3 Geological Frame Work of the Area

The major portion of the study area is occupied by the river alluvium consists of gravels, fine-to-coarse sand, clays and sandy clay of various shades of grey and brown and occasional gravels. The thickness of the deposits varies from 10 m in the south to 60 m in the north. The thickness is an increase in an easterly direction towards the coastline east of Minjur that is 60 m. The region between the Araniyar and Korattalaiyar rivers has a maximum thickness of alluvium, to the tune of 60 m. In the western part of Gummidipondi block laterites are observed. The alluvial deposits are underlain by the Tertiary and Gondwana formations, which are generally argillaceous in character. Regions are further underlain by crystalline rocks. However, in the study area, crystalline basement was not encountered to a depth of 764 mbgl in the exploratory borehole at Puduvoyal. However, UNDP studies carried out during 1967–1970 has reported occurrence of crystalline rocks in two locations, i.e., Vellaivayalchavdi at depth of 106 mbgl and at Athipattu villages of 60 mbgl. Figure 2 shows the geological map of the study area, and geological succession of the study area is given in Table 1.

3.1 Gondwanas

These deposits are comprised of massive pile of lacustrine fluviales as well as marine deposits. The Upper Gondwana Sediments consist of two stages, viz. the lower Sriperumbudur stage consisting of fluviatile clays, shales and feldspathic sandstones and the Satyavedu stage representing the marine sediments of ferruginous sandstones, conglomerates and boulders. The Sriperumbudur stage occurs as patches, with easterly dips at low angles. The age of the Sriperumbudur formation is not certain, but the impressions of the foraminifera and ammonites are suggestive of an age varying from Upper Jurassic to Lower Cretaceous.

The Satyavedu stage comprises beds of conglomerate mixed with a few beds of coarse mottled sandstone, beds of clayey sandstones and sandy shales. The conglomerates with a sandy clayey matrix are hard and compact, and exposures of it are invariably strewn with shingle, pebbles and boulders. The strata show, in general, a decreasing lateral gradation in the coarseness of component deposits. The conglomerates and boulder beds occur nearer to the crystalline rocks. The total thickness of the formation exceeds 30 m.

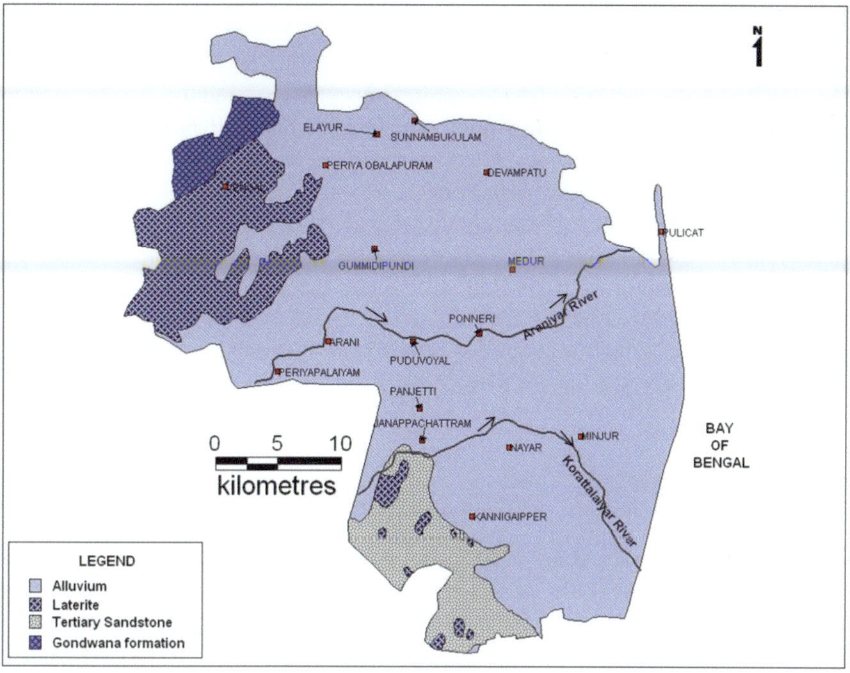

Fig. 2 Geological map of the study area

Table 1 Geological succession of the study area

Era	Period	Epoch	Formation	Lithology
Cenozoic	Quaternary	Holocene		Alluvium, clay, sand, gravel, pebbles and beach sand
		Pleistocene	Boulder bed	Boulders, cobbles, pebbles and gravel mainly quartzite in a clayey matrix
	Tertiary	Pliocene–Miocene		Sandstone, shale and clays
Unconformity				
Mesozoic	Cretaceous	Upper Gondwana	Satyavedu	Boulders, quarzitic sandstones, ferruginous conglomerate with clayey matrix and intercalations of sandstones
			Sriperumbudur	Clay, shale, feldspathic sandstone
Unconformity				
Proterozic	–	–	–	Basic and ultrabasic intrusives, gneiss and charnockites

3.2 Tertiary

The Tertiary beds comprise friable, white, speckled and reddish-brown mottled quartz grit and friable quartzose grits, which are white and brownish in colour becoming whiter at lower depths. Coarse laterite capping changes with depths into reticularly cellular, sandy clay grits. The latter again appear to grade into coarse, friable, mottled grits which become pure white with depth. Similar formations are traced north of the Araniyar river and along the southern bank of the Korattalaiyar River in a line running from the Red Hill lake to the north–north-west. The rocks belonging to this period have been assigned to the Miocene–Pliocene (Cuddalore) series but no fossil evidence of age has been found. It is only on stratigraphical and lithological evidence that they are separated from the upper Gondwanas.

3.3 Quaternary

Boulder Bed: Gondwana series is overlain by the deposits known as boulder bed in the eastern part of the study area. This bed consists of a mixture of rounded to subrounded boulders, cobbles, pebbles and gravel in a clayey sandy matrix and is partly compacted. These deposits represent a marginal facies of fluvial deposits worked out from Gondwana conglomerates. Further east, this bed abruptly thins out to a few feet in thickness, and as it is overlain by alluvium, it forms a good marker of transition to Gondwana deposits.

Laterite: The Tertiary friable sandstone and Gondwana series are commonly capped by scoriaceous and pisolitic laterite. It is noticed around Kannigapuram, Red Hills and Palaiya Erumaimattupalaiyam areas. The laterite of Red Hills is of a conglomeratic type comprising mainly fragments of subrounded quartz bound by cindered ferruginous cement. Its thickness ranges from 1.50 to 4.5 m.

Alluvium: The youngest formations in the area are the alluvium, which was deposited on the worn-down and eroded surface of Tertiary and Gondwana rocks by the major river. It is noted that the alluvial plains in the eastern part of the area entirely span the lower reaches of Araniyar and Korattalaiyar and branch off into two separate plains farther east. A cross section along the west–east direction near the Panjetty-Minjur area is represented in figure. The alluvium consists of gravel, fine-to-coarse sand, clay and sandy clay of various shades of grey and brown.

Commonly, the different types are intercalated (or) dovetailed in the form of lenses and pockets which point out the erratic geometry of the deposition, caused by the migration and varying flow velocities of old rivers. Exploratory drilling shows that the thickness of these deposits increases progressively in an easterly direction towards the coast line east of Minjur, where it is about 50–60 m thick. The wind deposited sand in the form of irregular, low flat dunes ranging in width from less than 0.1 km to about a kilometre that occurs all along the coast, except where they are interrupted by the river outlets. The most striking dunes are located near Pulicat, where they have grown by wind action into irregular mounds of 12–15 m high.

4 Methodology

To understand and decipher the freshwater/saline water interface and to assess the impact of the seawater intrusion, following studies were carried out: (1) detailed hydrogeological survey was carried out through, (2) well inventory survey was conducted in 124 wells, (3) 80 representative key wells have been established, and (4) groundwater level, EC, pH and temperature measured in the field and 45 samples were collected and analysed for bromide concentration. Representative wells included dugwells/openwells and shallow tubewells. Groundwater samples were analysed for bromide concentration in hydrogeochemical laboratory of the Central Ground Water Board, South Eastern Coastal Region, Chennai. Diverse methods of Br analysis exist. Bromide was analysed using the ion-selective electrodes. In a single-laboratory evaluation, a series of standards with known bromide concentrations was analysed with a bromide ISE (AHPA 1992). Measurements were conducted over three consecutive days using an Orion 9435 bromide ISE. Several point calibration (1.0, 4.00 and 40.0 mg/L bromide) was performed prior and in between the analysis. Once calibration was performed successfully, bromide concentrations in the groundwater samples of the study area were determined.

5 Results and Discussion

5.1 Hydrogeology

The major portion of the study area is occupied by recent alluvium, and groundwater is found to occur in unconfined/semi-confined conditions in these formations. Groundwater is developed by dugwells and tubewells. Dugwells are observed only in the eastern part/western and northern parts of the study area, i.e. Minjur, Athipattu and Nandiampakkam; in southern parts, i.e. Manali area; and in northern/western fringes, i.e. Gummidipondi. The depth of the dugwells ranges from 3 to 8 mbgl, while the depth of the tubewells ranges from 15 to 80 mbgl. Deepest tubewell of 80 mbgl is observed in western part of Gummidipondi block. The groundwater level in the dugwells ranges from 1.25 (village Tirupalavanam, northern eastern part of the study area) to 8.7 mbgl (village Uthandikandigal, north-western part of the study area) during the premonsoon period (Fig. 3). The groundwater level during the post-monsoon period ranges from 1.0 mbgl (Athipattu village) to 5.56 mbgl (Mandarapakkam village) (Fig. 3).

In the central portion of the study area, i.e. the area between the two rivers, Araniyar and Koratalaiyar river, groundwater is developed by tubewells ranging in depth from 30 to 50 mbgl, the groundwater level in these tubewells is reported to range from 21 to 25 mbgl (reported water level) during the premonsoon period, and it is developed by shallow tubewells ranging in depth from 30 to 50 mbgl. During the post-monsoon period, the groundwater level ranges from 11.64 (Ammarakulam

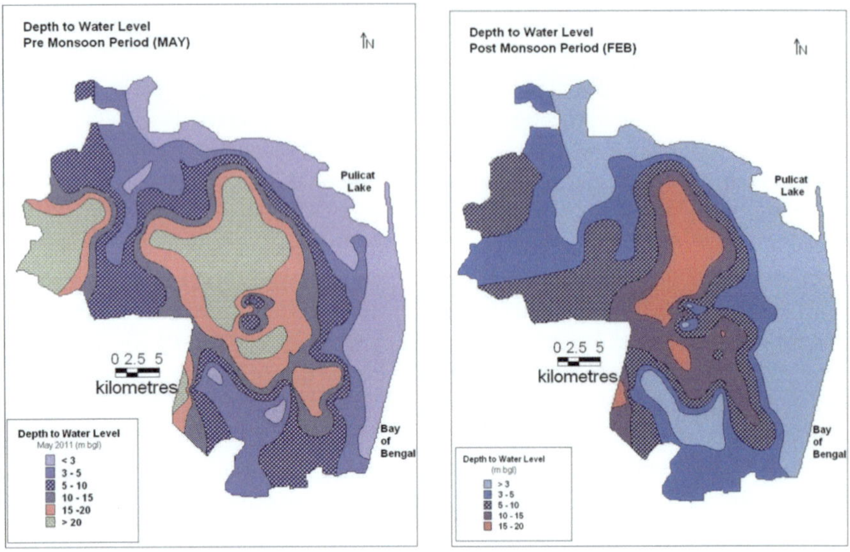

Fig. 3 Depth to water level (pre- and post-monsoon)

village) to 19.60 mbgl (Eliyambedu village). The spatial distribution of the groundwater levels also reveals deeper groundwater level in the central portion of the study area. Numerous pumping wells are observed between these two rivers and in the right-hand side bank of the Korattalaiyar River, supplying drinking water to various panchayats in the district. Groundwater level during the post-monsoon period ranges from 1.0 mbgl at Athipattu to 19.60 mbgl at Eliyambedu village. In the alluvial deposits occurring in the central portion between the Araniyar and Korattalaiyar rivers, the change in groundwater level from pre-post-monsoon shows substantial rise ranging from 3 to 10 m.

Several studies have been taken up ETO, UNDP and CGWB in the study area, and through long duration, pumping test aquifer parameters have been determined. The hydraulic conductivity value ranges from 81 m/day (Seerapalayam village) to 203 m/day (Nallur village). Transmissivity value ranges from 807 m^2/day (Vallur village) to 3974 m^2/day (Panjetty village). It is also observed that tubewells located in the palaeochannels have very high transmissivity values compared to the tubewells tapping alluvium/boulders. Two tubewells in Ponneri village were tested during UNDP studies which revealed that tubewell in palaeochannel has transmissivity of 1117 to 149 m^2/day in the other region.

The tubewells tapping the semi-confined/confined aquifer has a storage coefficient ranging from 1.85×10^{-3} in Vallur village to maximum of 6.2×10^{-4} determined in Duranallur village. The wells tapping the unconfined aquifer whose specific yield values ranges from 7% (Ponneri village) to 20% (Minjur village). Pockets of shallow water levels are observed in the central portion near village Krishnapuram. These might be due to the presence of perched aquifers in these areas. The dugwells are observed in these areas. A check dam was constructed across the Korattalaiyar River near the Siruvakkam village. The impact of the check is clearly seen by the dilution effect in the EC concentration of the groundwater in the upstream of the check dam. There are numerous pumping wells supplying water to various panchayats of the district. The groundwater yield has also increased considerably in the wells located in Kamaravapalayam, Vannipakkam and Siruvakkam villages.

5.2 *Groundwater Quality*

Groundwater was pumped from 6 well fields (Minjur, Kannigaipair, Thamaraipakkam, Panjetty, floodplain fields and Poondi well field) located in this alluvial river basin. In the last decade, 3 well fields had to be abandoned due to seawater intrusion. In order to decipher the extent of the seawater intrusion, EC values were measured along with temperature, pH and GPS locations. EC values range from 140 $\mu S/cm^2$ (village Andarkulam located in the central portion of the study area) to >6450 $\mu S/cm^2$ (village Urananmedu located in the eastern part of Minjur block). Pulicat town (NE corner) has very poor quality of groundwater. EC values range from 6000 to 20,000 $\mu S/cm^2$. The spatial distribution of EC values recorded during May 2011 is represented in Fig. 4a. TDS values below 50 mg/L are reported in few pockets in the central portion of the study area. High EC values are observed in the north-eastern part of the study area, which is due to the saline Pulicat Lake. Eastern part of the study area has high EC values ranging from 2000 to 6450 μS/cm. Reasons for high EC values are attributed to seawater intrusion. However, exact demarcation of freshwater/saline water interface is established with the help of bromide concentration of groundwater and is discussed below.

During post-monsoon period, EC values range from 250 $\mu S/cm^2$ (village Govindaraju Kandigai located in the north-western portion of the study area in between Tachur and Madarapakkam) to 7010 $\mu S/cm^2$ (village Urananmedu located in the eastern part of Minjur block). Pulicat town has very poor quality of groundwater. The spatial distribution of EC values recorded during Feb 2012 is represented in Fig. 4b. Quality of groundwater below 50 mbgl is reported to be poor in few pockets in the central portion of the study area. High EC values are observed in the north-eastern part of the study area, which is due to the saline Pulicat Lake. Eastern part of the study area has high EC values ranging from 1000 to 7010 μS/cm.

A detailed observation of Fig. 4 reveals few fresh groundwater pockets are observed in the eastern part of study area within the intruded region. These potable freshwater pockets are observed upstream of the tanks near Vayalur village. EC values range from 300 to 1100 μS/cm, and groundwater levels are shallow during the premonsoon, while in the villages downstream side of the tanks has saline groundwater (Urananmedu village). Reasons for the poor quality in the downstream side are mainly due to the limited quantum of freshwater. Fresh groundwater in

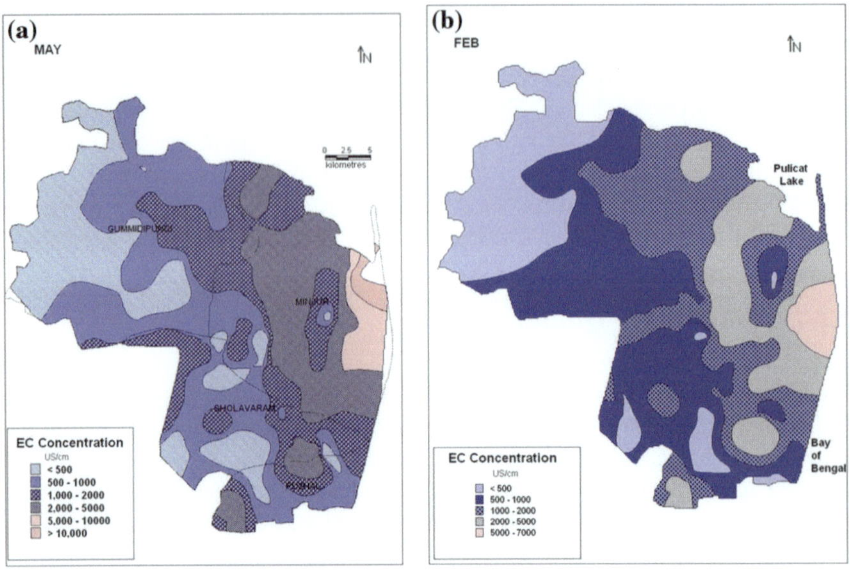

Fig. 4 Iso-conductivity map of the study area

these pockets may due to the two tanks located downstream of Voyalur village. The tanks get sufficient water after the monsoon, but the quantum is less sufficient to push the denser seawater intruded into the aquifer to the seaward side (eastern side); hence, it creates a freshwater mound in the upstream side of the tanks instead of the downstream side.

5.3 *Deciphering Freshwater/Seawater Intrusion*

Detailed hydrogeological survey in the study area revealed that denser saline water has intruded the freshwater aquifer in the Minjur-Mouthambedu (on the eastern side) to Vellanoda-Madaravaram village (on the central portion and east of Tachur). The groundwater level has gone below the mean sea level due to extensive pumping for agricultural and drinking water supply schemes. This has resulted inward movement of the denser seawater to maintain the hydrodynamic balance of the aquifer system. 45 groundwater samples were collected and analysed for cations and anion along with bromide concentration. Bromide is absent/minimal in freshwater, while seawater contains 51–71 mg/L of bromide (Schoonen et al. 1995; Vengosh and Pankratov 1998). Bromide concentration in the study area ranges from 0 to 22.1 mg/L. Spatial distribution of the bromide concentration (Fig. 5) clearly helps in demarcating the seawater intruded area. The bromide values reduce from east to west from 22.1 to 0 mg/L (Fig. 5), and with reference to the horizontal

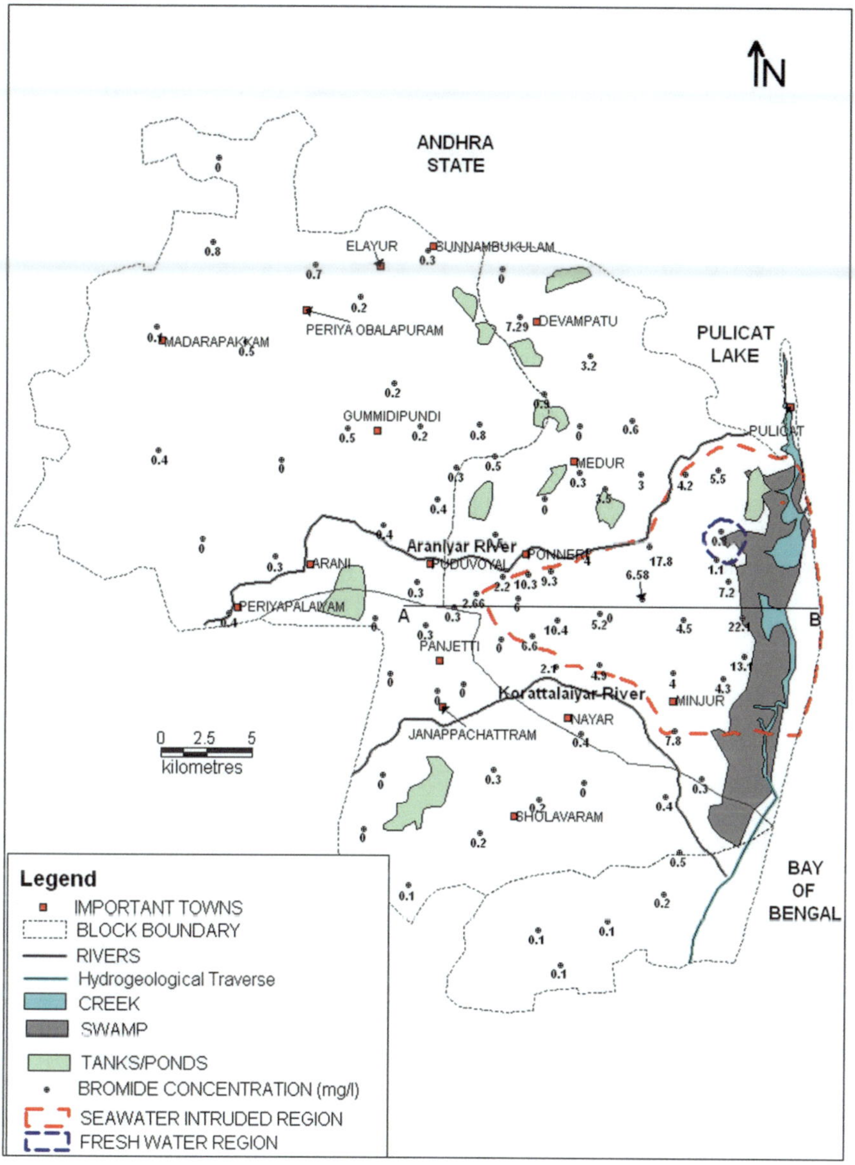

Fig. 5 Map showing seawater intruded region

distance from the coast, the freshwater/saline water interface is observed between 16.5 and 17.0 km from the coast.

The region about 1–1.5 km^2 upstream of Urnanmedu/Sanglibedu tanks covering Voyalur-Tirunelvoyal villages has fresh groundwater resources, and agriculture activity is being practiced in these regions. The EC of these villages ranges from 300

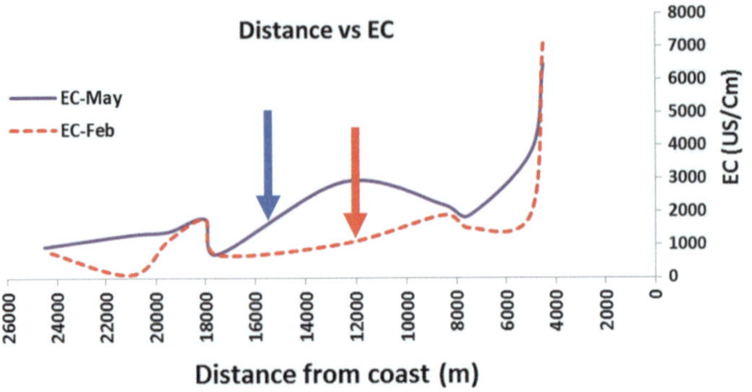

Fig. 6 EC variation along east–west direction

to 1000 μS/cm, and the bromide concentration is also very low (0.8–1.1 mg/L). Further, based on the field study along the traverse A-B (Fig. 5), EC, temperature and pH were monitored during pre- and post-monsoon periods (Fig. 6). It is observed that the seawater intrusion in the study area is not in a linear form and the interface is moving from pre- and post-monsoon periods. The movement of interface is about to 2–3 km towards the seaward side. The interface during the post-monsoon period is observed between 14 and 14.5 km. Heavy withdrawal of groundwater in the central portion of the study area for agricultural, industrial and drinking water supply is main cause for intrusion of denser saline water into the freshwater aquifer.

5.4 *Management Solutions*

(a) There is an urgent need to adopt suitable management plan in the area to control seawater ingress through the aquifer and built up freshwater resources.
(b) Seawater has intruded inland, and FW/SW interface is observed 16.5–17.0 km in the Minjur-Mouthambedu to Panjetty during the premonsoon period, and during post-monsoon period, the freshwater/saline water interface is observed 14–14.5 km inland. The natural process is capable of pushing the interface outward towards the sea which is 2–3 km. Artificial recharge structures like the check dams across the korattalaiyar River can be constructed for effective recharge.
(c) Groundwater sold by household to industries by tankers in the villages Minjur/Athipattu and Vallur needs to be stopped.
(d) The success stories of tanks/ponds in Urananmedu/Sanglikuppam and Tirunelvoyal areas can be taken in other seawater intruded villages. The existing tanks need to be desilted.
(e) Groundwater withdrawal needs to be regulated through participatory groundwater management.

6 Conclusion

(a) The summary findings of the research carried out in the area are pointed as below.
(b) Groundwater is developed by dugwells and tubewells. Dugwells (depth range 3–10 mbgl) are observed only in the eastern/western and northern parts of the study area, i.e. Minjur, Athipattu and Nandiampakkam; in southern parts, i.e. Manali area; and northern/western fringes, i.e. Gummidipondi. In the central portion between the Araniyar and korattalaiyar, groundwater is developed by tubewells with depth ranging from 30 to 50 mbgl tapping the alluvium.
(c) Groundwater level in the dugwells ranges from 1.25 to 8.71 mbgl during premonsoon period and from 1.0 to 5.6 mbgl during post-monsoon period. Groundwater level in the tubewells ranges from 21 to 25 mbgl during pre-monsoon period and 11.64–19.60 mbgl during post-monsoon period.
(d) EC of groundwater ranges from 140 to 6450 μS/cm during premonsoon period and from 250 to 7010 μS/cm during post-monsoon period.
(e) Groundwater is extensively pumped for drinking water supply to Chennai from 6 well fields—Tamaraipakkam, Panjetty, Kannigaipair, Minjur (abandoned), floodplains and Siruvakkam well fields.
(f) Bromide concentration is nil/minimal in freshwater and that of seawater it is in the range of 51–71 mg/L. Bromide concentration of the study area ranges from 0 to 22.1 mg/L.
(g) In detailed analysis of the groundwater level, EC and Bromide, it is observed that seawater has intruded inland and FW/SW interface is observed 16.5–17 km in the Minjur-Mouthambedu to Panjetty during the premonsoon period, and during post-monsoon period, the freshwater/saline water interface is observed 14–14.5 km inland.
(h) Artificial recharge structure in the Korattalaiyar River and in the field plains needs to be taken in large scale to push the FW/SW interface further seaward.

Acknowledgements The authors express their sincere gratitude to the Er. K.B. Biswas, Chairman of Central Ground Water Board, Faridabad. Authors express their heartfelt thanks to Dr. Dipankar Saha, Member (SAM) for his utiring efforts in shaping the manuscript and for his evaluable suggestion. Authors also thank Member (WQ & TT) and to Sh. D.S.C. Thambi, Retd. Member, CGWB, Faridabad, for their support and guidance. Thanks are also expressed to Head of Office, SECR, CGWB, and all the officers and officials of CGWB, SECR, Chennai, for their support during the study.

References

APHA (1992) Standard methods for the examination of water and wastewater (Franson, Mary Ann H, editor), 18th edn. American Public Health Association, Washington
CGWB (1981) Puduvoyal basic data report of Puduvoyal well site, p 20

Davis SN, Whittemore DO, Fabryka-Martin J (1996) Uses of chloride/bromide ratios in studies of potable water. Ground Water 36(2):338–350

Gnanasundar D, Elango L (1999) Ground water quality assessment of a coastal aquifer using geoelectrical techniques. Int J Envi Hydrol 6:21–33

Kim Y, Lee K-S, Koh D-C, Lee D-H, Lee S-G, Park W-B, Koh G-W, Woo NC (2003) Hydrogeochemical and isotopic evidence of groundwater salinization in a coastal aquifer: a case study in Jeju volcanic island, Korea. J Hydrol 270:282–294

Mukherjee A, Saha D, Harvey CF, Taylor RG, Ahmed KM (2015) Groundwater systems of the Indian Sub-Continent (Soumendra N. Bhanja). J Hydrol Reg Stud. http://dx.doi.org/10.1016/j.ejrh.2015.03.005

Saha D (2009) Arsenic groundwater contamination in parts of middle Ganga Plain, Bihar. Curr Sci 97(6)

Sathish S, Elango L, Rajesh R, Sarma VS (2011) Assessment of seawater mixing in a coastal aquifer by high resolution electrical resistivity tomography. Int J Env Sci Tech 8(3):483–492

Schoonen MAA, Devoe V, Brown CJ (1995) Bromide in Long Island ground waters and surface waters. Geology of Long Island and Metropolitan New York, April 22 1995, Program with Abstracts. Stony Brook, New York, Long Island Geologists, pp 118–126

Vengosh A, Pankratov I (1998) Chloride/bromide and chloride/fluoride ratios of domestic sewage effluents and associated contaminated ground water. Ground Water 36(5):815–824

Geogenic Fluoride Contamination in Two Diverse Geological Settings in West Bengal

S. Brahma

Abstract Fluoride contamination in groundwater has been studied in two important geological settings of West Bengal, viz. (i) alluvium of Rampurhat–Nalhati area of Birbhum district, and (ii) consolidated formation of Barabazar block of Purulia district. Rampurhat–Nalhati area in Birbhum district is located in 'basin margin' area represented by Tertiary sediments overlain by Quaternary deposits. A two aquifer system exists in the area: (i) shallow aquifer in the Pleistocene Older Alluvium in Quaternaries, and (ii) deeper aquifer within the Tertiary sediments. The shallow aquifer has high groundwater potentiality with low fluoride content, while the deeper aquifer system is characterized by comparatively low yield and high fluoride concentration. Maximum concentration of fluoride of 10.7 mg/L has been encountered at Junudpur in Birbhum district. At places, only one continuous aquifer system exists where the groundwater potentiality is high with low fluoride content. The fluoride content of groundwater has, in general, an inverse relationship with calcium and magnesium content and positive relationship with sodium concentration. Barabazar block in Purulia district is situated in the eastern fringe of Chhotanagpur Gneissic Complex, occupied mostly by meta-sedimentaries and a few thin linear patches of alluvium along narrow courses of stream. Groundwater prospect is better in northern parts underlain by gneissic rocks than in the south, which is mainly underlain by phyllites and schists. No relationship could be established between the fluoride content of groundwater with the concentrations of bicarbonate, calcium and sodium. Groundwater is generally potable, except at places near 'south shear zone', where groundwater with high concentration of fluoride and iron has been encountered. The concentration of fluoride increases with depth as well as in proximity to shear zone. In both cases, fluoride concentration in groundwater increases with depth.

Keywords Rampurhat–Nalhati · Barabazar · Purulia south shear zone
Aquifer

S. Brahma (✉)
Central Ground Water Board, Eastern Region, C. P. Block-6, Sector-V,
Salt Lake, Kolkata 700091, India
e-mail: supriyo_brahma@yahoo.co.in

D. Saha et al. (eds.), *Clean and Sustainable Groundwater in India*,
Springer Hydrogeology, https://doi.org/10.1007/978-981-10-4552-3_4

1 Introduction

Fluoride is essential for human beings. Drinking water is one of the major pathways for ingestion of fluoride in human body. Fluoride in drinking waters might help prevent tooth decay if the level of fluoride intake is within the permissible limit of 0.5 and 1.5 mg/L (WHO 2004). Considering the climatic condition and other related factors, Bureau of Indian Standards has prescribed 1 and 1.50 mg/L as desirable and permissible limits of fluoride, respectively, for drinking water. Excessive intake of fluoride causes fluorosis, which may lead to serious health hazards. Around 200 million people from 25 nations have health risks because of high fluoride in groundwater (Ayoob and Gupta 2006). Most of the people affected by high fluoride concentration in groundwater live in the tropical countries where the per capita consumption of water is more because of the prevailing climate (Brindha and Elango 2011).

Due to consumption of fluoride-contaminated water in many parts of India, villagers have been found to be suffering from dental, skeletal and non-skeletal fluorosis and various other morphological deformities, and even permanent crippling. Sporadic higher concentrations of fluoride have been reported in groundwater in many parts of the country in both consolidated and unconsolidated deposits. Dental fluorosis is endemic in 14 states and 150,000 villages in India with the problem most pronounced in the states of Andhra Pradesh, Bihar, Gujarat, Madhya Pradesh, Punjab, Rajasthan, Tamil Nadu and Uttar Pradesh (Pillai and Stanley 2002).

Fluoride contamination of groundwater in West Bengal was first detected during 1997 in Nalhati I block in Birbhum district. The occurrence of fluoride in groundwater beyond 1.50 mg/L is in a rather sporadic manner than following a definite pattern in 43 blocks in seven districts in West Bengal. In Purulia district, out of a total 20 blocks, 17 are fluoride infested where out of a total 3304 habitations, 233 are in risk zone. In Birbhum district out of a total 19, seven blocks, viz. Nalhati I, Rampurhat I, Mayureswar I, Sainthia, Siuri II, Rajnagar and Khayrasol have been reported to be fluoride affected by 'Fluoride Task Force', set up by Govt. of West Bengal; in this district, 134 out of a total of 1493 habitations are in the risk zone of fluoride infestation.

In present paper, geogenic fluoride contamination in groundwater has been dealt in two diverse geologic settings: (a) in parts of Rampurhat–Nalhati area of Birbhum district, and (b) in Barabazar block of Purulia district.

Rampurhat–Nalhati area of Birbhum district lies in the north-western part of West Bengal; it represents the marginal part of the 'Bengal Basin' and is fluoride infested (PHE Directorate, Govt. of West Bengal 2008). This terrain is flat and mainly covered by alluvium. Normal annual rainfall is 1601 mm; but, the area faces shortage of drinking water due to topographic undulation and sporadic poor quality of its groundwater due to high fluoride concentration in it.

Barabazar block is situated in the south-eastern part of Purulia district. It covers an area of 418.06 km^2 and falls in the eastern fringe of Chhotanagpur Gneissic Complex. Barabazar is characterized by undulating topography with denudation hills, mostly in the southern part, the maximum elevation of which is 331 m above mean sea level (amsl). The present area has a subtropical climate with annual rainfall varying within 1100–1500 mm. Topographic characteristics, hard formation with poor aquifer and erratic rainfall create drought situation.

2 Materials and Methodology

The present study has been carried out in two field seasons. Exploration by rotary rigs in Rampurhat–Nalhati area of Birbhum district has been carried out primarily for delineation of aquifers based on lithological and their characteristics, e.g. groundwater potentiality, its quality including fluoride. The target of exploration by this rotary rig is fixed to 225 m. Electrical logging in boreholes by UPTRON logger has been conducted at three sites, viz. Margram, Tarapur and Nowapara. By e-logging, continuous array of resistivity data, both long-normal (64″) and short-normal (16″), and spontaneous potential data of formations in down-log and up-log have been recorded. Clean drill-cutting samples of individual borehole have been kept at 3 m interval and after study of those samples, litho-logs have been prepared. These litho-logs have been compared with respective e-logs; after through correlation, delineation of aquifers in the present area has been made. Then, design of shallow and deep tube wells tapping shallow aquifer and deep aquifer, respectively, has been recommended at different site. After construction of tube wells at different locations, yield tests of tube wells are conducted and water samples from exploratory tube wells are collected for chemical analysis. As the number of exploration is limited in the area, a few other hand pump-fitted and submersible pump-fitted tube wells in and around the Rampurhat–Nalhati area have also been inventoried and water samples from these wells have also been collected for laboratory analysis.

At two places, viz. Margram and Tarapur, both in Rampurhat II block, two-aquifer system occurs; so, tube wells tapping deeper and shallow aquifers, respectively, have been constructed. In Nalhati II block, only one aquifer exists. Groundwater samples collected from deep, shallow and single continuous, from exploratory and other private and government wells representing deep/shallow/single aquifer, are analysed in the laboratory of Central Ground Water Board. Selected drill-cut litho-samples have been sent to Geological Survey of India; those samples are powdered and analysed in their chemical laboratory for determination of F-content in those drill-cut samples.

In other field season, hydrogeological investigation has been carried out in fluoride infested parts of Purulia district. Investigations have been conducted both in pre-monsoon and post-monsoon. Sampling has been done following the standard procedure (Central Ground Water Board, Lucknow 1996). Attention was given to collect fresh samples from open wells and hand pump-fitted tube wells so that the samples collected represented the quality of groundwater in the aquifer concerned.

Groundwater samples have been analysed following standard methods (American Public Health association 2012) in the laboratory of Central Ground

Water Board, Kolkata. pH and electrical conductivity in groundwater have been measured, respectively, by potentiometric method using pH metre and electrometric method by a conductivity metre. Among major ions, carbonate (CO_3^{2-}) and bicarbonate (HCO_3^-) in groundwater have been analysed by acid–base titration using H_2SO_4, and chloride (Cl^-) by titrimetric method using $AgNO_3$. UV-spectrophotometer has been used for determination of nitrate (NO_3^-); turbidimetric and molybdophosphoric acid methods by spectrophotometer have been applied, respectively, for determination of sulphate (SO_4^{2-}) and phosphate (PO_4^{3-}) ions. Spectrophotometer has also been used by eriochrome cyanin R as reagent (SPADNS method) for determination of fluoride (F^-) content and ammonium molybdate as reagent for SiO_2 in groundwater. EDTA-titrimetric method has been applied for determination of total hardness (TH), calcium (Ca^{2+}) and magnesium (Mg^{2+}) in water, whereas both sodium (Na^+) and potassium (K^+) in it have been analysed using flame emission method by flame photometer. For determination of total iron, spectrophotometry method has been carried out by using 1,10-phenanthroline as reagent.

3 Rampurhat–Nalhati Area, Birbhum District

Exploration has been carried out at four places: viz. (i) Margram (24° 09′ 02″; 87° 51′ 54), about 6 km ESE of Rampurhat, (ii) Tarapur (24° 06′ 26″; 87° 49′ 12″), 8 km ESE of Rampurhat, in Rampurhat II block, (iii) Lohapur (24° 17′ 31″; 87° 57′ 45″), 30 km N of Rampurhat and (iv) Nowapara (24° 16′ 56″; 88° 27″), about 6 km ESE of Lohapur, in Nalhati II block (Fig. 1). Tertiaries and overlying

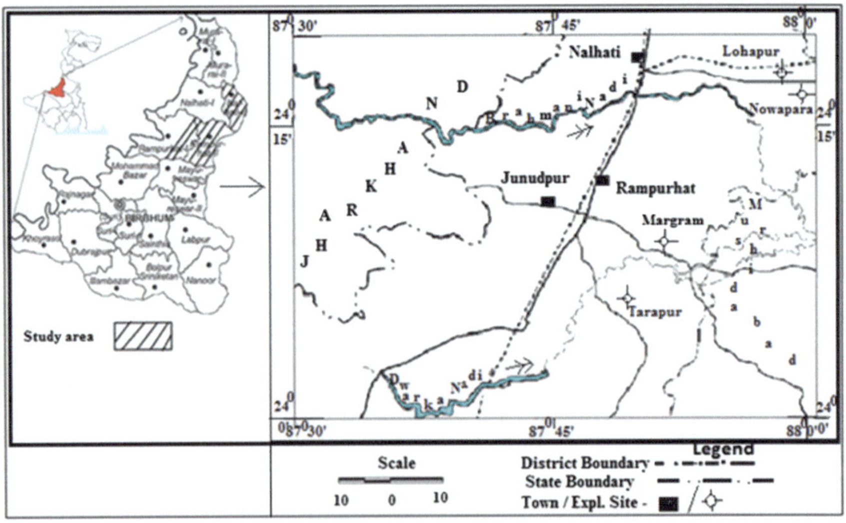

Fig. 1 Rampurhat–Nalhati area, Birbhum district

unconsolidated Quaternaries have been encountered in three boreholes, viz. Margram, Tarapur and Lohapur. At Nowapara, Rajmahal basalts (?) have been encountered beneath Tertiary Formation. Around Rampurhat, the thickness of Older Alluvium, consisting mainly of yellowish to brownish yellow sands, varies between 24 and 40 m. Older Alluvium is underlain by the Tertiaries represented mainly by whitish grey to grey mixtures of sand and clay; greyish clay bed separates the overlying Older Alluvium from the Tertiaries. The thickness of alluvium increases from west to east. In Nalhati II Block, topmost Older Alluvium occurs within 50–60 m. Alluvium is yellowish white to pale yellow. Tertiaries include mixture of grey sands and finer clastics.

3.1 *Hydrogeology*

At Margram and Tarapur, two-aquifer system has been established. Groundwater under unconfined condition occurs in shallow aquifers of Older Alluvium. The Tertiaries form the deeper aquifer, where groundwater occurs under semi-confined to confined condition. The shallow aquifer corresponds to the thickness of the Older Alluvium. In Nalhati area, however, there is only one continuous aquifer running through both Tertiaries and Quaternaries. Depth to water level (DTW) and discharge of different abstraction structures have been shown in Table 3.1. At Tarapur in Rampurhat II block, historical data of water level in shallow piezometer show falling trend in water level both in pre- and post-monsoon to the tune of 0.49 and 0.43 m/yr, respectively. The depth of submersible pump-fitted tube wells used for

Table 1 Details of tube wells in Rampurhat II and Nalhati II blocks, Birbhum district

S. No.	Location (block)	Well type	Depth to water level (mbgl)	Discharge (lps)	Depth (m)
1	Margram (Rampurhat II)	Deep expl. well	12.40	6.2	160
2	Margram (Rampurhat II)	Shallow expl. well	17.40	26	67
3	Rampurhat (Rampurhat II)	Mark I hand pump-fitted tube well	–	2.5	30
4.	Tarapur (Rampurhat II)	Deep expl. well	14.65	2.6	148
5	Tarapur (Rampurhat II)	Shallow expl. well	17.95	2.1	82
6	Nowapara (Nalhati II)	Expl. well	11.12	35	90
7	Lohapur (Nalhati II)	Expl. well	16.45	22	100
8	Junudpur (Rampurhat I), 9 km <S> of Rampurhat	Mark II hand pump-fitted tube well	–	2.5	61
9	Junudpur (Rampurhat I), 9 km <S> of Rampurhat	Submersible pump-fitted tube well	–	10	21
10	Tarapur	Piezometer	18.12	15.46	50

irrigation generally remains within depth range between 10 and 50 m. The depth of Mark I and Mark II hand pump-fitted tube wells around the area is generally within 30 m and more than 50 m, respectively. The average yield of tube wells varies from 18 to 22.7 m^3/hr (Datta and Bandyopadhyay 2006). A perusal of Table 1 indicates that water level in deep exploratory well (DEW) both at Margram and Tarapur is shallower than those in shallow exploratory well (SEW). In Rampurhat area, the discharge of shallow wells including exploratory wells at Margram and Tarapur, shallow tube well of 21 m depth at Junudpur, 9 km SW of Rampurhat, varies within 10–26 lps excepting at Tarapur, where SEW shows discharge of 2.1 lps due to technical reason. But, the discharge of DEWs is low and it ranges between 2.1 and 6.2 lps. So, the potentiality of Older Alluvium is more than that in Tertiaries. In Nalhati II area, water level in Lohapur is deeper than that in Nowapara; the discharge of the well at Lohapur is relatively lower than that at Nowapara.

3.1.1 Analytical Data

Groundwater

The quality of groundwater tapped by selected abstraction structures and exploratory wells in Rampurhat II and Nalhati II blocks has been summarized in Table 2. Groundwater in shallow aquifer is potable, but in deeper aquifer is not suitable for drinking. Fluoride and sodium in groundwater of deeper aquifer range between 5.9 and 10.7 mg/L and between 185 and 267 mg/L, respectively; in shallow wells, they range within 0.40–0.84 and 23–46 mg/L. Data show that Ca content is high in groundwater of shallow wells than in deeper wells, and its concentration is indirectly proportional to concentration of both Na and F. The fluoride content of groundwater has an inverse relationship with calcium and magnesium content. Overall high pH, high carbonate plus bicarbonate and low calcium plus magnesium in groundwater lead to leaching of fluoride which results in the concentration of fluoride in groundwater (Brindha et al. 2011).

Deeper aquifer in Tertiaries especially in Rampurhat II block is not potable; but the quality in the shallow aquifer is potable. In Nalhati II area, fluoride content in groundwater is low; it ranges between 0.72 and 0.76 mg/L and it is potable. Chemical facies of groundwater has been studied by plotting chemical parameters in Piper diagram (Fig. 2), which shows that groundwater in shallow aquifer is Ca–HCO_3 type, but in deeper aquifer it is more akin to Na–HCO_3 type. Most of the studies indicate the increase in fluoride composition in groundwater with increase in depth from ground surface (Hudak and Sanmanee 2003; Valenzuela-Va'squez et al. 2006).

Drill-Cut Samples

In Rampurhat area, drill-cut litho-samples up to 64.30 m are yellow sand and are quartzo-feldspathic. Older Alluvium forming the topmost unconfined shallow

Table 2 Quality of groundwater in wells of Rampurhat II and Nalhati II blocks, Birbhum district

Chemical constituents (EC in μS/cm at 25 °C, others in mg/L)

Location	Well type	Type of aquifer	pH	EC	Fe	HCO_3	CO_3	Ca	Mg	Cl	SO_4	F	SiO_2	NO_3	Na	K	TH	PO_4
Margram	Shallow expl. well	Shallow aquifer	8.42	409	0.13	195	12	42	9.72	21	–	0.4	48	0	23	0.9	145	0.18
Rampurhat	Mark I hand pump-fitted tube well	Shallow aquifer	7.76	1210	0.5	348	0	142	45	199	52	0.54	33	25	54	1.6	540	0.01
Tarapur	Shallow expl. well	Shallow aquifer	7.55	397	0.09	226	0	46	12.16	14	–	0.62	36	0	23	1.3	165	0.24
Junudpur	Submersible pump-fitted tube well	Shallow aquifer	7.7	368	3.36	134	0	20	3.64	53	–	0.84	27	2	46	0.8	65	0.01
Margram	Deep expl. well	Deeper aquifer	8.78	1180	0.39	573	36	6	2.43	92	–	6.15	14	–	267	4.9	25	0.65
Tarapur	Deep expl. well	Deeper aquifer	8.8	881	0.67	360	27	6	1.21	71	–	5.9	12	0	203	5	20	4.46
Nowapara	Expl. well	Deeper aquifer	7.19	447	0.06	226	0	20	17	18	–	0.76	32	1	41	1.9	120	–
Lohapur	Expl. well	Deeper aquifer	8.26	586	0.11	366	0	60	24	50	–	0.72	33	–	54	2.9	250	0.13
Junudpur	Deep tube well	Deeper aquifer	8.2	1060	0.12	348	0	30	7.29	209	–	10.7	17	4	185	4.8	105	–

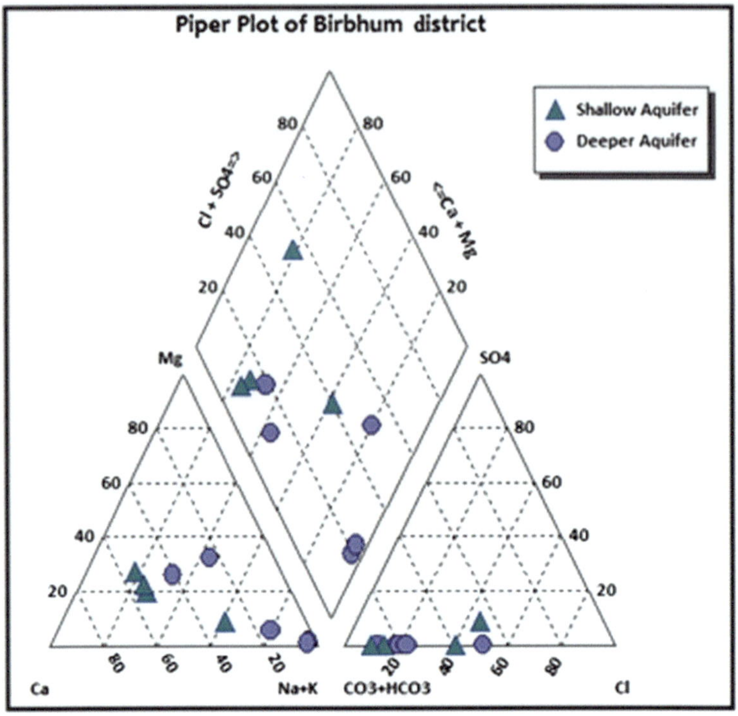

Fig. 2 Chemical facies of ground water of shallow and deeper aquifers

aquifer, analytical data of sample within 51.60–64.30 m contain less than 100 ppm fluoride. Finer clastics are prevalent in deeper aquifer of Tertiaries; sample between 64.30 and 82.20 m contains 315 ppm of fluoride ions (Fig. 3) and below 82.20 m, litho-units contain fluoride ranging from less than 100–164 ppm.

In Nalhati area, Older Alluvium occurs within 88.0 m and is represented by yellow to white, fine to coarse grained material; at Lohapur, sample between 43.90 and 54.10 m contains 120 ppm of fluoride (Fig. 4). Below Older Alluvium, Tertiaries consist mostly of greyish clayey rock fragments and silts; samples at 94.90–98.30, 98.30–101.70 and 162.40–183.00 m litho-samples contain fluorides ranging from 168 to 284 ppm.

4 Barabazar Block, Purulia District

Detailed hydrogeological investigations have been carried out in perennially drought prone Barabazar block (Fig. 5) of Purulia district. 'South shear zone' trending almost NW-SE lies in the southern part of study area. The area belongs to the Precambrian Chhotanagpur Gneissic Complex (CGC) comprising mainly of

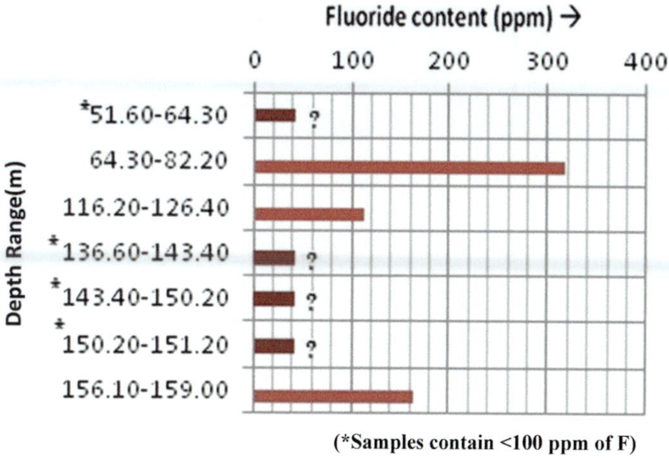

Fig. 3 Fluoride concentration in drill-cut soil samples at Margram, Rampurhat II block

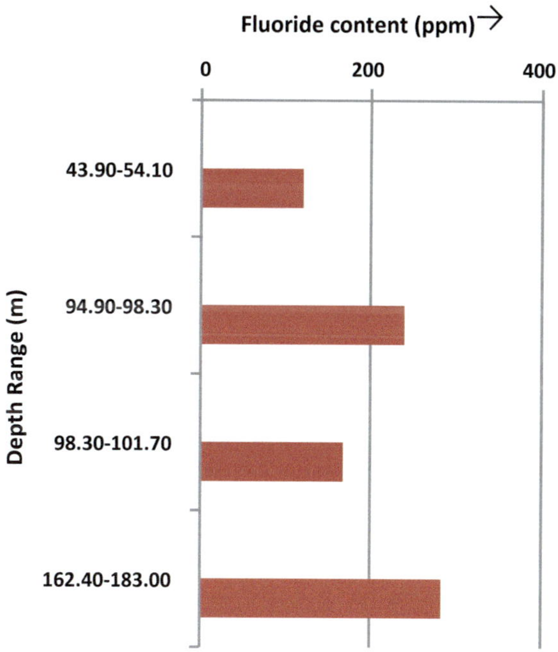

Fig. 4 Fluoride concentration in drill-cut soil samples at Lohapur, Nalhati II block

metamorphics: (i) granite gneisses, with intrusive bodies of granites, in the northern part and (ii) phyllites, slates and mica schists, with enclosures of metamorphosed basic rocks in the southern part of Barabazar block; alluvium is restricted along narrow courses of streams. Mainly three sets of joint, viz. NNE–SSW, NW–SE and

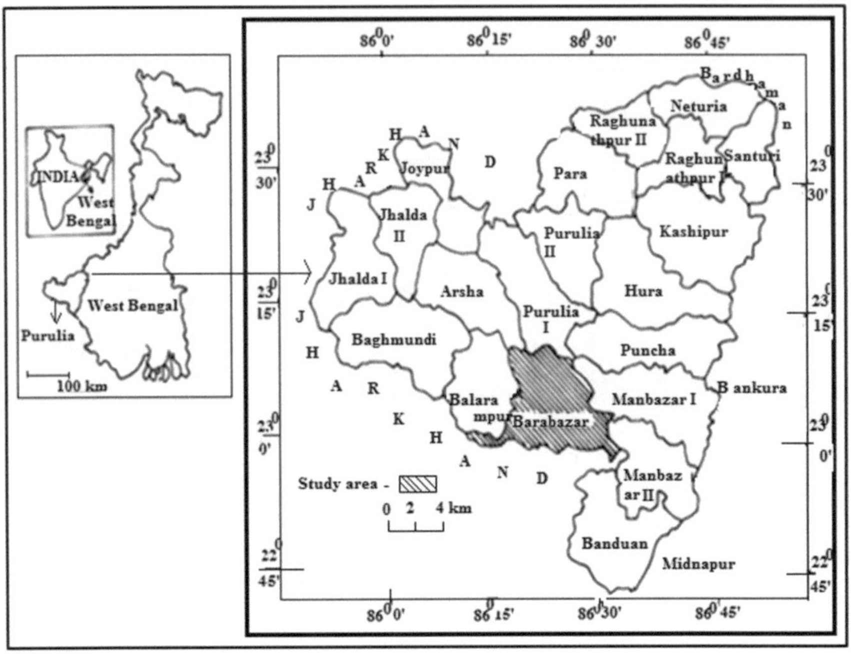

Fig. 5 Barabazar block, Purulia district, West Bengal, India

NE–SW are predominant in the northern granitic part. In the southern part, phyllite and mica schist dominate and NW–SE trending lineaments are relatively persistent in nature.

4.1 Hydrogeology

Groundwater occurs in top weathered mantle ranging up to about 15 m and in the interconnected fractures/joints. The depth to water level in open wells ranges within 5.01–9.82 m below ground level in pre-monsoon and is given in Table 3. The saturated fractures in hard rocks are generally encountered down to 100 m depth (Ghosh and Talukdar 1996); in consolidated formations, groundwater occurs under semi-confined to confined condition (Guha 2009). Deeper potential aquifers in hard rocks are controlled by lineaments (Brahma 2001) of structural elements.

4.2 Analytical Data

The concentration of chemical parameters of groundwater of observation wells, viz. open wells and hand pump-fitted bore wells in villages of eight blocks of Jhalda I,

Table 3 Depth to water level in Barabazar block, Purulia district

Village	Location (coordinates in degree)	Well depth (m)	Depth to water level (m below ground level)	
			(Pre-monsoon)	(Post-monsoon)
Takriya	(23.1693; 86.2688)	15.00	5.98	2.50
Puriara	(23.1562; 86.3564)	9.82	6.98	2.77
Manpur	(23.09194; 86.36556)	7.70	5.01	2.11
Banjora	(23.093889; 86.458338)	9.35	9.19	4.56
Bhabanipur	(23.05139; 86.29444)	8.10	7.85	3.85
Barabazar	(23.01917; 86.347219)	8.40	6.90	4.31
Hizla	(23.03694; 86.41972)	9.00	5.80	2.00
Sindri	(23.03889; 86.48499)	14.00	7.26	2.90
Kukurchanti	(23.58333; 86.29583)	11.22	9.82	5.98

Jhalda II, Joypur, Arsha, Baghmundi, Barabazar, Purulia I and Purulia II falling in western part of Purulia district, has been tabulated in Table 4.

The quality parameters of the groundwater samples of open and bore wells of above mentioned eight blocks have been plotted in Piper diagram (Fig. 6); it has been observed: (i) facies characteristics of groundwater in open wells remain Ca-HCO_3–Cl type to Ca–Na–HCO_3–Cl type in both seasons; while the same in bore wells exhibits Ca–HCO_3 to Ca–HCO_3-Cl type facies characteristics in post-monsoon. The quality of groundwater in selected villages of Barabazar block, viz. Beldih, Puriara, Bhabanipur, Roladih, Hizla, Sindri and Murgakocha, has been marked bold in Table 4 in Barabazar block, and groundwater in open wells and bore wells is characterized generally by Ca–HCO_3–Cl type facies both in pre-monsoon and post-monsoon (Fig. 7).

The concentration of chemical parameters in Barabazar block and western part of Purulia district is mostly within BIS permissible limits, excepting occasional cases, when one or two parameters cross the limit. EC of groundwater has often been found high and this may be due to stagnant condition of the concerned wells. The concentrations of fluoride and iron, which are geogenic in origin, in ground water have been found exceptionally high at certain locations. This has resulted contamination of the vital natural resource in the drought prone part of the district and causes serious threat to lives of human beings and cattle of the area.

Analytical data show that fluoride contents in ground water in selected villages of Barabazar block of Purulia district range up to 0.82 mg/L in open well and 1.62 mg/L in hand pump-fitted bore well in pre-monsoon and the same ranges up to 2.6 mg/L in both abstraction structures during post-monsoon; maximum Fe content in groundwater in open well and bore well goes up to 0.39 and 7.03 mg/L, respectively, in pre-monsoon and the same ranges up to 0.07 and 1.97 mg/L in post-monsoon. No relationship could be established between the fluoride content of groundwater with the concentrations of bicarbonate, calcium and sodium.

In seven blocks covering western part of Purulia district excluding Barabazar block, fluoride concentration in groundwater of open well ranges up to 1.4 mg/L in

Table 4 Details of chemical analysis of groundwater samples collected in pre- and post-monsoon period in Purulia district

Chemical constituents (EC in μS/cm at 25 °C, other parameters in mg/L)

S No.	Location (coordinates in degrees)	Type	pH	EC	Fe	HCO_3	CO_3	Ca	Mg	Cl	SO_4	F	SiO_2	NO_3	Na	K	TH	PO_4
1	**Beldih (23.05087; 86.29683)**	**Pre-monsoon_Open Well**	**8.54**	**393**	**0.39**	**85**	**6**	**40**	**16**	**60**	**14**	**0.3**	**0**	**33**	**20**	**1.4**	**165**	**0**
2	Ghagra (23.50111; 86.13417)	Pre-monsoon_Open Well	8.2	806	0.08	244	–	–	29	74	42	0.6	28	23	38	1.7	280	0
3	Chakra (23.3806; 86.3113)	Pre-monsoon_Open Well	8.05	498	0.15	128	–	40	13	64	20	0.55	35	5	26	1.5	155	0
4	**Puriara Panarda (23.10694; 86.28417)**	**Pre-monsoon_Open Well**	**7.69**	**1410**	**0.09**	**366**	–	**128**	**23**	**213**	**63**	**0.51**	**50**	**40**	**128**	**24**	**415**	**0.19**
5	**Bhabanipur (23.05139; 86.29444)**	**Pre-monsoon_Open Well**	**8.1**	**890**	**0.09**	**201**	–	**90**	**16**	**96**	**40**	**0.75**	**35**	**30**	**32**	**6.7**	**290**	**0.33**
6	**Hizla (23.03694; 86.41972)**	**Pre-monsoon_Open Well**	**7.86**	**1710**	**0.09**	**464**	–	**122**	**29**	**213**	**57**	**0.85**	**42**	**39**	**96**	**99**	**425**	**0.34**
7	**Sindri (23.03889; 86.48499)**	**Pre-monsoon_Open Well**	**8.24**	**261**	**0.37**	**91**	–	**18**	**3.64**	**18**	**1**	**1.42**	**59**	**4**	**19**	**1.6**	**60**	**0**
8	Chipida (23.2337; 86.3674)	Pre-monsoon_Open Well	7.69	2000	0.12	512	–	74	81	319	88	0.64	59	38	207	9.6	520	0.002
9	Thumba Jhalda (23.33444; 86.23472)	Pre-monsoon_Open Well	7.95	1410	0.008	433	–	130	24	121	40	0.31	42	38	80	5	425	0

(continued)

Table 4 (continued)

Chemical constituents (EC in μS/cm at 25 °C, other parameters in mg/L)																	
S No.	Location (coordinates in degrees)	Type	pH	EC	Fe	HCO_3	CO_3	Ca	Mg	Cl	SO_4	F	SiO_2	NO_3	Na	K	TH	PO_4
10	Lohadungri (23.23472; 86.16778)	Pre-monsoon_Open Well	8.2	768	0.08	220	–	68	30	74	32	0.6	34	39	13	2.6	295	0
11	Birgram (23.17194; 85.99745)	Pre-monsoon_Open Well	7.7	3870	0.03	1098	–	348	122	574	107	0.32	42	41	198	2.5	1370	0.01
12	Suisa (23.19527; 85.91861)	Pre-monsoon_Open Well	8 2	448	0.1	140	–	52	9.72	64	7	0.48	41	18	21	1.4	170	0
13	Chakkeyari (23.27444; 85.84499)	Pre-monsoon_Open Well	8 2	236	0.02	73	–	18	7.29	35	1	0.34	11	4	17	2.2	75	0
14	Kolabund (23.4088; 86.3629)	Pre-monsoon_Open Well	8 17	163	0.12	91	–	12	9.72	21	2	1.4	34	1	14	0.9	70	0
15	Hutmura (23.3486; 86.48181)	Pre-monsoon_Open Well	7 78	434	0.16	140	–	34	8.51	60	15	0.34	42	16	46	2.4	120	0.09
16	Talmu (23.54833; 86.1)	Pre-monsoon_Open Well	7.68	2120	0.1	823	–	140	90	181	81	0.52	16	39	101	71	720	0.22
17	Talmu (23.55; 86.10417)	Pre-monsoon_Bore Well	8.34	905	2.75	189	–	90	40	145	76	3.7	39	50	44	2.6	390	0
18	**Beldih (23.05078; 86.29697)**	**Pre-monsoon_Bore Well**	**8.55**	**518**	**0.88**	**146**	–	**78**	**13**	**71**	**16**	**1.62**	**20**	**25**	**8**	**1.8**	**250**	**0**
19	**Beldih (23.05082; 86.29699)**	**Pre-monsoon_Bore Well**	**8.3**	**1080**	**7.03**	**122**	–	**88**	**34**	**266**	**37**	**0.63**	**31**	**38**	**87**	**6.2**	**360**	**0**
20	**Rupapatia (23.082126; 86.37981)**	**Post-monsoon_Open Well**	**7.64**	**1450**	–	**250**	–	**92**	**17**	**191**	**56**	**0.76**	**56**	**80**	**120**	**7.7**	**300**	**0.47**

(continued)

Table 4 (continued)

S No.	Location	Type	pH	EC	Fe	HCO_3	CO_3	Ca	Mg	Cl	SO_4	F	SiO_2	NO_3	Na	K	TH	PO_4
Chemical constituents (EC in µS/cm at 25^0 C, others in mg/L)																		
21	**Roladih (23.060751; 86.37561)**	**Post-monsoon_Open Well**	**7.34**	**788**	–	**171**	–	**70**	**15**	**152**	**43**	**1.55**	**48**	**8**	**72**	**2.5**	**235**	**0.15**
22	**Sargo (23.04194; 86.39739)**	**Post-monsoon_Open Well**	**7.15**	**1680**	–	**268**	–	**210**	**49**	**421**	**79**	**0.93**	**48**	**42**	**84**	**14.5**	**725**	**0.17**
23	**Pathariaghata (23.102337; 86.47673)**	**Post-monsoon_Open Well**	**7.51**	**145**	–	**67**	–	**10**	**6.08**	**14**	**1**	**0.43**	**56**	**3**	**13**	**0.2**	**50**	**0.11**
24	**Fatehpur (23.04313; 86.26950)**	**Post-monsoon_Open Well**	**6.7**	**347**	**0.66**	**73**	–	**32**	**11**	**21**	**11**	**0.26**	**45**	**80**	**17**	**1.5**	**125**	**0.23**
25	**Beldih (23.05087; 86.29683)**	**Post-monsoon_Open Well**	**6.95**	**240**	**0.07**	**61**	–	**28**	**4.86**	**35**	**7**	**1.05**	**19**	**12**	**10**	**0.6**	**90**	**0.18**
26	Gunja (23.42387; 86.22280)	Post-monsoon_Open Well	7.57	612	–	195	–	64	21	82	25	2.4	3.8	42	36	1.7	245	0.42
27	Ranibandh (22.86528; 86.78297)	Post-monsoon_Open Well	7.77	175	0.008	61	–	14	3.64	18	2	0.9	38	15	14	0.5	50	0.06
28	Dindiha (23.38199; 86.26326)	Post-monsoon_Open Well	7.26	290	0.05	104	–	26	8.51	35	22	0.26	4	5	25	8.7	100	0.03
29	Talmu (23.54833; 86.1)	Post-monsoon_Open Well	7.37	1220	0.02	598	–	166	64	85	136	0.86	28	1	63	7.3	680	0.06
30	Kalyanpur(Kasangi) (23.51044; 86.05884)	Post-monsoon_Open Well	7.5	227	–	73	–	14	6.08	21	13	4.5	36	30	29	1.2	60	0.15
31	Jaipur (23.16861; 86.16111)	Post-monsoon_Open Well	7.27	868	0.004	293	–	98	15	117	30	0.73	26	4	59	1.6	564	0.09
32	Murusaram (23.43897; 86.86.04788)	Post-monsoon_Open Well	7.27	618	–	153	–	64	17	78	41	0.6	33	28	19	5.2	230	0.11

(continued)

Table 4 (continued)

Chemical constituents (EC in µS/cm at 25⁰ C, others in mg/L)																	
S No.	Location	Type	pH	EC	Fe	HCO_3	CO_3	Ca	Mg	Cl	SO_4	F	SiO_2	NO_3	Na	K	TH	PO_4
33	Sarjemahatu (23.46788; 86.08423)	Post-monsoon_Open Well	7.6	562	–	244	–	88	16	53	45	3.9	39	2	7	1.1	285	0.15
34	Mukundpur (23.39776; 86.14534)	Post-monsoon_Open Well	7.53	654	–	262	–	70	18	43	34	0.74	11	4	35	12	250	0.6
35	Kotshila (23.40694; 86.06667)	Post-monsoon_Open Well	7.38	616	–	110	–	54	12	39	49	0.5	19	4	62	1.4	185	0.1
36	Kulajanga (23.39291; 85.95238)	Post-monsoon_Open Well	7.52	406	–	128	–	38	16	32	32	0.4	28	28	18	1.9	160	0.15
37	Hesalbera (23.42287; 85.95659)	Post-monsoon_Open Well	7.28	910	–	256	–	120	11	96	50	0.64	28	39	34	2.9	345	0.23
38	Murgakocha (23.55478; 86.04623)	Post-monsoon_Open Well	7.01	852	–	256	–	60	17	89	48	2.6	16	28	67	2.7	220	0.53
39	Rupapatia (23.08209; 86.37939)	Post-monsoon_Bore Well	7.29	779	1.97	195	–	64	16	113	38	0.7	42	26	65	3.1	225	0.05
40	Roladih (23.060875; 86.37601)	Post-monsoon_Bore Well	7.44	686	0.83	195	–	76	11	99	29	2.3	45	16	44	1.9	235	0.05
Sl No.	Location	Chemical constituents (EC in µS/cm at 25 °C, others in mg/l)																
		Type	pH	EC	Fe	HCO_3	CO_3	Ca	Mg	Cl	SO_4	F	SiO_2	NO_3	Na	K	TH	PO_4
41	**Sargo (23.29722; 86.2125)**	**Post-monsoon_Bore Well**	**7.63**	**270**	**0.16**	**110**	–	**26**	**3.64**	**18**	**2**	**0.5**	**48**	**4**	**17**	**0.7**	**80**	**0.09**
42	**Joramohul (23.; 86.)**	**Post-monsoon_Bore Well**	**7.31**	**670**	**1.04**	**153**	–	**44**	**18**	**85**	**32**	**1.02**	**38**	**27**	**46**	**2.6**	**185**	**0.05**
43	**Pathariaghata (23.12223; 86.47833)**	**Post-monsoon_Bore Well**	**7.61**	**233**	–	**22**	–	**14**	**7.29**	**18**	**3**	**0.45**	**56**	**13**	**20**	**1.3**	**65**	**0.06**
44	**Fatehpur (23.04487; 86.25789)**	**Post-monsoon_Bore Well**	**7.02**	**1490**	**0.44**	**275**	–	**184**	**64**	**362**	**53**	**0.24**	**30**	**45**	**36**	**5.5**	**725**	**0.05**
45	**Beldih (23.05078; 86.29697)**	**Post-monsoon_Bore Well**	**7.18**	**444**	**0.12**	**159**	–	**68**	**9.72**	**67**	**11**	**1.85**	**33**	**15**	**15**	**1.5**	**210**	**0.05**

(continued)

Table 4 (continued)

Sl No.	Location	Chemical constituents (EC in μS/cm at 25 °C, others in mg/l)															
		Type	pH	EC	Fe	HCO_3	CO_3	Ca	Mg	Cl	SO_4	F	SiO_2	NO_3	Na	K	TH	PO_4
46	Gunja (23.42455; 86.22434)	Post-monsoon_Bore Well	7.25	595	0.07	268	–	54	16	53	22	1.06	33	26	60	1.1	200	0.06
47	Alkusha (23.46855; 86.13979)	Post-monsoon_Bore Well	7.18	363	0.95	128	–	34	11	50	37	0.41	45	1	27	1	130	0.09
48	Talmu (23.55; 86.10417)	Post-monsoon_Bore Well	7.67	686	0.57	120	-	96	6.08	14	108	2.6	28	1	1	2	265	0.04
49	Kalyanpur (Kasangi) (23.51165; 86.0657)	Post-monsoon_Bore Well	7.33	389	0.01	226	–	50	8.51	14	9	5.8	28	1	25	1.6	160	0.04
50	Jaipur (23.16888; 86.16222)	Post-monsoon_Bore Well	7.25	645	0.15	146	–	82	11	92	38	0.76	28	31	29	2.6	250	0.06
51	Murusaram (23.43279; 86.04644)	Post-monsoon_Bore Well	7.32	796	–	207	–	98	18	99	44	0.5	28	34	22	3.2	320	0.06
52	Sarjemahatu (23.46834; 86.08444)	Post-monsoon_Bore Well	7.51	322	0.1	146	–	44	7.29	11	13	5.8	36	1	11	0.9	125	0.06
53	Mukundpur (23.39605; 86.14611)	Post-monsoon_Bore Well	7.41	449	0.004	183	–	48	11	32	31	2.1	21	4	27	2.2	165	0.18
54	Kotshila (23.40446; 86.06638)	Post-monsoon_Bore Well	7.6	339	0.29	73	–	30	2.43	117	11	0.49	36	8	19	1	85	0.1
55	Kulajanga (23.39344; 85.95678)	Post-monsoon_Bore Well	7.2	399	1.2	152	–	36	19	131	26	0.66	23	11	20	8.5	170	0.07
56	Hesalbera (23.42567; 85.95688)	Post-monsoon_Bore Well	6.96	1030	0.1	238	–	150	15	131	64	0.43	28	38	33	5.7	435	0.1
57	**Murgakocha (23.55435; 86.046642)**	**Post-monsoon_Bore Well**	**6.75**	**2000**	**1.58**	**342**	–	**206**	**34**	**404**	**82**	**2.6**	**21**	**37**	**140**	**5.4**	**655**	**0.09**

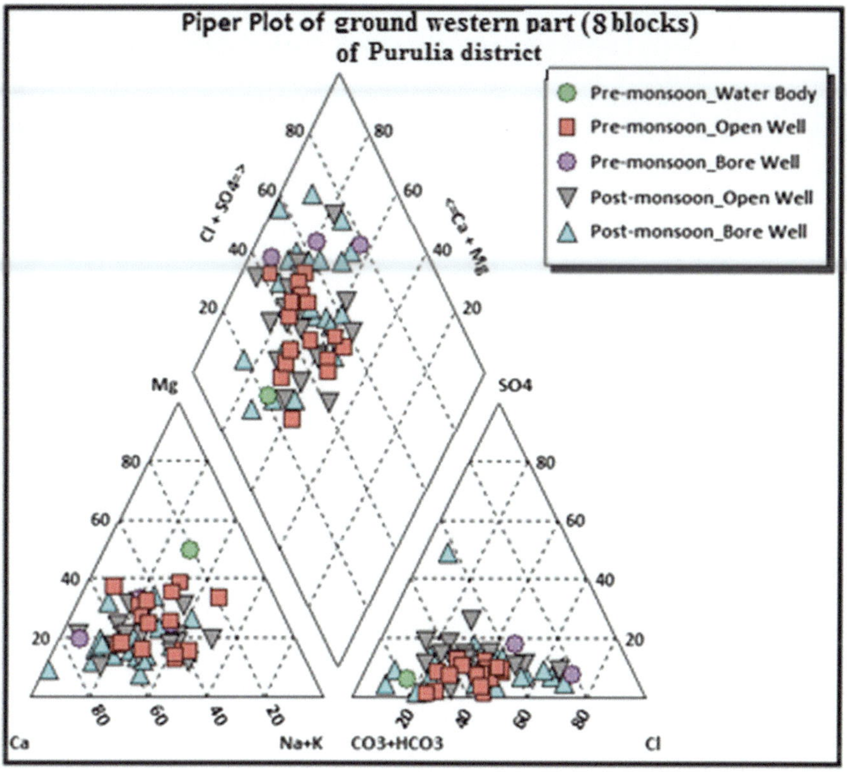

Fig. 6 Facies plot of chemical parameters of groundwater in eight blocks of Purulia district

pre-monsoon and 4.5 mg/L in post-monsoon, whereas in bore well it ranges up to 3.7 mg/L in pre-monsoon and 5.8 mg/L in post-monsoon; NE–SW trending 'north shear zone' of Purulia district passes through the northern part of this area.

A perusal of data (**Annexure I**) shows that in Barabazar block, high fluoride and iron concentration in groundwater have been encountered, in general, at villages, e.g. Beldih, Murgakocha, Roladih and Rupapatia, which are located at the proximity of 'south shear zone' (Fig. 10). Also, in this area the concentration of fluoride increases with the depth of the abstraction structures (Figs. 8 and 9).

5 Discussion and Conclusion

5.1 Rampurhat–Nalhati Area, Birbhum District

In Rampurhat area, presence of two-aquifer system has been established in Rampurhat area: shallow aquifer, conforming to the thickness of Older Alluvium of the Quaternaries, occurs under unconfined condition and the deeper aquifer in the

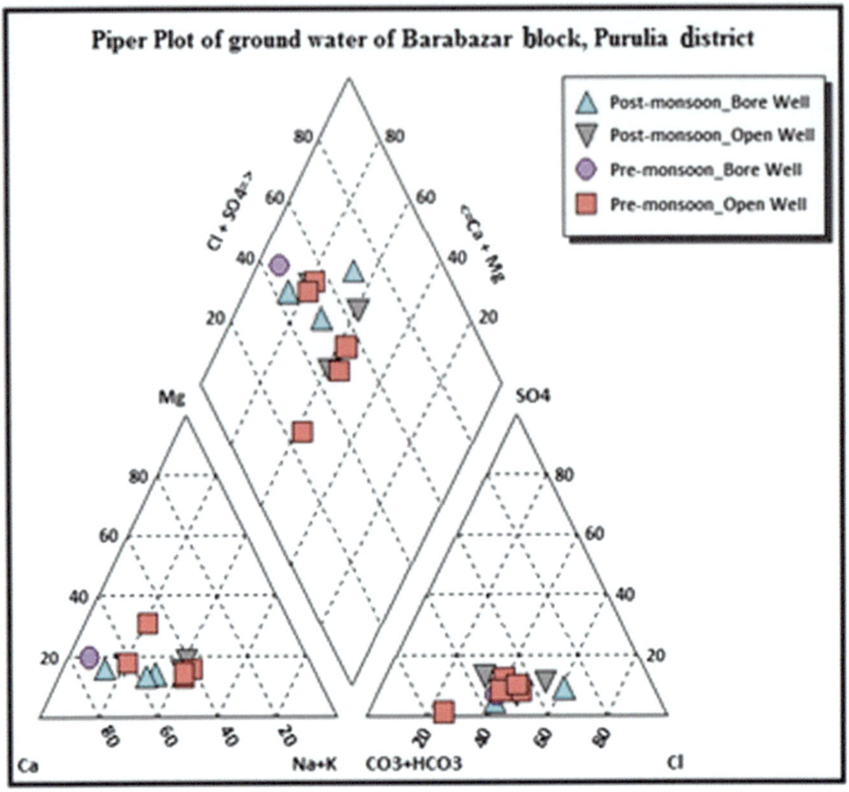

Fig. 7 Facies plot of groundwater in Barabazar block

Tertiaries, under semi-confined to confined condition. Hydrogeologically, these two aquifers are quite different from each other and are separated by a thick clay. Depth to water level (DTW) in deep exploratory well (DEW), both at Margram and Tarapur, is shallower than that in the shallow exploratory well (SEW). Potentiality of Older Allvium, mainly tapped in SEW, is more than that of the Tertiaries, tapped in DEW. Groundwater in deeper aquifer is characterized by high fluoride and high sodium content, whereas these constituents are quite low in groundwater of shallow aquifer. In Nalhati II block, however, there is only one aquifer continuously from the Tertiaries at the bottom to the overlying Quaternaries; fluoride and sodium content in groundwater is low and groundwater in this part is potable.

5.2 *Barabazar Block, Purulia District*

In the northern, granite gneisses predominate, whereas in the southern part slates, phyllites and mica schists are prevalent. The effect of deformation activities on

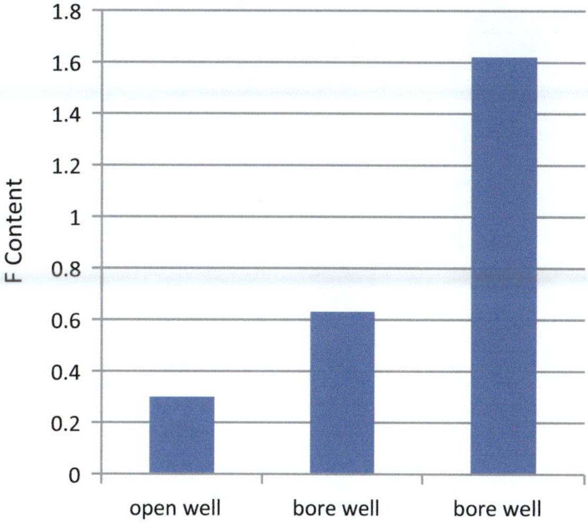

Fig. 8 Increase in fluoride content in groundwater in pre-monsoon at Beldih, Barabazar block

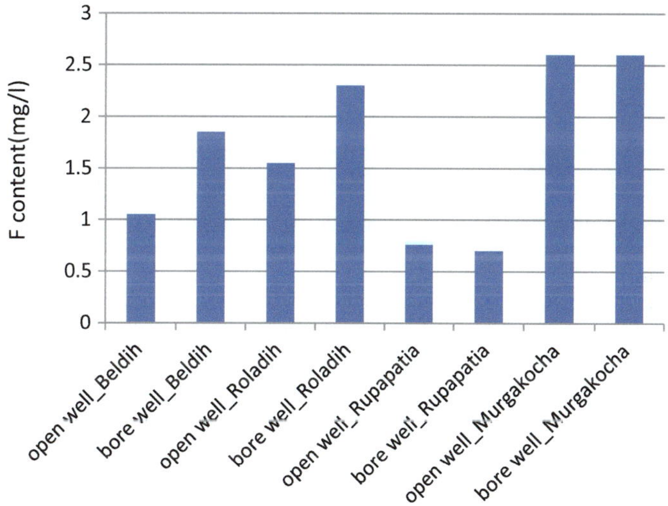

Fig. 9 Increase in fluoride content in groundwater in post-monsoon in different villages, Barabazar block

granite gneisses of the north is more pronounced than that in slates and phyllites. As a result, former type is more permeable than the latter. Hence, groundwater prospect is relatively better in the northern part.

Many granitic rocks have elevated fluoride concentration (World Health Organization 1970). The main source of fluoride in groundwater is considered to be

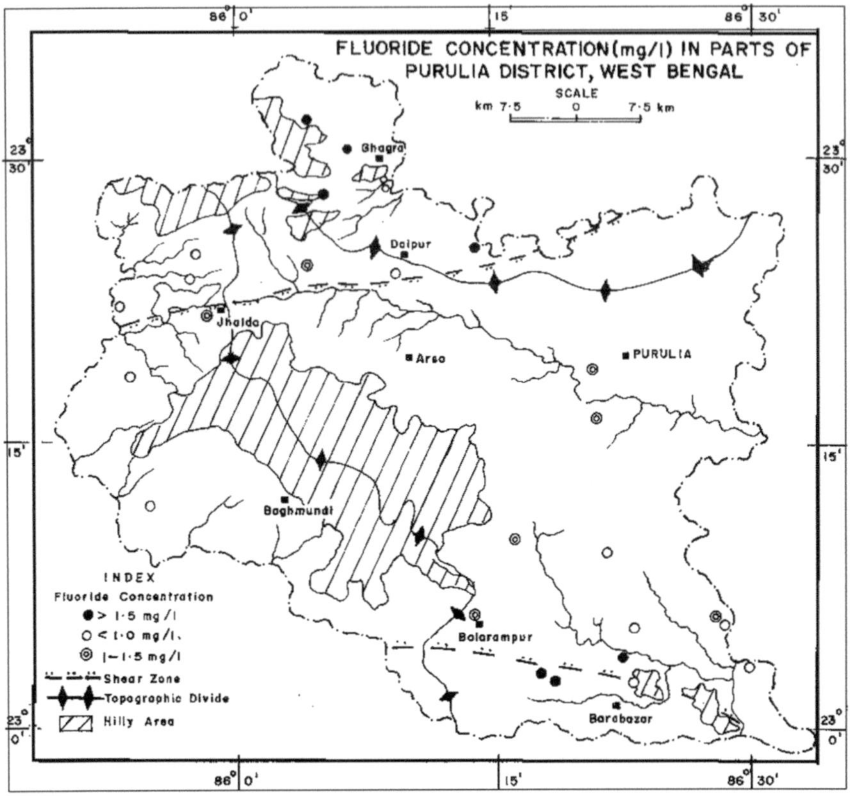

Fig. 10 Fluoride concentration in groundwater in parts of Purulia district

fluoride-bearing minerals such as fluorspar (CaF_2), fluorapatite [$Ca_5(PO_4)_3F$], cryolite, and hydroxylapatite in rocks (Farooqi et al. 2007). Overall, the natural concentration of fluoride in groundwater depends on the geological, chemical and physical characteristics of the aquifer, the porosity and acidity of the soil and rocks, the surrounding temperature, the action of other chemical elements, depth of the aquifer and intensity of weathering (Feenstra et al. 2007).

The chemical parameters of groundwater are mostly within BIS limits, but at places in the southern part of Barabazar block especially near the 'south shear zone', its quality is bad due to high fluoride and iron concentration. Both iron and fluoride contents in groundwater increase with depth and towards the proximity to shear zone. Saxena and Ahmed (2001) put forth that alkaline conditions with pH ranging between 7.6 and 8.6 are favourable for dissolution of fluorite mineral from the host rocks. Very deep aquifers are safer than shallow aquifers. (Shashank et al. 2012). The fluoride content of groundwater has no relation with the concentrations of bicarbonate, calcium and sodium.

In both these terrains, though diverse in nature, fluoride concentration in groundwater increases with depth.

Acknowledgements The author expresses deep sense of gratitude to Shri K.B. Biswas, Chairman, Central Ground Water Board for rendering permission to present this paper in one day workshop titled 'Bhujal Manthan' at Kurukshetra, Haryana. The author would like to thank Dr. D. Saha, Member (SAM) for critically scrutinizing the paper. Analysis of drill-cut samples by Geological Survey of India is deeply acknowledged. The author is indebted to Shri Atlanto Chowdhury, Asst. Chemist for his assistance and cooperation for facies analysis of groundwater.

References

American Public Health Association (2012) Standard methods for the examination of water and waste water, 22nd edn

Ayoob SK, Gupta AK (2006) Fluoride in drinking water: a review on the status and stress effects. Crit Rev Environ Sci Technol 36:433–487

Brahma S (2001) Ground water exploration aided by study of lineaments for tapping the deeper potential aquifers in hard rock terrain in parts of Subarnapur and Sambalpur districts, Orissa. In: Bhu H, Agrawal V (eds) Ground water resources, evaluation & management, Proc. Sem., XII IGC, vol. I, pp 106–112

Brindha K, Elango L (2011) Fluoride in groundwater: causes, implications and mitigation measures. In: Monroy, S.D. (ed) Fluoride properties, applications and environmental management, pp 111–136

Central Ground Water Board, Lucknow (1996) Guidelines for ground water pollution studies, p 21

Dutta S, Bandyopadhyay S (2006) Hydrogeology and ground water development of Birbhum District, West Bengal; Central ground water board, ER Technical Report, Series 'D', No. 188, 23 p

Farooqi A et al (2007) Toxic fluoride and arsenic contaminated groundwater in the Lahore and Kasur districts, Punjab, Pakistan and possible contaminant sources. J. Environ Pollut. 145:839–849

Feenstra L, Vasak L, Griffioen J (2007) Fluoride in groundwater: overview and evaluation of removal methods. International Groundwater Resources Assessment Centre Report nr. SP 2007-1, 1–21

Ghosh AK, Talukdar T (1996) Hydrogeology of the hard rock aquifers in Purulia District, West Bengal—scope for sustainable management. In: Aquifer characteristics and groundwater management in eastern India, Proc. Sem., Department of Geological Sciences, Jadavpur University, Kolkata, March, pp 61–67

Guha RK (2009) Ground water exploration in West Bengal, Central Ground Water Board, ER Technical Report, Series 'D', No. 219

Hudak PF, Sanmanee S (2003) Spatial patterns of nitrate, chloride, sulfate, and fluoride concentrations in the woodbine aquifer of north-central Texas. Environ Monit Assess 82:311–320

PHE Directorate, Govt. of West Bengal (2008) Detailed project report on piped water supply scheme for Atla & adjoining Moujas, Block—Rampurhat—I, District—Birbhum

Pillai KS, Stanley VA (2002) Implications of fluoride—an endless uncertainty. J Environ Biol 23:81–97

World Health Organisation (1970) Fluoride and human health. WHO monograph series no. 59

Brindha K et al (2011) Fluoride contamination in groundwater in parts of Nalgonda district, Andhra Pradesh, India. Environ Monit Assess 172:481–49

World Health Organization (2004) Guidelines for drinking-water quality, 3rd edn. World health organization report, Geneva, 1, Recommendations

Shashank S. et al (2012) Evaluation of fluoride contamination in groundwater sources in hard rock terrainin Garhwa district, Jharkhand, India. Int J Environ Sci 3(3):1022

Saxena VK, Ahmed S (2001) Dissolution of fluoride in groundwater: a water–rock interaction study. Environ Geol 40(9):1084–1087

Valenzuela-Va'squez L, Ramı'rez-Herna'ndez J, Reyes-Lo'pez J, Sol-Uribe A, La'zaro-Mancilla O (2006) The origin of fluoride in groundwater supply to Hermosillo city, Sonora, Mexico. Environ Geol 51:17–27

Petro-Geochemical Analysis and Their Correlation for Genesis of Fluoride Contamination in Groundwater of District Sonbhadra, U.P., India

H.K. Pandey, S.K. Srivastava and Prashant Pandey

Abstract In many parts of India, high concentration of fluoride in drinking water has been causing severe health problems in human, such as dental and skeletal fluorosis. The present study has been aimed to assess the role of hydrogeological characteristics on fluoride contamination of groundwater. This includes the monitoring of parameters such as lithological compositions, groundwater level behavior, and different chemical parameters to understand their impact on fluoride concentration in the study area of Sonbhadra district, U.P. Chemical parameters such as pH, F^-, EC, TDS, chloride, total hardness, alkalinity, phosphate, Ca^{2+}, Mg^{2+}, Na^+, and K^+ have been determined. A petrological study of rocks reveals that mainly granite, granite gneiss, and pegmatite types of rocks are found in area. In general, high fluoride content has been observed in shallow aquifer due to erosion of rock. Therefore, the most dominant reason for leaching of fluoride in groundwater is due to the weathering of various fluoride-bearing minerals in the area.

Keywords Fluoride · Groundwater · Health hazard · Sonbhadra
Uttar Pradesh

1 Introduction

Vast stretches of peninsular India are occupied by granitic and gneissic rocks of Precambrian age (older than 540 million years) in which F-bearing minerals commonly occur as accessory phases. Prolonged chemical weathering of these

H.K. Pandey (✉)
Department of Civil Engineering, Motilal Nehru National Institute of Technology, Allahabad 211004, India
e-mail: hkp@mnnit.ac.in; pandey_hk@rediffmail.com

S.K. Srivastava
Central Ground Water Board, Northern Region, Lucknow 226021, India

P. Pandey
Department of Geology, Benaras Hindu University, Varanasi 221005, India

D. Saha et al. (eds.), *Clean and Sustainable Groundwater in India*,
Springer Hydrogeology, https://doi.org/10.1007/978-981-10-4552-3_5

rocks resulted in elevated contents of F^- in groundwater in several parts of India (Patel et al. 2014). WHO has prescribed a range of fluoride from 0.6 to 1.5 mg/L in drinking water as suitable for human consumption. In study conducted by WHO, it has been estimated that more than 200 million people living all over the world are consuming water with fluoride content more than 1 mg/L (WHO 1984). In many parts of India, high concentration of fluoride in drinking water has been causing severe health problems in human, such as dental and skeletal fluorosis. It is the 13th most abundant element in the earth's crust, with abundance as 0.06–0.09% (Tebutt 1983) with an average crustal abundance as 300 mg/kg (Fawell et. al 2006). Fluoride is the most electronegative element in the earth crust and also forms both organic and inorganic compounds. When dissolved in water, fluoride does not show any change in color, taste, or odor; hence, it is impossible to detect it through physical examination. It can only be estimated by the chemical analysis of the water samples.

2 Study Area

The study area lies in Sonbhadra district of Uttar Pradesh, (83° 30′N to 24° 30′E) and falls under Survey of India toposheet 63P/7. It is southernmost district of U.P. (Fig. 1). Its geographical area is 6788 Km^2. The climate of the area is subhumid and is characterized by hot summer, pleasant monsoon, and cold winter season with an average annual rainfall of 1115 mm. However, 90% of the precipitation takes place during the months of June to September. May is the hottest month with the mean daily maximum temperature as 45.5 °C. The average temperature in a whole year ranges from 16.15 to 39.80 °C. The relative humidity ranges from 25 to 81% with a mean of 70%. The district is characterized by hard rocks with undulating and hilly topographic features. Very limited valley fill deposits are observed along the river course. River Son divides the district into two distinct topographic divisions. The thickness of alluvium or weathered zone is more in north of river Son, while it is negligible in south of the river. The district can again be divided into three distinct units; such as (i) table land stretching from Vindhyan scrap to Kaimur ranges, (ii) valleys of Son River, (iii) hills and ravines. The elevation of the area varies from 150 to 400 m amsl. The general slope of the tract is from north to south.

2.1 Geology of the Area

The rock types in Sonbhadra Distt. range from Archean granitoid complex to recent deposits along the Son river and its tributaries. The stratigraphic sequence with the broad lithological composition of different units is given in Table 1. The oldest rocks of the Archean granitoid complex have been found to be exposed in the southeast area of the district which covers the Babhani block and few areas of Muirpur and Dudhi blocks. Bijawar Group of rocks covering parts of Chopan,

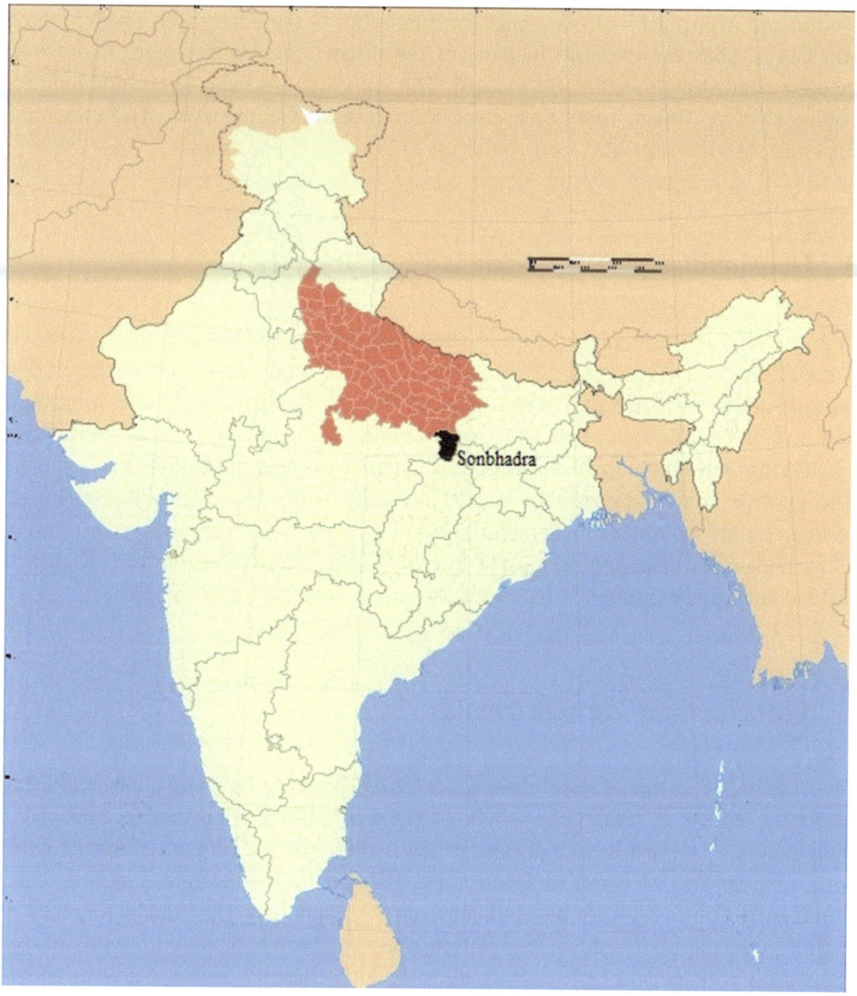

Fig. 1 Study area in U.P. of India

Table 1 Stratigraphic sequences in Sonbhadra district, U.P.

Stratigraphic period	Formation	Lithology
Recent	Alluvium	Clay, silt, and medium-grained sand.
Lower cretaceous–Jurassic	Gondwana	Shale, sandstone, and clay
Meso to Neoproterozok	Vindhyan	Shale, sandstone, limestone, and pyroclastic rocks.
Palaeo-proterozoic	Bijawar	Phyllites, quartzites, and basic and ultra basic rocks.
Archean	Granitoid complex	Granite gneiss, schist, and pegmatite

Dudhi, and Muirpur blocks occupies the area south of the River Son. The Vindhaya and Gondwanas are exposed in parts of the district. The laterites are found over quartize sand stone in 400–500 m amsl. The alluvium has been observed along the course of river Belan, river Son, river Rihand, and Kanhar river. The maximum thickness of alluvium has been found up to 30 m in these areas.

3 Geochemistry of Fluoride

Fluorine (F_2) is a greenish diatomic gas and highly reactive such that it is never encountered in its elemental gaseous state except in some industrial processes. The fluoride occurs in different minerals such as fluorite, fluorapatite, biotite, muscovite and hornblende, Sellaite, Fluorspar, and Cryolite. Probable sources of fluoride are weathering and leaching of fluoride-bearing minerals under alkaline environment. The fluoride content in groundwater is increased with time mainly due to longer residence time of waters in the aquifer zone in contact with fluoride-bearing minerals, high rates of evaporation, and intensive and long-term irrigation (Khandare 2013). The various minerals having fluoride are depicted in Table 2.

4 Fluoride and Human Health

Fluoride is one of the necessary elements for the growth of animal bones and teeth. However, excessive intake of F^{-} over a long period of time has negative effects on human health and results in a disease called fluorosis, which is known for the worst form of bone disorder and body crippling (Khandare 2013). Ingestion of F^{-} predominantly comes from drinking water sources, especially groundwater. Intake of high F^{-} concentrated water may result in acute to chronic skeletal fluorosis if used for a prolonged period of time. Crippling skeletal fluorosis might occur in people who have ingested 10–20 mg of fluoride per day for over 10–20 years. Other health disorders that occur due to consumption of high fluoride in drinking water are muscle fiber degeneration, low hemoglobin levels, deformities in RBCs, excessive

Table 2 Fluoride-bearing mineral (after)

Minerals	Chemical formula	wt% of fluoride (%)
Sellaite	MgF_2	61
Villiumite	NaF	55
Fluorite (Fluorspar)	CaF_2	49
Cryolite	Na_3AlF_6	45
Bastnaesite	(Ce, La) $(CO_3)F$	9
Fluorapatite	$Ca3(PO_3)3F$	3–4

thirst, headache, skin rashes, nervousness, neurological manifestations, depression, gastrointestinal problems, urinary tract malfunctioning, nausea, abdominal pain, tingling sensation in fingers and toes, reduced immunity, repeated abortions or still births, male sterility, etc.

5 Methodology

The fresh rock samples were collected from six nearby areas from where groundwater samples were collected (The rock samples were collected after excavate 1 feet below ground surface where the outcrops are not exposed). Only six representative rock samples were collected (wherever changes in lithological characteristics). The representative six rock samples with their numbering and locations are given below in the Table 3. The rock samples were sent to Department of Geology, BHU for making thin sections of rock and microphotograph (CN & PPL) with the help of petrological microscope.

The proper purging has been done before the water samples were collected from the hand pumps. In case of open dug wells, water was drawn from after proper agitating the water at least 30 cm below the surface of water level column. Samples were collected in the good quality HDPE bottles. The bottles were then sealed in such a manner that collected water samples did not contact with air. Samples of groundwater were collected during the entire period of February 2014 from 22 villages from both the groundwater abstraction structures, viz., hand pumps and open wells (Fig. 2). The open wells depth varies from 5.45 to 20 m below ground level. Groundwater level in the wells was measured using a measuring tape, and fluoride content of groundwater samples was measured using a fluoride field kit in field.

The water samples were analyzed in the Laboratory of Environmental Engineering, MNNIT Allahabad, and Regional Chemical Laboratory of Central Ground Water Board, Ministry of Water Resources, Govt. of India follows the standard methods for the examination of water and waste water by APHA (2000). The water samples were analyzed for pH, TDS, total hardness, alkalinity, chloride, phosphate calcium, magnesium, sodium, and potassium. The SPADNS colorimetric method is used which is based on the reaction between fluoride and a zirconium-dye. Fluoride reacts with the dyelake, dissociating a portion of it into

Table 3 Locations of rock samples

S.No.	Village	Cordinates	Rock sample
1	Rohiniadamar	24° 15′ 53.7″N and 83° 22′ 38.7″E	R-7
2	Charakpathali	24° 19′ 23.8″N and 83° 22′ 02.1″E	R-9
3	Pandubiya	24° 20′ 35.4″N and 83° 21′ 30.8″E	R-14
4	Gobardaha	24° 19′ 02.3″N and 83° 22′ 21.1″E	R-19
5	Mithniya	24° 23′ 58.4″N and 83° 21′ 43.3″E	R-23
6	Barwakhand	24° 26′ 24.6″N and 83° 22′ 17.2″E	R-28

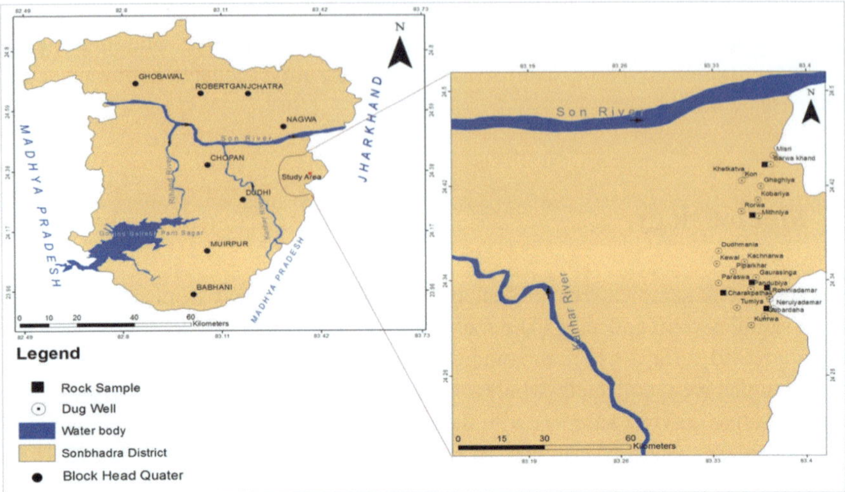

Fig. 2 Location map of groundwater and rock samples, district Sonbhadra (U.P.)

colorless complex anion (ZrF_6^{2-}) and the dye. As the amount of fluoride increases, the color produced becomes progressively lighter.

First the standard curve was prepared; the standard range of fluoride concentration of 0–1.4 mg/L was prepared by diluting appropriate quantities of standard fluoride solution to 50 ml with distilled water, 5 ml each of SPADNS solution. Zirconyl-acid reagent, or 10 ml mixed acid—zirconyl SPADNS reagent was mixed in each standard. The spectrophotometer was set to zero absorbance with the reference solution, and absorbance readings of standards were taken immediately.

Sample pre-treatment: If the sample contains residual chloride, 1 drop of (0.05 ml) $NaAsO_2$ solution per 0.1 $MgCl_2$ is added and thoroughly mixed.

Color development: A 50-ml sample or a portion diluted to 50 ml is used. The temperature of sample is adjusted to that used for standard curve. Then, 5 ml each of SPADNS solution and zirconyl-acid reagent or 10 ml acid-zirconyl SPADNS reagent are added and mixed thoroughly Then, absorbance reading is taken immediately or at any subsequent time, after setting the reference point. If the absorbance falls beyond the range of the standard curve, the process is repeated using a smaller sample volume.

6 Results and Discussion

6.1 Optical Mineralogy of Rock

Optical mineralogy of rock of different thin sections has been carried out particularly for the location where high concentration of fluoride in groundwater has been analyzed and reported. Salient observations regarding the minerals composition of rocks have been studied since fluorite, apatite, mica, and various other minerals take part in rock–water interaction and release **fluoride** into the groundwater, it is imperative to know the presence of minerals in the rock specimen microscopically. Microscopic analysis of the rock samples shows quartz, microcline, fluorapatite ($Ca_{10}(PO_4)_6(F)_2$), Biotite ($K(Mg, Fe)_3$ $AlSi_3O_{10}(F, OH)_2$), hornblende $(Ca, Na)_{2-3}(Mg, Fe, Al)_5(Al, Si)_8O_{22}(OH, F)_2$, Muscovite ($KAl_2$ $(Si_3Al)O_{10}(OH, F)_2$), and minor chlorite, in which apatite and biotite are the fluorine-bearing minerals. Feldspars occasionally sericitized. Apatites with variable grain size are closely associated with the hydrous minerals at the boundary of the coarse-grained feldspar and quartz. As far as immediate association is concerned, apatite grains are closely related to biotite and at places chlorite and myrmekites. The close association of apatite grains with the hydrous phases and myrmekites at the grain boundary of feldspar is probably caused by the late fluid activity from crystallization of the granite body. Photomicrograph of six number of rock samples has been narrated and explained below (Figs. 3–9).

Alteration of the mineral as well as fractures of rock is quite extensive which may also responsible for higher concentration of fluoride. Rock's nature which is

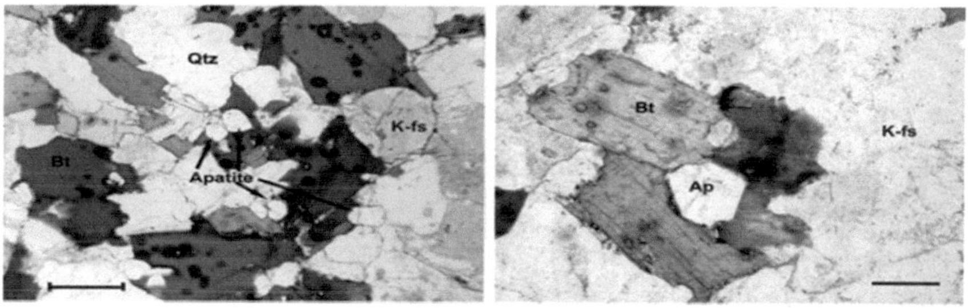

Fig. 3 Minerals in the rock samples of microscopic nature

Fig. 4 Photomicrograph of rock (PPL) near village—Mithniya

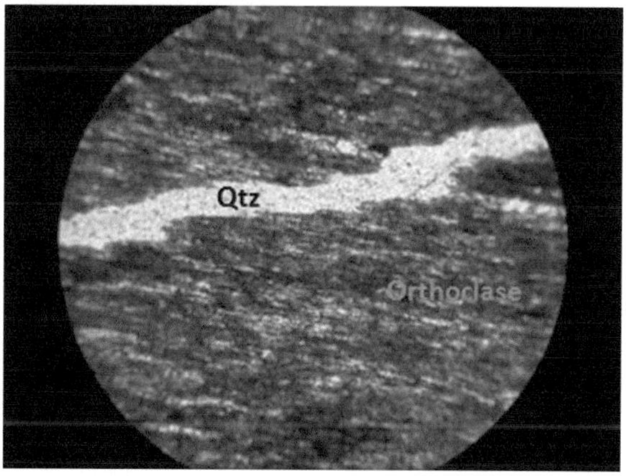

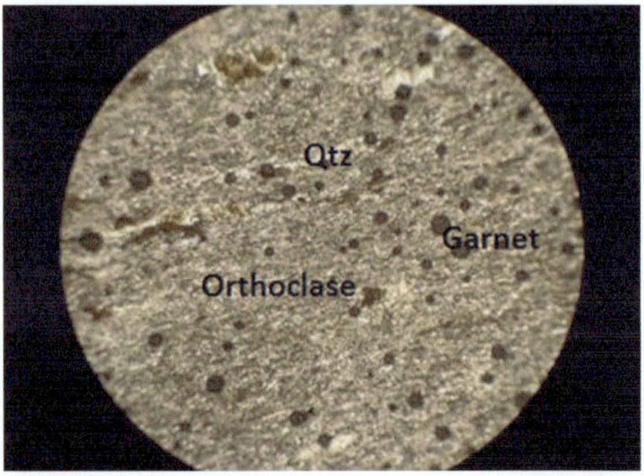

Fig. 5 Photomicrograph of rock (PPL) near village—Pandubiya

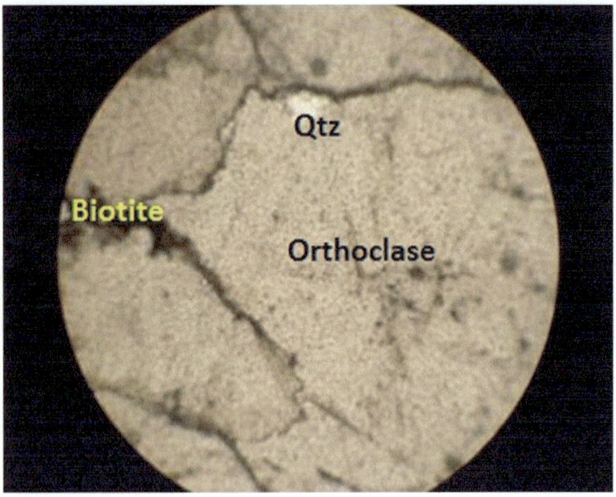

Fig. 6 Photomicrograph of rock (PPL) near village—Gobardaha

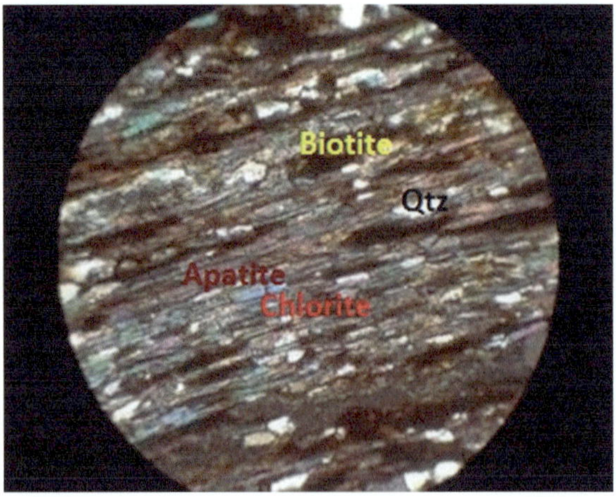

Fig. 7 Photomicrograph of rock (CN) near village—Charakpathali

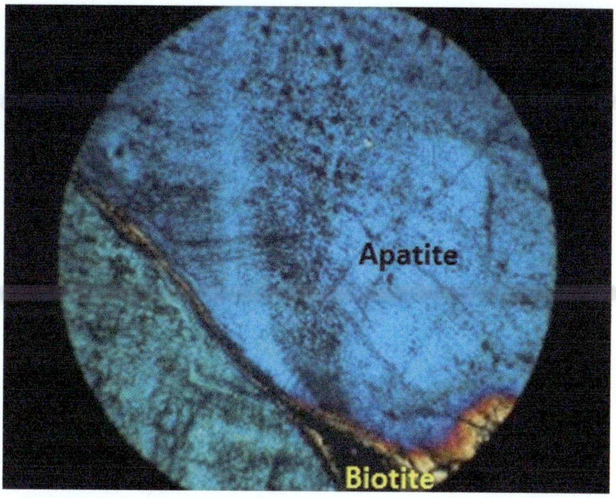

Fig. 8 Photomicrograph of rock (CN) near village

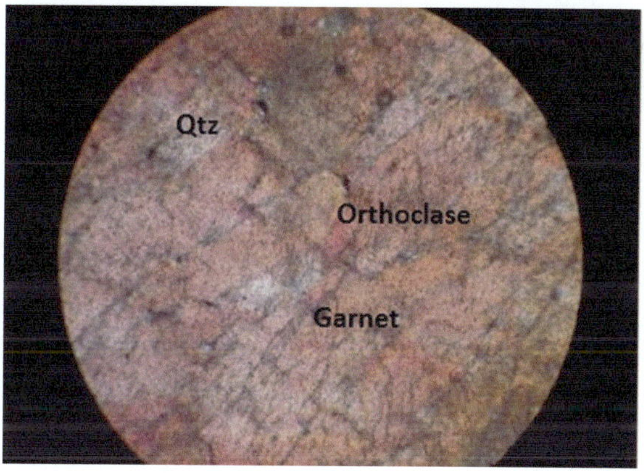

Fig. 9 Photomicrograph of rock (PPL) near village—Barwakha

occurring up to the depth of 198 m reflects that the natures of rock are granitic and granite gneiss. Water level varies from 6.7 to 11.25 mbgl. In a shallow water level, more concentration of fluoride has been observed and as the water level goes down, then there is low fluoride concentration; these are due to erosion of rocks and more contact time for water–rock interaction. These fluoride-rich salts in the soil get leached from the soil along with the percolating rain water during subsequent rains and get mixed with the groundwater.

6.2 *Exploratory Bore well Study*

The data obtained for four number of bore wells drilled in the area by CGWB (CGWB, 2011) have been studied. The drilled depth, fractures zones uncounted,

Table 4 Lithological log of exploratory well at district Sonbhadra, U.P. (CGWB 2011)

S. No.	Location and cordinates	Drilled depth/bedrock (mbgl)	Fractures zones (mbgl)	Water level (mbgl)	Yield (lpm)	Over burden (mbgl)	Geology
1	Dudhi (Tumiya) 24° 19′ 31.5″N and 83° 20′ 27.2″E	122.4	35.00–38.00	4.56	4125	9.95	Granite and granitic gneiss
2	Piparkhar 24° 21′ 07.8″N and 83° 20′ 16.4″E	190.35	50.00–50.40	20.3	6100	12.4	Granite and granitic gneiss
3	Kachnarwa 24° 21′ 29.8″N and 83° 21′ 06.2″E	386.55	19.75	9.95	2296	12.25	Granite, pegmatite, and granitic gneiss
4	Kunrwa 24° 19′ 04.5″N and 83° 22′ 32.6″E	198.2	13.00–15.00 and 154.00	7.6	1125	13	Sticky clay yellow in color, weathered gray color granite with quartz and feldspar and granitoid

water level, yield, overburden thickness, and geology are provided in Table 4. The nature of rocks which are occurring down to the depth of 198 bgl are granite and granite gneiss. The well yield range from 40 to 896 L/min (1 pm). The fractures encountered in depth range from 19.75 mbgl as deep as 154.0 mbgl. Generally only one set of fracture is found in the area. The yield of the fracture varies from 100 up to 290 lpm shows poor-to-moderate potential.

6.3 *Groundwater Regime*

The groundwater level is monitored through a network of wells from dug well or open well during the month of May to June (Pre-monsoon) as well as November (Post-monsoon). During pre-monsoon, water level varies from contrast to earlier statement. The western part of the study area shows deeper levels (Fig. 10). The groundwater level during post-monsoon period varies from 2.00 to 14.00 mbgl in the area (Fig. 11).

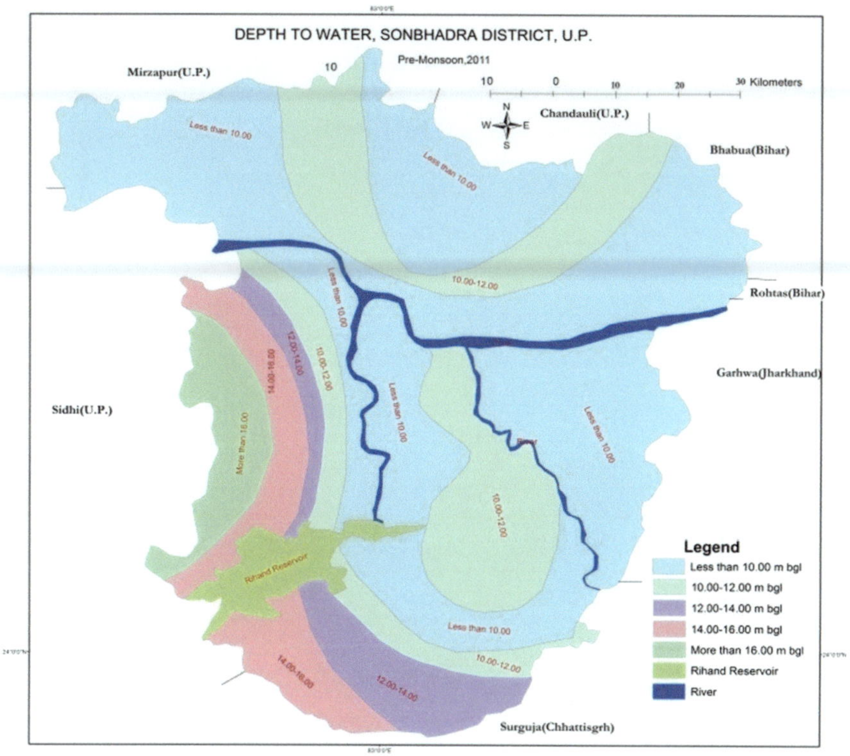

Fig. 10 Pre-monsoon water level of Sonbhadra district

6.4 Depth to Water Level Versus Fluoride Concentration

The fluoride concentration in groundwater and the water levels of the wells are compared in 16 places and presented in Table 5. The F^{-} conc. versus the water level depths is presented in Figs. 12 and 13. The graphs indicate that as the water level goes deeper, F^{-} concentration also decreases (Fig. 13). This may be due to higher rock–water interaction at shallow levels. High conc. >1.6 mg/L is confined where water levels are within 10 mbgl or so. At places with deeper water levels, the concentration has gone down.

6.5 Hydrochemistry of Groundwater

Among all the villages from where water sample are collected, eight villages having fluoride content more than 1.5 mg/L and seven villages below 0.5 mg/L. The analytical results and their concentration of each parameter are given in Table 6. The graphical representations of the chemical parameters are produced in Fig. 14.

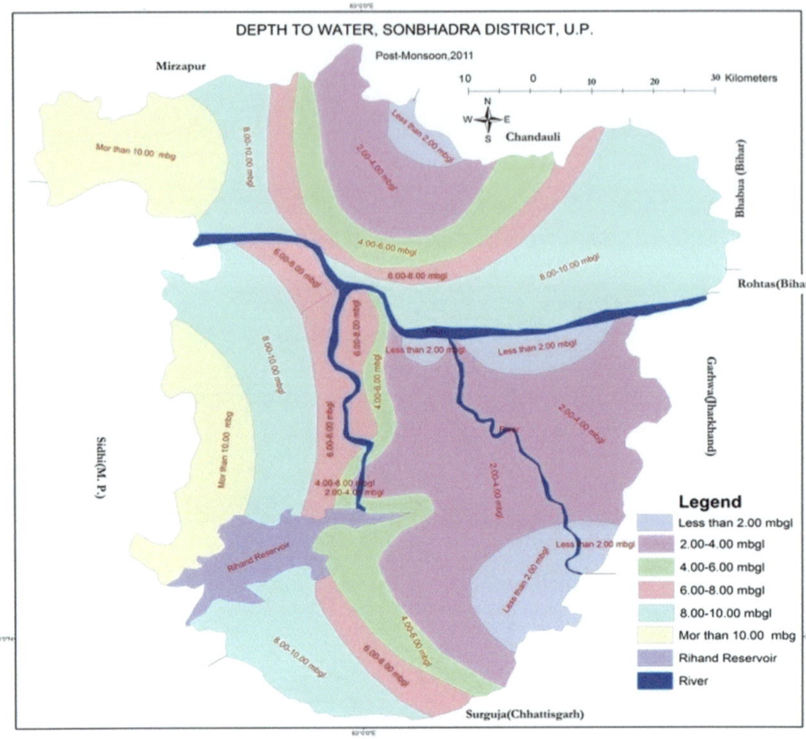

Fig. 11 Post-monsoon water level of Sonbhadra district

Table 5 Variation of fluoride with groundwater level

S. No.	Name of village	Water level (mbgl)	Fluoride (mg/L)
1	Neruiyadamar	9.3	1.91
2	Charakpathali	8.3	1.90
3	Rohiniadamar	6.7	1.86
4	Gobardaha	6.7	1.80
5	Pandubiya	10.1	1.83
6	Tumiya	7.9	1.82
7	Barwakhar	8.5	1.68
8	Gaurasinga	9.5	1.63
9	Kewal	10.7	0.68
10	Dudhmania	11.7	0.68
11	Mithniya	10.1	0.68
12	Kachnarwa	11.25	0.52
13	Rorwa	10.2	0.36
14	Kon	10.1	0.32
15	Lalbijora	11.5	0.30
16	Khetkatwa	13.3	0.23

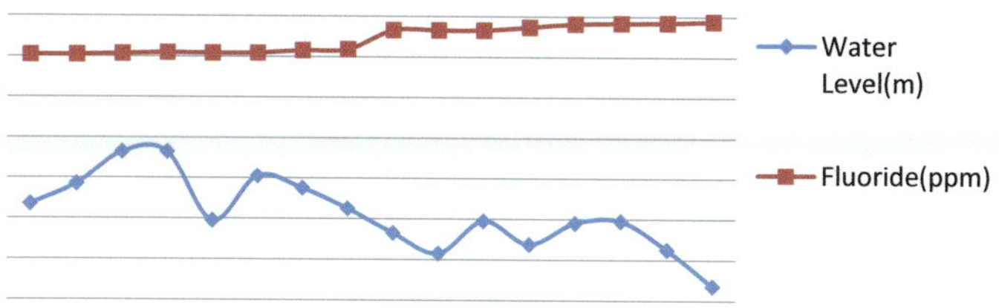

Fig. 12 Comparison of fluoride concentration and water level

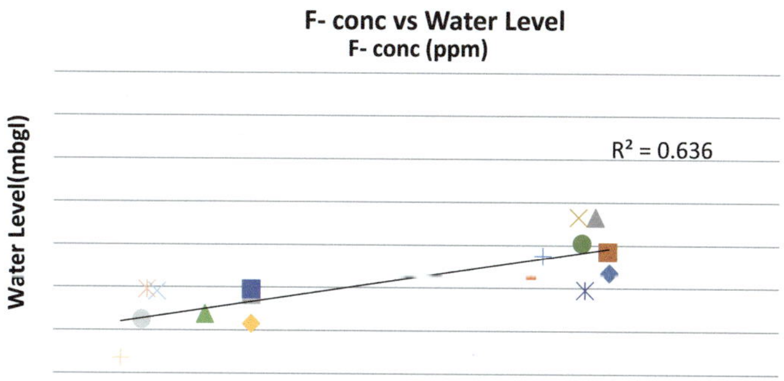

Fig. 13 Fluoride concentration versus water level of the villages in Sonbhadra district

6.5.1 Correlation Among Parameters

Karl Pearson correlation matrix among different chemical parameters is produced in Table 7. The correlation coefficient (r) lies between +1 and −1. When correlation coefficient is in the range of +0.8 to 1.0 and −0.8 to −1.0, then the correlation between the parameters is characterized as strong. In case value remains in the range of +0.5 to 0.8 and −0.5 to −0.8, it is characterized as moderate and considered weak when it is in the range of +0.0 to 0.5 and−0.0 to −0.5. Strong correlations between fluoride and sodium (0.94) and fluoride and alkalinity (0.81). Moderate correlation exists between fluoride and pH (0.63), fluoride and TDS (0.70), and fluoride and potassium (−0.53). The parameters have week correlation among each other are fluoride and hardness (−0.18), fluoride and calcium (−0.40), fluoride and magnesium (0.10), fluoride and chloride (−0.18), and fluoride and phosphate (−0.47). F^- is the most electronegative element, and thus it reacts immediately to form fluoride compounds and therefore making presence of free F^- an obsolete possibility but under favorable physicochemical conditions with long residence time it may occur as dissolved from groundwater (Handa 1975; Salve et al 2008; Gupta and Mondal 2012). The HCO_3^-/Ca^{2+} ratio at all the places is found to be greater than one, suggesting favorability of dissolution processes. F^- is adsorbed on the surface of the clay at acidic pH. When pH value is high, it indicates the elevated concentration of F^- in groundwater. Replacement of the exchangeable F^- of fluoride-containing minerals (muscovite) by the OH^- in groundwater with

Table 6 Chemical analysis of groundwater sample in Sonbhadra district

S.No.	Village	F^- mg\L	pH	Hardness mg\L	Alkalinity mg\L	TDS mg\L	Ca mg\L	Mg mg\L	Na mg\L	Cl mg\L	K\ mg\L	Phosphate(µg/L)
1	Neruiyadamar	1.92	7.1	140	330	412	42	8.5	108	28.49	0.7	15
2	Charakpathali	1.91	7.57	135	300	421	22	19.5	104	24.49	0.7	28
3	Rohiniadamar	1.87	7.17	135	330	351	34	12	98	17.99	0.6	23
4	Gobardaha	1.81	6.88	90	210	297	22	8.5	88	12.49	0.6	42
5	Pandubiya	1.83	7.24	155	240	294	50	7.3	52	12.99	0.4	93
6	Tumiya	1.82	7.66	120	308	450	27	13	109	30.99	0.9	60
7	Barwakhar	1.69	6.68	170	250	293	53	9.1	75	31.49	0.74	72
8	Gaurasinga	1.63	6.97	150	280	342	37	14	73	22.99	0.4	24
9	Kunrwa	1.29	6.26	130	220	275	30	13	54	12.49	0.7	25
10	Paraswa	0.93	6.84	120	150	286	42	3.6	52	15.49	0.98	62
11	Kewal	0.69	6.46	155	188	235	36	16	42	18.49	0.8	57
12	Dudhmania	0.69	7.46	115	184	247	38	4.9	43	14.49	1.2	72
13	Mithniya	0.69	7.13	150	226	343	50	6.1	52	20.49	1	84
14	Kobariya	0.6	6.5	130	204	284	50	1.2	36	17.99	0.8	90
15	Kachnarwa	0.53	6.76	195	194	235	70	4.9	41	62.48	1.4	11
16	Piparkhar	0.46	6.34	100	180	157	32	4.9	31	15.99	0.9	69
17	Rorwa	0.36	6.36	110	166	241	38	3.6	28	15.99	0.5	99
18	Ghaghiya	0.35	6.26	140	182	231	40	9.7	49	21.49	2.09	66
19	Kon	0.33	6.58	150	160	265	53	4.3	21	30.99	1.55	63
20	Misri	0.33	6.75	255	240	255	42	36	39	18.99	1.77	91
21	Lalbijora	0.31	6.24	110	176	205	30	8.5	21	15.99	0.96	48
22	Khetkatwa	0.23	6.6	185	160	328	67	4.3	32	62.98	1.7	86

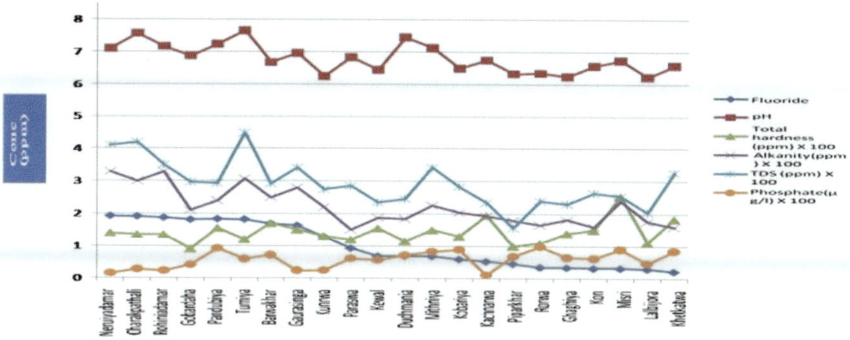

Fig. 14 Graphical representation of chemical parameter of groundwater

high value of pH increases the concentration of fluoride in groundwater (Gupta and Mondal 2012).

F^- in muscovite can be replaced by hydroxyl ions as shown below:

Muscovite

$$KAl_2[AlSi_3O_{10}]F_2 + 2OH^- \rightarrow KAl_2[AlSi_3O_{10}][OH]_2 + 2F^-$$

Most of the high-fluoride groundwater samples contain Ca^{2+} and Mg^{2+} as the predominant cation. HCO_3^- and Cl^- are the predominant anions for the different kinds of high-fluoride groundwater samples (Fig. 15). Result of the samples shows that the nature of the groundwater is HCO_3–CO_3 type and Na and K^+ type with Cl domination (Fig. 16).

Gibbs (1970) suggested a diagram to evaluate the relationship of chemical components of water from their respective aquifer lithologies. Three distinct fields, namely precipitation dominance, evaporation dominance, and rock dominance areas are shown in the Gibbs diagram (Fig. 17) Gibb's ratios are calculated with the formulae given below:

$$\text{Gibb's Ratio I(for anion)} = Cl/(Cl + HCO_3)$$
$$\text{Gibb's Ratio II(for cation)} = Na + K/(Na + K + Ca)$$

where all ions are expressed in mg\L.

Almost all the samples fall at the margin of rock weathering and not in Fig. 17 dominance area suggesting that majority of high-fluoride groundwater samples are mainly attributed to rock weathering with minor contribution by evaporation.

Table 7 Correlation matrix for chemical constituents/parameter of groundwater sample

	TDS	pH	Hardness	F-	PO_4^{2-}	Alkanity	EC	Ca	Mg	Na	Cl	K
TDS	1											
pH	0.74	1										
Hardness	0.06	−0.02	1									
F^-	0.70	0.63	−0.18	1								
PO_4^{2-}	−0.29	−0.17	−0.02	−0.47	1							
Alkalinity	0.76	0.64	0.09	0.81	−0.47	1						
EC	0.89	0.67	0.19	0.73	−0.26	0.87	1					
Ca	−0.15	−0.17	0.58	−0.40	0.25	−0.32	−0.15	1				
Mg	0.07	0.05	0.50	0.10	−0.08	0.34	0.22	−0.31	1			
Na	0.78	0.72	−0.10	0.95	−0.40	0.84	0.81	−0.35	0.16	1		
Cl	0.17	0.02	0.58	−0.18	−0.16	−0.05	0.14	0.71	−0.16	−0.10	1	
K	−0.27	−0.28	0.57	−0.53	0.26	−0.35	−0.14	0.49	0.10	−0.4	0.46	1

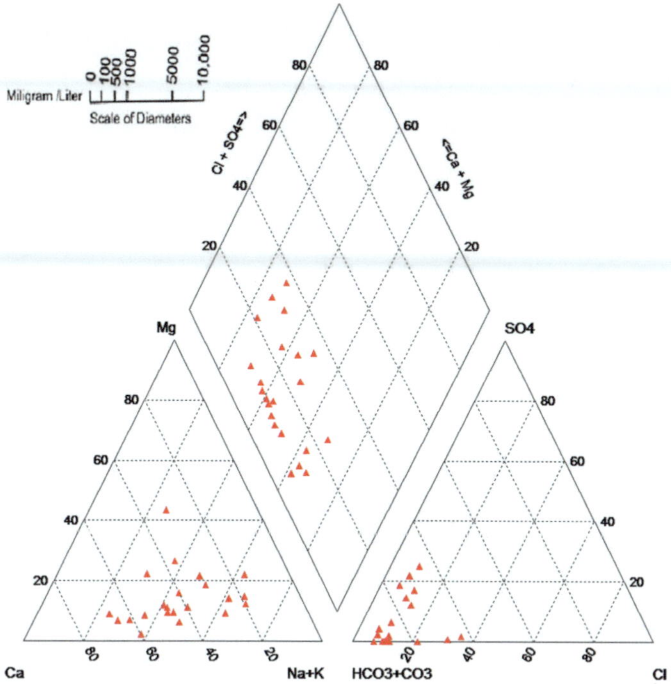

Fig. 15 Groundwater sample plotted in piper trilinear diagram (Piper 1944)

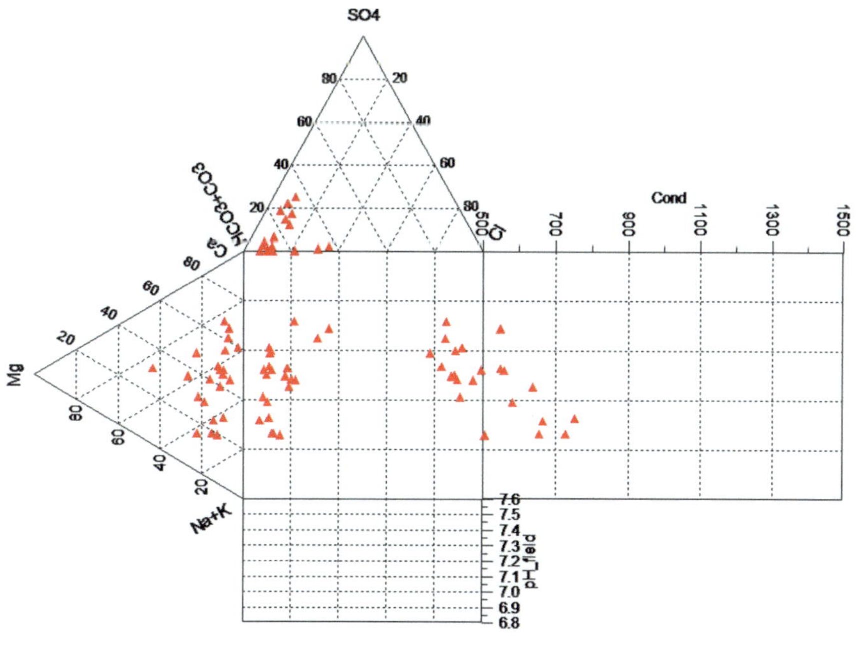

Fig. 16 Durov diagram for representing the analysis of groundwater quality

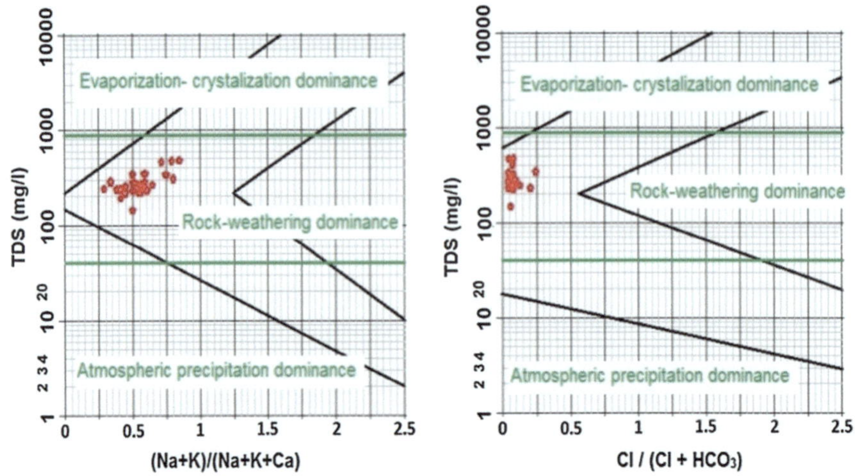

Fig. 17 Mechanism controlling the chemistry of groundwater as recorded by Gibbs diagram (1970)

7 Conclusion

Thin sections of rock samples reveal that the area is characterized by granite, pegmatite, and granite gneiss. Minerals which are present in above motioned rocks are fluoro apatite ($Ca_{10}(PO_4)_6(2)$, biotite($K(Mg, Fe)_3AlSi_3O_{10}(F, OH)_2$), muscovite ($KAl_2(Si_3Al)O_{10}(OH, F)_2$), hornblende ($(Ca, Na)_2$ 3$(Mg, Fe, Al)_5(Al, Si)_8O_{22}(OH, F)_2$), and other accessory silicate minerals. Fluoride ion in the water is mainly due to leaching of fluoro apatite ($Ca_{10}(PO_4)_6(2)$ and biotite($K(Mg, Fe)_3AlSi_3O_{10}(F, OH)_2$) minerals. It is also observed that at certain locations, the nature of rock as observed is granitic and granite gneiss up to a depth of 198 mbgl, with intermittent metapelites and clastic sedimentary rocks. In above research area, groundwater level ranges from 4.5 to 20.3 mbgl during pre-monsoon with the yield from 40 to 896 lpm. As per the data of exploratory wells of CGWB, the fractures encountered in different rocks are observed mostly at shallow depth (25–78 mbgl).It is also important to mention that shallow groundwater indicates higher concentrations of fluoride because the contact time with fluoride-bearing minerals is more. The litho-log of bore holes indicates that fractured rocks are suitable locales for transport of fluoride-bearing minerals than massive rocks.

Results indicate that 36.4% of (8 samples out of 22) groundwater samples have higher concentration of fluoride than the maximum permissible limit (1.5 mg/L) for drinking water. The fluoride concentration in groundwater is higher (>1.5 ppm) at Neruiyadamar, Charakpathali, Rohiniadamar, Gobaradaha, Pandubiya, Tumiya, pandubiya, and Gaurasinga villages of district Sonbhadra. Calcium concentration in groundwater is low under high alkalinity and high pH, which permits free mobility of fluoride ion into the groundwater. Better linear regression between sodium

(R2 = 0.94) and fluoride concentration is observed. Nonlinear dependency of fluoride is also observed with other parameters such as hardness, Ca, Mg, K, Cl, and PO4. Strong correlation of pH, sodium, TDS, EC, and alkalinity is observed with fluoride, whereas hardness, K, Mg, Ca, Cl, and phosphate show a weak correlation. As per Gibbs diagram, concentration of all important parameter is accelerated due to rock weathering. In Wilcox diagram, most of the point falls under the category of S1–C2, which shows that water quality is potable from the salinity point of view.

Acknowledgements Authors are thankful to Prof. P. Chakraborty, Director, MNNIT Allahabad for sanctioning the Institute Project under which the present study was carried out. We are also thankful to the Central Ground Water Board, Ministry, Govt. of India, Lucknow, for chemical analysis of groundwater samples. Authors are grateful to U.P. Jal Nigam for providing the baseline information regarding the fluoride-affected villages and Department of Geology, Banaras Hindu University, Varanasi, for preparation of thin section of rocks. We are thankful to M/S Hindalco Industries, Renukoot of Sonbhadra, for providing the logistic support during study. Authors are also thankful to Mr. Saurabh Kumar, Ph.D., Department of Civil Engineering, MNNIT Allahabad and Mr. Vivek Tiwari, M. Tech student, Institute of Agriculture Sciences, BHU, Varanasi, for their continuous involvement in preparation of maps and arrangement of the manuscript.

References

APHA (2000) Standard methods for the examination of water and waste water, 20th edn. American Public Health Association, Washington DC, pp 4–129

Fawell J, Bailey K, Chilton J, Dahi, E, Fewtrell L, Magara Y (2006) Fluoride in drinking water, WHO, IWA publishing, 1-144.Federation Dentaire International, 1990. FDI Basic Facts 1990: Dentistry around the world, London, FDI

Gibbs RJ (1970) Mechanisms controlling World's water chemistry. Science 170:1088–1090, 170, 795–840

Gupta S, Mondal D (2012) Geochemical provenance and spatial distribution in groundwater in parts of Raniganj coal field, West Bengal, India. Arch Appl Sci Res 4(1):296–306

Handa BK (1975) Geochemistry and genesis of fluoride containing groundwater in India. Groundwater 13:275–281

Khandare HW (2013) Fluoride contaminated water and its implications on human health—a review, department of geology, Armori, dist. Gadchiroli, India. Int J Chem Tech Res 5(1): 502–511

Patel SC et al (2014) Fluoride contamination of groundwater in parts of eastern India and a preliminary experimental study of fluoride adsorption by natural haematite iron ore and synthetic magnetite, Jharkhand, India. Environ Earth Sci 14(1):12–18

Piper AM (1944) A Graphic Procedure in the Geochemical Interpretation of Water Analysis. Am Geophys Union Trans 25, 914–923

Salve PR, Maurya A, Kumbhare PS, Ramteke DS, Wate SR (2008) Assessment of groundwater quality with respect to fluoride. Bul Environ Contam Toxicol 81:289–293

Tebutt THY (1983) Relationship between natural water quality and health, United Nations Educational, Scientific and Cultural Organization, Paris

WHO (1984) Guidelines for drinking water quality. In: Health criteria and other supporting information, 2 edn. World Health Organization, Geneva

Hydrochemical Evaluation of Fluoride-Rich Groundwater in Cherlapalli Watershed, a Fluorosis Endemic Area, Nalgonda District, Telangana State

Pandith Madhnure, P.N. Rao, K.M. Prasad, A.D. Rao and J.S. Kumar

Abstract Nalgonda District is widely known as one of the fluorosis endemic districts of India. Groundwater quality from Cherlapalli watershed, Nalgonda District, is evaluated with reference to fluoride. In the present study, concentration of fluoride is reported as high as 7.1 mg/L, while NO_3^- reaches up to 490 mg/L. Higher F concentrations are detected in highly weathered and fractured zones in structurally controlled central part of the area. Vertical variation in fluoride concentrations is observed down to 120 m depth, indicating local geological influence. F^- and Ca^{2+} show moderate degree of negative correlation, indicating F^- enrichment by removal of Ca^{2+} during rock-water interaction. Positive correlation between Na^+ and F^- and pH and F^- reveals that Na^+ is released into groundwater under alkaline conditions. Groundwater is mainly of Na–Mg–HCO_3- and Na–HCO_3 types. It is recommended that for drinking use surface water be blended with groundwater in fluoride-affected areas under the proposed Water Grid Project by local Government. Artificial recharge measures through percolation tanks and check dams are recommended in the central part where thick de-saturated weathered zone is available. De-silting of existing tanks under the ongoing Govt.-sponsored Mission Kakatiya may be taken on priority basis in the area.

Keywords Watershed · Granite · Groundwater · Fluoride · Contamination Telangana

---

P. Madhnure (✉) · P.N. Rao · K.M. Prasad · A.D. Rao · J.S. Kumar
Central Ground Water Board, Southern Region, Hyderabad, India
e-mail: pandithmadhnure@gmail.com

D. Saha et al. (eds.), *Clean and Sustainable Groundwater in India*,
Springer Hydrogeology, https://doi.org/10.1007/978-981-10-4552-3_6

1 Introduction

Groundwater being a natural solvent never occurs in pure form. It always contains some amount of solids and gases as a result of continuous interaction with porous rock media as it passes through it. Groundwater, thus enriched in different constituents through complex water-rock interaction, ion exchange may result in enrichment of certain elements and become unfit for human consumption. For healthy human growth, some elements are essential in trace amount, but their higher concentrations can cause toxic effects and F^- is one of such toxic elements (CGWB 1999). To protect tooth decay and enhance bone development, F^- concentration in the range of 0.6–1.2 mg/L in drinking water is essential, and ingestion of high-fluoride-containing drinking water can create severe health problem like fluorosis (Wood 1974). According to Susheela (2003), fluorosis is an endemic health problem affecting >67 million people all over the world.

Fluorosis was detected in India as early as the 1930s in Nellore District (Andhra Pradesh) (Shrott et al. 1937). Presently, it is reported that 20 states consisting of ~66.62 million people are affected (including 6 million children) (UNICEF 1999; RGNDWM 1993; Susheela 2000). In India, maximum concentration of F^- in groundwater reported is 48 mg/L (CGWB 1999), which is much higher than the maximum permissible limit for drinking use 1.5 mg/L (BIS 2003). Occurrences of F^- in groundwater, its genesis, and possible remedial measures have been studied by many workers in India (Ramesam and Rajagopalan 1985; Jacks et al. 2005; Ramamohan Rao et al. 1993; Wodeyar and Sreenivasan 1996; Agrawalet al. 1997; Subba Rao et al. 1998; Sujatha 2003; Saxena and Ahmed 2001; Madhnure et al. 2007; Reddy 2014). This study is carried out in order to understand the hydrochemistry in relation to hydrogeology and dwell upon geochemical evaluation of groundwater and reasons for elevated fluoride contamination in groundwater.

2 Study Area

The Cherlapalli watershed having a geographical area of 128 km^2 lies between East Longitudes 79° 08′ 50″–79°19′ 45″ and North Latitudes 17° 04′ 36″–17° 12′ 15″ (Fig. 1). The watershed-covering 18 villages have a population of ~40000 (2011 census). The climate of the area is tropical to semiarid with 750-mm normal annual rainfall. The study area is part of the Halia river sub-basin of the river Krishna basin and crisscrossed by 3 sets of lineaments. The area is underlain by granite rocks of Archaean-Paleoproterozoic age with basic intrusive rocks (dolerite) at places. The major landforms observed are pediplain (both medium and shallow), followed by buried pediplain. During kharif season, cotton and paddy are mainly grown, and during rabi season, paddy is mainly grown.

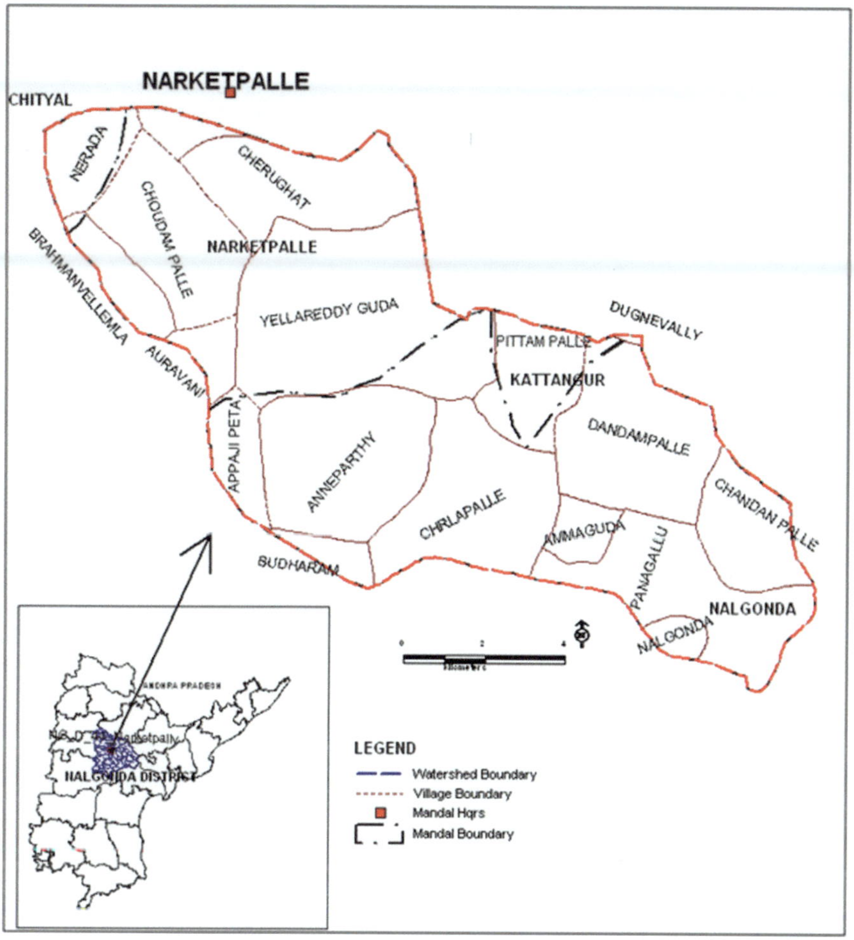

Fig. 1 Location map of study area

3 Hydrogeological Setup

In the weathered portion of granite, groundwater occurs under unconfined conditions, and in fractured zone of granite, it occurs under semiconfined to confined conditions. The weathered zone thickness goes up to about 30 m depth, followed by fractured/fissured zone which generally extends up to 100 m depth. However, in central part, fractures are encountered up to 187 m below ground. Most of the dug wells which tap weathered zone have gone dry in the recharge and in the intermediate areas of the watershed. In the discharge areas of the watershed, the water remains in the dug wells only seasonally. Presently, groundwater is extracted through bore wells of 60–100 m depth, with yield in the range of 0.1 to 6.5 lps (liter per second). The hydrogeology of the area is presented in Fig. 2. The depth of

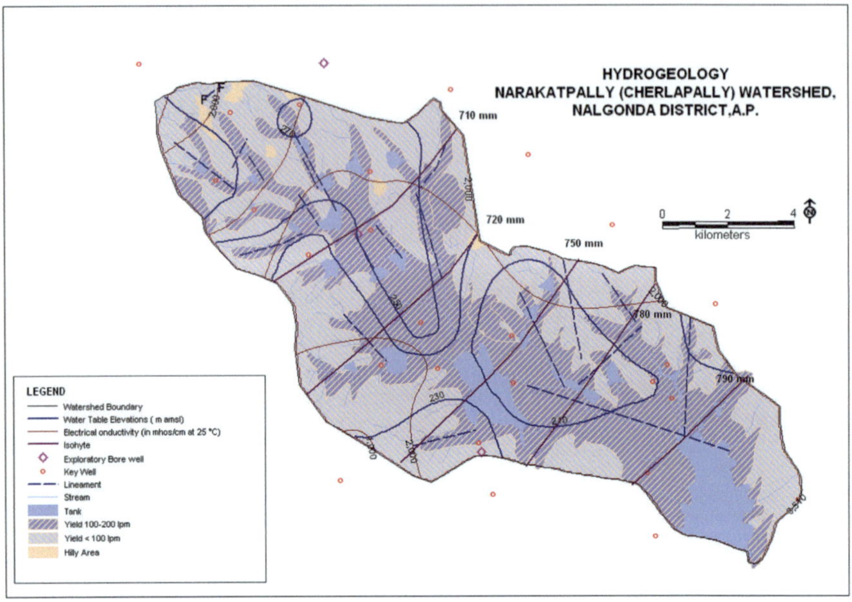

Fig. 2 Hydrogeology map

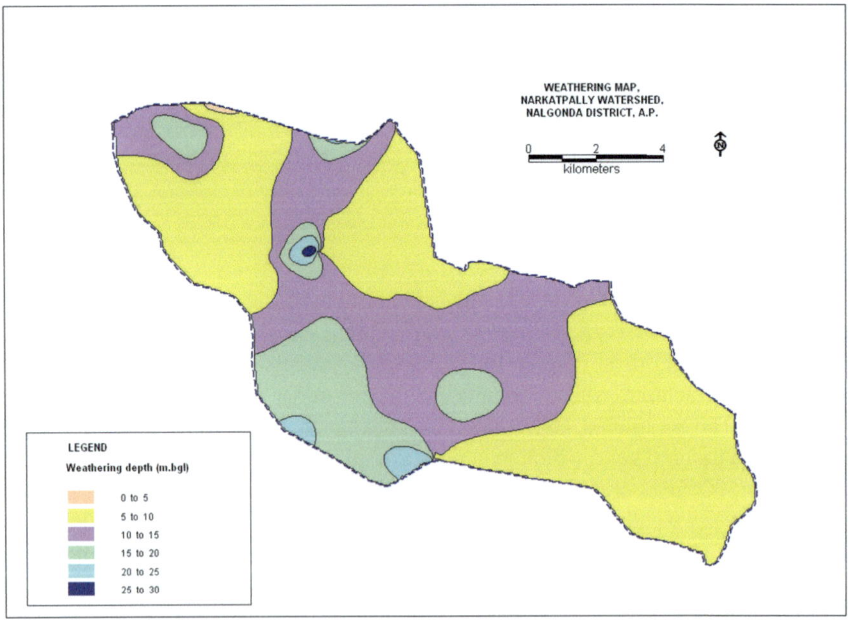

Fig. 3 Depth of weathering map

weathering varies from 4 to 27.3 m below ground level (mbgl) with an average of 10.3 m. However, spatially 5–10 m weathering depth is more predominant, followed by 10–15 m depths (Fig. 3). The fractures are encountered between 13 mbgl at Sherubaigudem and 187 mbgl at Yellareddigudem villages, but majority of them occur between 30 and 50 mbgl. During the month of May (pre-monsoon season) and November (post-monsoon season), depth to water levels varies from 3.2 to 51.7 and 3.0 to 47.7 mbgl, respectively (Fig. 4a, b). In major part of the study area, the water levels are in the range of 10 to 20 mbgl in both the seasons. The net annual groundwater availability is 1791 hectare meter (ha m), while the extraction is 1017 ha m. The watershed falls under safe category with the stage of groundwater development of 57% (SGWD and CGWB 2012).

4 Methodology

Total 17 samples, tapping both weathered (dug wells) and fractured zones (hand pumps), were collected during pre-monsoon season (May). From hand pumps, samples were collected after 10 min of pumping to get fresh water from aquifer, and from dug wells, samples were collected from 30 cm below the water levels. The samples were collected in good-quality plastic bottles and stored properly before transferring to chemical laboratory. The samples were analyzed for 13 constituents (pH, TDS, TH, Ca^{2+}, Mg^{2+}, Na^+, K^+, CO_3^{2-}, HCO_3^-, Cl^-, SO_4^{2-}, NO_3^-, and F^-) in CGWB Laboratory at Hyderabad (NABL Accredited) by following standard methods (APHA 1998). The cation/anion balance for the analyzed data is within the permissible range of ±5% (APHA 1998) (Table 1). Hydrochemical evaluation of groundwater has been studied based on trilinear Piper diagram (Hill 1940; Piper 1944). Drinking and irrigation suitability of groundwater is assessed based on Bureau of Indian Standard (BIS 2003) and Wilcox diagram (Wilcox 1955).

5 Results and Discussions

Groundwater in general is of neutral to mildly alkaline in nature as the pH ranges from 7.27 to 8.4 (average: 7.73). Electrical conductivity (EC) ranges from 718 to 3600 μS/cm (average: 1532) (Table 1), and total hardness (TH) varies from 85 to 820 mg/L (average: 369). The calcium (Ca^{2+}) and magnesium (Mg^{2+}) concentrations vary from 6 to 257 mg/L (average: 75 mg/L) and 17 to 112 mg/L (average: 44 mg/L), respectively, which is commonly observed in groundwater. The sodium (Na^+) concentration ranges between 35 and 483 mg/L (average: 172 mg/L) and is the predominant cation. Potassium ranges between 1 and 294 mg/L (average: 26 mg/L). High K^+ (>200 mg/L) is noticed at Cheruvugattu, which could be due to the use of chemicals containing potash in blasting of exposed granite rocks. Contribution of Ca^{2+}, Mg^{2+}, Na^+, and K^+ to total cations is about 25, 25, 47, and 3% respectively.

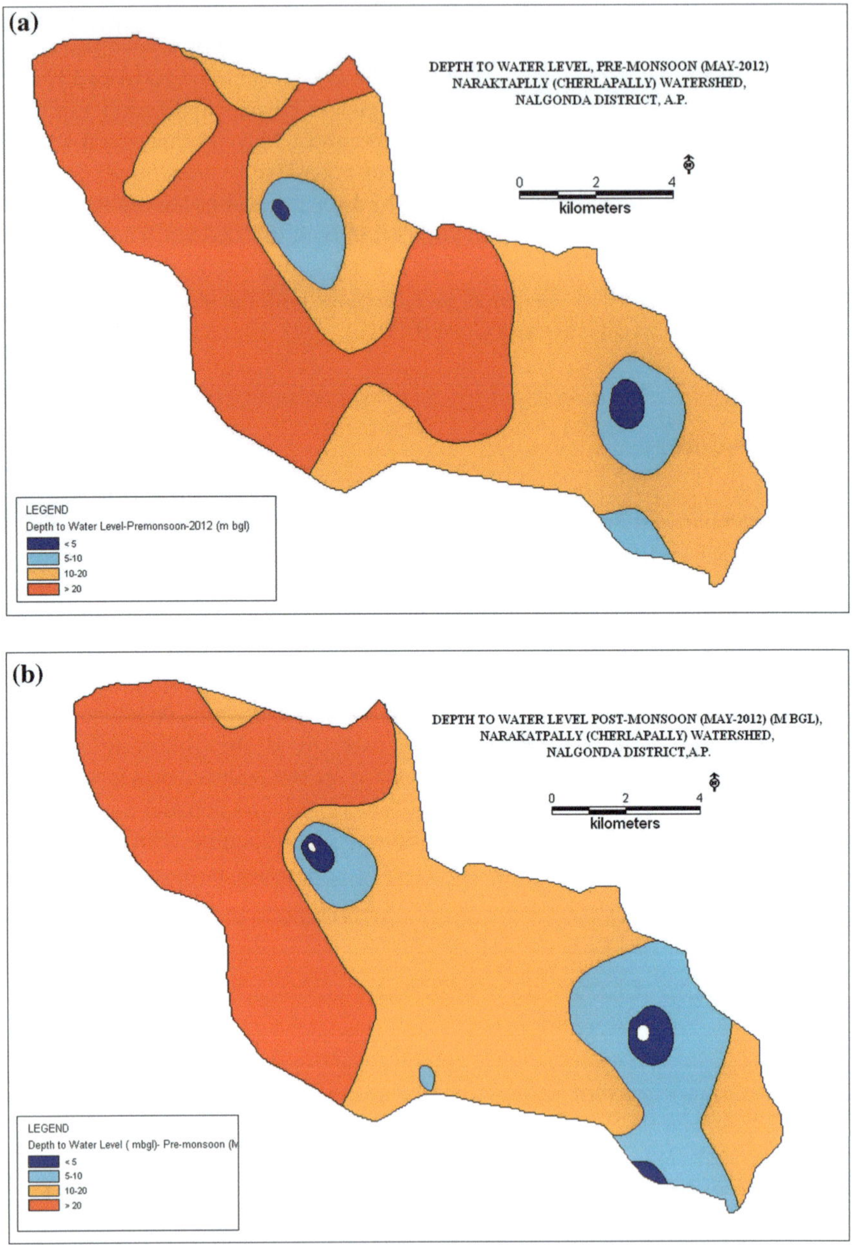

Fig. 4 **a** DTW (Pre-monsoon-May, 2012) **b** DTW (Post-monsoon-November, 2012)

Table 1 Summarized results of groundwater quality data (Pre-monsoon, 2012)

S. No.	Location	X	Y	pH	EC	TH	Ca	Mg	Na	K	HCO_3	Cl	SO_4	NO_3	F	Water type
					µS/cm	mg/L										
1	Anaparthy	79.2142	17.1256	7.79	1340	330	36	58	142	19	515	113	34	**50**	**2.2**	Na–Mg–HCO_3–Cl
2	Badagudem	79.2514	17.1208	8.25	1200	335	46	53	152	8	671	53	27	20	1.1	Na–Mg–HCO_3
3	Battapotulagudem	79.2944	17.1256	7.36	1180	380	86	40	92	2	336	142	67	40	0.2	Ca–Na–Mg–HCO_3–Cl
4	Chadanpalli	79.2958	17.1167	8.16	1105	200	34	28	189	1	488	99	67	0	0.7	Na–HCO_3–Cl
5	Chadanpalli	79.3317	17.0903	7.42	718	300	90	18	35	2	244	64	58	30	0.6	Ca–Na–HCO_3–Cl
6	Cherlapalli	79.2416	17.1052	7.5	2100	550	172	29	229	3	366	269	137	**280**	0.6	Na–Ca–Cl–HCO_3
7	Cheruvugattu	79.2111	17.1765	7.29	3600	**820**	**257**	44	274	294	451	461	356	**490**	1.1	Ca–Na–K–Cl–SO_4–HCO_3
8	ChinnaNarayanpur	79.1911	17.1941	7.27	1300	420	100	41	80	48	287	163	56	**160**	1.5	Ca–Na–Mg–HCO_3–Cl
9	Dandempalli	79.2903	17.1211	7.62	1539	370	96	32	175	3	415	227	48	43	0.3	Na–Ca–HCO_3–Cl
10	Kammagudem	79.2254	17.1369	8.4	2490	190	16	36	483	1	616	213	127	19	**6.4**	Na–HCO_3–Cl–CO_3
11	Kondapakagudem	79.178	17.1665	7.45	1640	**615**	62	**112**	93	1	372	277	68	**65**	0.5	Mg–Na–Cl–HCO_3
12	Panagal	79.2889	17.0972	8.06	1266	320	48	49	154	30	561	103	66	10	0.3	Na–Mg–HCO_3–Cl
13	Pittampalli	79.2508	17.1333	7.84	1393	410	54	67	129	4	543	149	5	43	0.4	Na–Mg–HCO_3–Cl
14	Serubaigudem	79.1935	17.1547	7.78	1600	325	38	56	216	4	732	64	43	**70**	**3.4**	Na–Mg–HCO_3
15	Sitarampuram colony	79.2302	17.1249	7.49	1360	245	36	38	190	15	592	85	41	28	**2.6**	Na–Mg–HCO_3
16	Yellareddigudem	79.2112	17.1612	8.36	1230	85	6	17	250	11	360	78	90	13	**7.1**	Na–HCO_3
17	Yenuguladhari	79.1713	17.1921	7.39	980	380	94	35	45	1	268	103	49	**70**	**2.7**	Ca–Mg–Na–HCO_3–Cl
			Minimum	7.27	718	85	6	17	35	1	244	53	5	0	0.24	
			Maximum	8.4	3600	820	257	112	483	294	732	461	356	490	7.1	
			Average	7.73	1532	369	75	44	172	26	460	157	79	84	1.86	
	Permissible limits in the absence of alternate source			**6.5–8.5**	–	**600**	**200**	**100**	–	–	–	**1000**	**400**	**45**	**1.5**	
	No of samples falling beyond permissible limits			**0**	–	**2**	**1**	**1**	–	–	–	**0**	**0**	**7**	**6**	

Note CO_3 below detectable limits

Among the anions, HCO_3^- dominates and its concentration varies from 244 to 732 mg/L (average: 460 mg/L). The other dominant anion is Cl^- which varies from 53 to 461 mg/L (average: 157 mg/L) followed by SO_4^{2-} ranging from 5 to 356 mg/L (average: 79 mg/L). Though average NO_3^-concentration is 84 mg/L, it ranges from nil to 490 mg/L.

For identification of different water facies of groundwater, Piper diagram is widely used as it gives best graphical representation (Hill 1940; Piper 1944; Saha et al. 2008, 2009). Groundwater is mainly Ca–Na–HCO_3 (5 nos) and Ca–Na–HCO_3–Cl (5 nos) type, and other types are Na–HCO_3–Cl (4 nos) and Ca–HCO_3–Cl (3 nos) (Fig. 5). Groundwater quality evaluation has shown F^--rich groundwater mainly belongs to Na–Mg–HCO_3 type.

Correlation between F^- and major parameters, viz. pH, Ca^{2+}, Na^+, and HCO_3^-, is studied (Fig. 6a–d). The scatter plot between F^- and Ca^{2+} shows moderate negative correlation with a correlation coefficient of 0.44, which is in agreement with proven hypothesis, i.e., F^- enrichment is facilitated by removal of Ca^{2+} through precipitations of calcite during water-rock interaction (Reddy 2014).

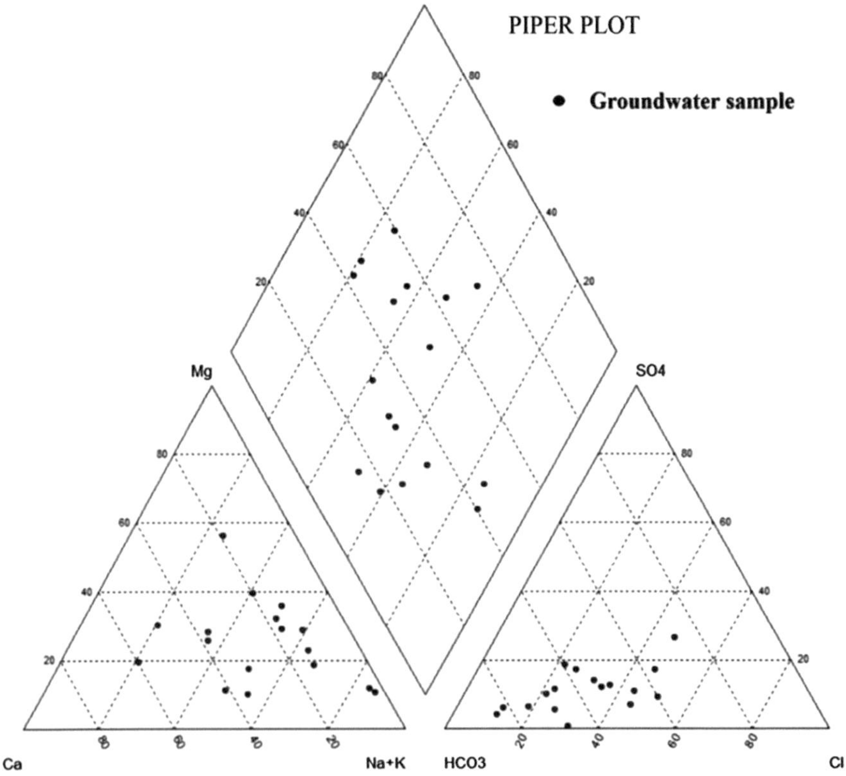

Fig. 5 Groundwater facies (Piper plot)

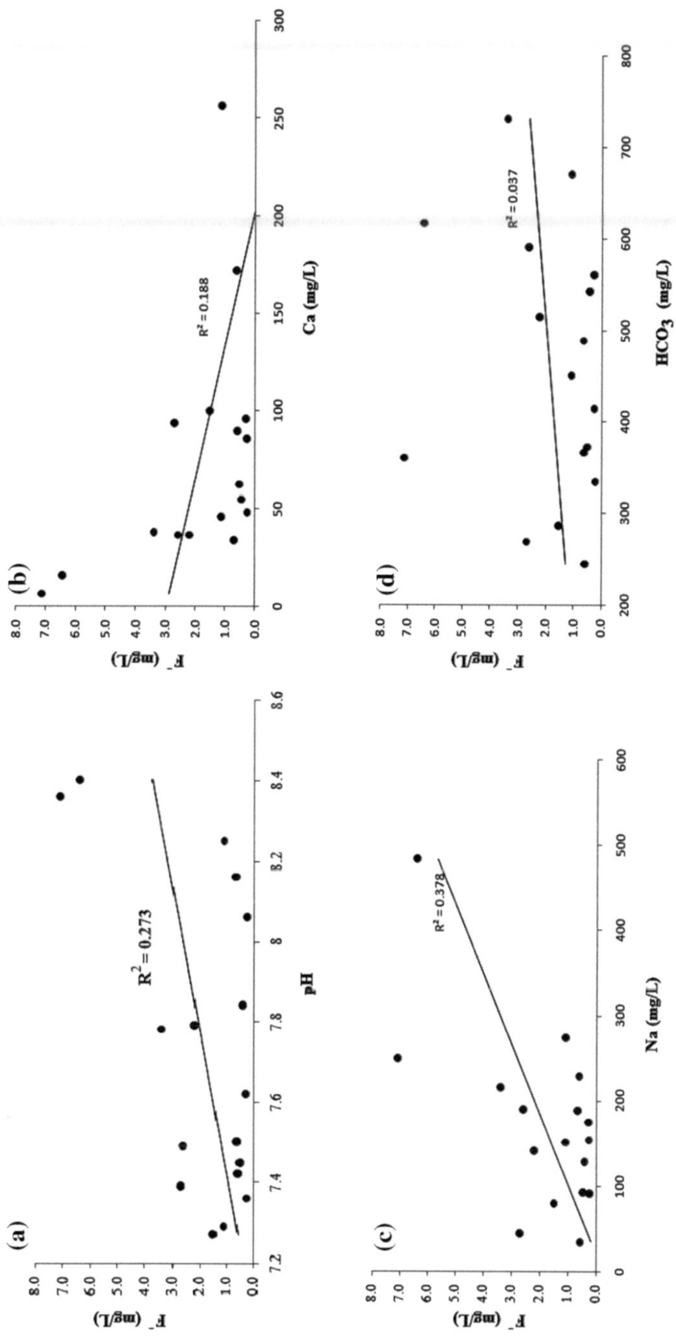

Fig. 6 a–d Interrelationships between variables

Apambire et al. (1997) has also observed that Na^+ concentration increases with F^-, thereby increasing the solubility of fluorite mineral in water. Fluoride and HCO_3^- pair has also shown a negative correlation (−0.11).

5.1 Fluoride Distribution

In the area, F^- concentration ranges from 0.24 to 7.1 mg/L (average: 1.86 mg/L). From the results, it is observed that wells located in highly weathered and proximal areas of lineaments in the watershed, particularly in and around Anaparthy, Yellareddigudem, and Cheruvugattu villages, have higher concentrations of F^- (Fig. 7). Depth-wise variation in fluoride concentrations has been established at Yellareddigudem village by systematic sampling of groundwater from respective fracture zones by sealing of other zones during drilling operations by Central Ground Water Board (Fig. 8). It is observed that fluoride concentration at the depth of 20 m is 2.8 mg/L and the discharge of fracture is 0.9 lps; the concentration decreases to 2.2 mg/L with increase in discharge 1.0 lps at the depth of 80 mbgl. However, at the depth of 120 m, discharge increased to 1.8 lps and concentration of F increased to 2.3 mg/L. Below this depth in spite of increase in discharge, F^-concentrations remains constant.

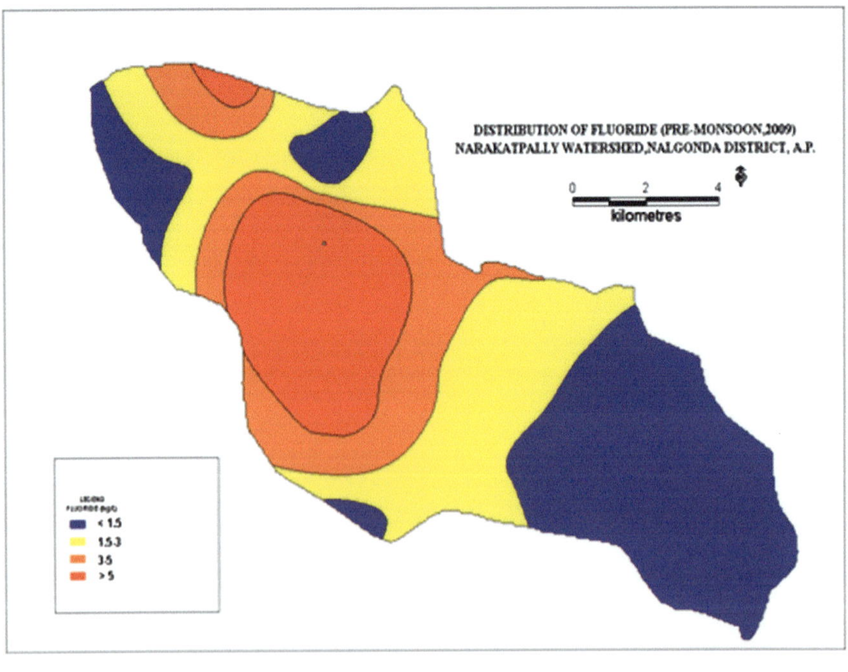

Fig. 7 Distribution of fluoride concentrations

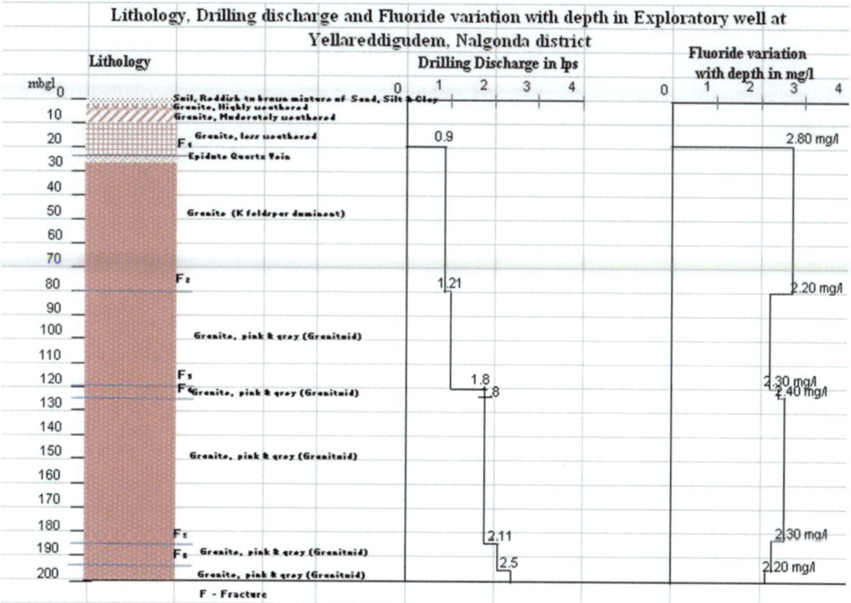

Fig. 8 Depth-wise variations in drilling discharge and fluoride concentrations

5.2 *Mobilization of Fluoride in Groundwater*

The host rock is granite, which is fractured and weathered. Analyses of rock samples have shown F^- content in the range of 325–3200 mg/kg (average: 1440 mg/kg) which is more than the world average (810 mg/kg) and 1.5 times higher than the average of adjoining Hyderabad granites (Rammohan Rao et al. 1993). In the study area, there are neither active volcanoes nor fluorine-based industries; therefore, it can be concluded that the F^- in groundwater is mainly come from fluoride-bearing minerals presents in country rock. The main fluoride-bearing minerals that contribute to the groundwater are fluorite, fluoroapatite, cryolite, apophyllite, and ferromagnesian silicates such as amphiboles and micas (Subba Rao et al. 1998). During the downward movement of groundwater from weathered zone to fractured zone, F^- is released into the groundwater through geochemical reactions and this has been observed by many authors in their research work (Ramesam and Rajgopalan 1985; Agrawal et al. 1997 and Subba Rao et al. 1998). The other causes of high fluoride in groundwater locally are semiarid to arid climate, ion exchange, etc. (Madhnure et al. 2007). Saxena and Ahmed (2001) have observed that alkaline pH (7.6–8.6) with high HCO_3^- (350–450 mg/L) and moderate EC (>1200) favors dissolution of fluorite in the groundwater.

5.3 Suitability of Groundwater for Different Uses

Groundwater of area is generally fit except for drinking problem of fluoride and nitrate. The pH, Na, K, Cl, and SO_4 ions remain within the maximum permissible limits of BIS (2003) for drinking water use. Fluoride concentration exceeds maximum permissible limit in 6 samples (35% of total) and nitrate in 7 samples (41% of total). The concentration of total hardness, Ca^{2+} and Mg^{2+}, exceeds maximum limits as prescribed by BIS 2003 in 2 nos, 1 no., 1 no. of samples, respectively. Overall 41% of samples are unfit for drinking purposes (BIS 2003).

The quality of groundwater decides agricultural productivity and quality of crops and thereby enhancing its price. Groundwater from the area mainly falls in C_3S_1 (9 nos) and C_3S_2 (4 nos) classes and 1 each in C_2S_1, C_3S_3, C_4S_2, and C_4S_4 classes indicating high to very high salinity hazard and low to medium category sodium hazard (Wilcox 1955) (Fig. 9). The groundwater with salinity hazard high to very high requires treatment before it is applied for irrigation, in case if applied, it reduces the crop growth (Figs. 5 and 6).

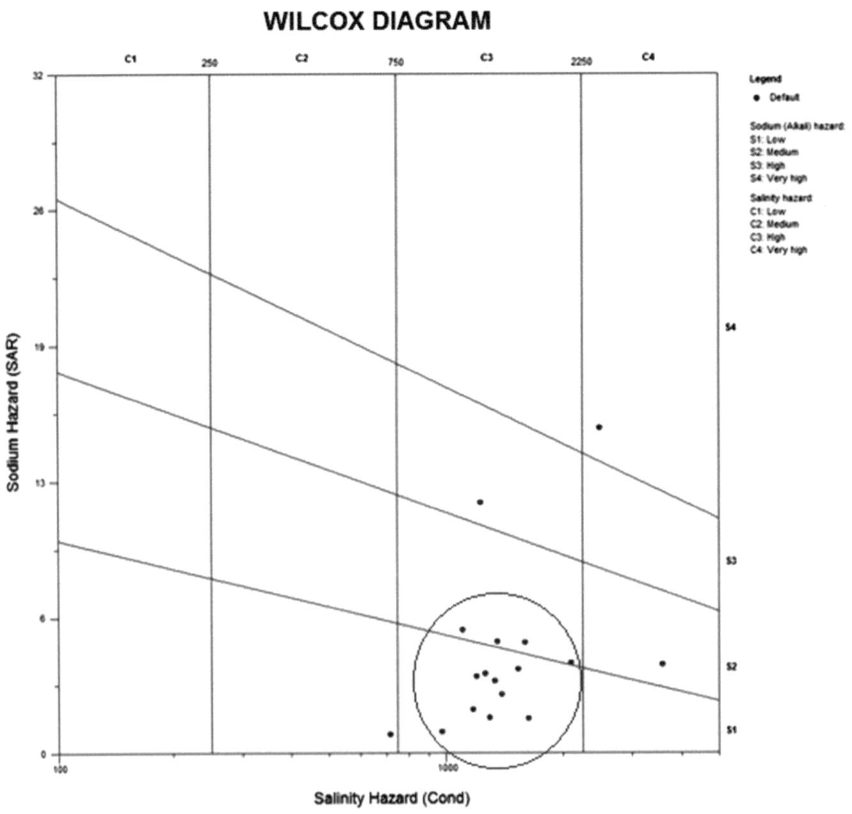

Fig. 9 Wilcox diagram (1955) showing suitability for irrigation

6 Conclusions

There is an overall decline in water levels in the area because of overexploitation, and shallow aquifers made up of weathered zone has mostly gone dry in recharge and intermediate zones. In this area, groundwater extraction is through bore wells tapping fractures at 60–100 m depth. Deeper fractures in central part can be exploited for other domestic/irrigation needs. The anthropogenic contaminant such as K^+ and NO_3^- is observed in some areas. The main cause for high F^- in groundwater is due to leaching of fluoride from fluoride-rich minerals present in the area, which is abetted by alkaline nature of groundwater, semiarid climate, and higher residence time of slow moving groundwater allowing prolong rock-water interaction. The study reveals that the major process of F^- enrichment in groundwater is by removal of Ca^{2+} due to rock-water interaction. During the chemical weathering of rock, clay (phyllosilicate group of minerals) is formed as residue and sodium present in mineral is released into groundwater and this is supported by positive correlation between and also by the fact that majority of fluoride-rich groundwater is of Na–Mg–HCO_3 type.

As more than 40% of samples are unfit for human consumptions, surface water supply under Mission Bhagiratha (Project conceived by Govt. of Telangana) should be provided to all the affected villages and fluoride-rich groundwater may be used for other purposes (including irrigation). Latest purification, separation, and decontamination techniques such as ion exchange and reverse osmosis may be adopted along with providing supplementary calcium, phosphorous-rich food, and multivitamins through midday meals particularly to the children below the ages of 14 years. The de-saturated weathered zone with a thickness of $\sim$10 m indicates good scope for further groundwater augmentation through artificial recharge, which will also reduce F^- load in groundwater by dilution. Creating awareness on impact of excessive intake of fluoride- and nitrate-rich groundwater on human health and regular health checkup are the other measures recommended.

Acknowledgements The authors thank Chairman, Shri K.B. Biswas, Member (TT & WQ) Sri K. C. Naik and Member (SAM), Dr. Dipankar Saha from Central Ground Water Board, Govt. of India, for their encouragement. Authors also extend their thanks to Regional Director, CGWB Southern Region for his valuable guidance. Authors thank Dr. Sudheer Kumar, Sc-D of CGWB for his help and support. The data provided by agencies like Ground Water Department, Govt. of Telangana Sate, is duly acknowledged.

References

Agrawal V, Vaish AK, Vaish P (1997) Groundwater quality: focus on fluoride and fluorosis in Rajasthan. Curr Sci 73(9):743–746

Apambire WB, Boyle DR, Michel FA (1997) Geochemistry, genesis, and health implications of fluoriferous groundwater in the upper regions of Ghana. Environ Geol 33(1):13–24

APHA (1998) Standard methods for the examination of water and waste water, 19th edn., American Public Health Association, Washington, DC, 20th Edition, p 10–161

BIS (2003) Drinking water-specification IS: 10500; 1991, Edition 2.1 (1993-01) Bureau of Indian Standards, New Delhi. p 11

CGWB (1999) High fluoride groundwater in India: Occurrences, genesis and remedies. Unpub. Report Central Ground Water Board, Ministry of Water Resources, Govt. of India, p 29

Hill RA (1940) Geochemical patterns in Coachella Valley. Trans Am Geophys Union 21:46–49

Jacks G, Bhattacharya P, Chaudhary V, Singh KP (2005) Controls on the genesis of some high-fluoride ground waters in India. Appl Geochem 20:221–228

Madhnure P, Sirsikar DY, Tiwari AN, Ranjan B and Malpe DB (2007) Occurrence of fluoride in the groundwater's of Pandharkawada area, Yavatmal district, Maharashtra, India. Curr Sci 95 (5):675–679

Piper AM (1944) A graphic procedure in the geochemical interpretation of water analysis. Trans Am Geophys Union 25:914–923

Ramamohan Rao NV, Rao N, Surya Prakash Rao K, Schuiling RD (1993) Fluorine distribution in waters of Nalgonda District, Andhra Pradesh, India. Environ Geol 21(1):84–89. doi:10.1007/BF00775055

Ramesam V, Rajagopalan K (1985) Fluoride ingestion into the natural waters of hard-rock areas Peninsular India. J Geol Soc India 26:125–132

Reddy AGS (2014) Geochemical evaluation of nitrate and fluoride contamination in varied hydrogeological environs of Prakasam district, southern India. Environ Earth Sci 71:4473–4495. doi:10.1007/s12665-013-2841x

RGNDWM (1993) Prevention and control of fluorosis in India: Health aspects. Rajiv Gandhi National Drinking Water Mission, Ministry of Rural Development, Govt. of India, New Delhi, v1

Saha D, Dhar YR, Sikdar PK (2008) Geochemical evolution of groundwater in the pleistocene aquifers of South Ganga Plain, Bihar. J Geol Soc India 71:473–482

Saha D, Sarangam H, Diwedi SN, Bhartariya KJ (2009) Evaluation of hydrochemical processes in the arsenic contaminated alluvial aquifers in parts of Mid-Ganga basib, Bihar, India. J Environ Earth Sci. doi:10.1007/s12665-009-0392-y

Saxena V, Ahmed S (2001) Dissolution of fluoride in groundwater: a water-rock interaction study. Environ Geol J 40(9):1084–1087. doi:10.1007/s002540100290

SGWD and CGWB (2012) Dynamic Ground Water Resources of Andhra Pradesh. State Ground Water Department, Govt. of Andhra Pradesh, Hyderabad and Central Ground Water Board, Southern Region, Govt. of India, Hyderabad. vol 1 & vol 2. p 598 & p 512

Shrott HE, Merobert GR, Barnald TW, Nayar ASM (1937) Endemeic fluorosis in the Madras presidency. Indian J Med Res 25:553–561

Subba Rao N, Krishna Rao G and John Devadas D (1998) Variation of fluoride in groundwater of crystalline terrain. J Environ Hydrol v.6, paper 3, 1–5

Sujatha D (2003) Fluoride levels in the groundwater of the south-eastern part of Rangareddy district, Andhra Pradesh. India. Environ Geol. 44(5):587–591. doi:10.1007/s00254-003-0795-0

Susheela AK (2000) Early detection of fluorosis in children Up-date series 5. Centre for Research on Nutrition Support System, Nutrition Foundation of India, New Delhi

Susheela AK (2003) A treaties on fluorosis. revised second edition, p 137

UNICEF (1999) State of art report on the extent of fluoride in drinking water and the resulting endemicity in India. UNICEF, New Delhi

Wilcox LV (1955) Classification and use of irrigation waters, circular no 969. United States Department of Agriculture, Washington DC

Wodeyar BK, Sreenivasan G (1996) Occurrence of fluoride in the ground waters and its impact in Peddavankahalla Basin, Bellary District, Karnataka- a preliminary study. J Curr Sci 70(1):71–74

Wood JM (1974) Biological cycles for toxic elements in the environment. Science 183:1049–1052

Uranium Contamination of Groundwater in Southwest Parts of Punjab State, India, with Special Reference to Role of Basement Granite

K.P. Singh, Naval Kishore, Naresh Tuli, R.S. Loyal, Mandeep Kaur and Jasbir Kaur Taak

Abstract The state of Punjab is known for potential aquifer system and also experiencing heavy groundwater extraction mainly to cater the irrigation sector from the shallow aquifers (<50 m depth). Uranium concentration in the groundwater from shallow tubewells (<50 m depth) ranges from 1.78 to 261 μg/L. Uranium concentration shows spatial variations even within a short distance of 200 m. In majority of samples, uranium concentration exceeds the permissible limit of 15 μg/L prescribed by WHO (2004) and 60 μg/L prescribed by AERB, DAE (2004). Most workers link the presence of uranium in groundwater to basement granite. The results of uranium analysis of basement granite sample collected earlier from Phulka (29° 29′; 75° 08′) during exploratory drilling carried out under UNDP-CGWB project reveal a normal concentration of 3.686 μg/g. Further, the granite core sample shows negligible (0.01 mR/h) radioactivity. In addition, granitic basement occurring at the depth of 242 m or more at other places could not have contributed uranium to groundwater occurring at shallow depth (up to 50 m). Uranium concentration in sediments/soils was found to range between 1 and 3 μg/g. Mobilization of uranium from sediments/soils under favorable geochemical conditions seems to be the possible explanation for uranium contamination in groundwater.

Keywords Uranium · Groundwater · Basement granite · Mobilization Sediments · Punjab

1 Introduction

Uranium content of groundwater varies markedly depending upon bedrock type and on proximity to uranium deposits as well as chemical composition of the water. Average uranium content in groundwater from the igneous terrains is 4.5 μg/L; in

---

K.P. Singh (✉) · N. Kishore · N. Tuli · R.S. Loyal · M. Kaur · J.K. Taak
Centre of Advanced Study in Geology, Punjab University, Chandigarh, India
e-mail: kpsingh1948@rediffmail.com

D. Saha et al. (eds.), *Clean and Sustainable Groundwater in India*,
Springer Hydrogeology, https://doi.org/10.1007/978-981-10-4552-3_7

sedimentary terrains about 26.2 μg/L; in metamorphics 4.4 μg/L; and in groundwater from sand and gravels about 2.5 μg/L though it can range from 0 to 74 μg/L (Scott and Barker 1962).

The contamination of uranium in groundwater is a matter of concern because of its chemical toxicity as a dissolved metal and radiotoxicity. Uranium is known to be nephrotoxic, and its toxic effects on kidney functions are known (Domingo 1995). Under some circumstances, the chemical toxicity of soluble uranium can even surpass its potential radiotoxic effects (Kumar et al. 2014). According to ATSDR (1999), the toxicity of uranium is a function of route of exposure, particle solubility, contact time, and route of elimination. In 1991, the US Environmental Protection Agency (US EPA) classified uranium as a confirmed human carcinogen (Group A). In view of the above, various regulatory bodies have proposed maximum contaminant levels of uranium in drinking water. US EPA has proposed maximum contaminated level (MCD) of 30 μg/L (USEPA 1991). Canada proposed interim maximum concentration of 20 μg/L. WHO (2004) recommends a reference level of 15 μg/L, whereas in India AERB, DAE (2004) proposed to maximum level of 60 μg/L based on radiotoxicity.

In view of large number of cancer cases being reported in the southwest parts of Punjab in India, the authors carried out uranium concentration level studies in the study area with an aim to know the extent of contamination and the possible mobilization process. This is the first report from Muktsar-Malout region of southwest Punjab, though the groundwater contamination by uranium from adjoining areas has been earlier reported by many workers (Singh 2010; Singh and Kishore 2010; Kumar et al. 2011a, b, 2014; Singh et al. 2012; Bajwa et al. 2015). The results of the study are discussed in the present paper.

2 Study Area

The study area forms a part of Indo-Gangetic alluvium plains and lies in Muktsar and Bhatinda districts of Punjab state (Fig. 1). The study focused on Muktsar-Malout blocks of Muktsar district and Talwandi Sabo block of Bhatinda district where large number of persons suffering from various health problems including cancer. The area has arid and semiarid climate with an average rainfall of about 300 mm. The Bhakra Canal (main branch) and the Kotla branch are the major canals flowing through the Talwandi Sabo area. In Muktsar-Malout belt, Sirhind feeder and Rajasthan feeder are the major canals. The area is also characterized by sand dunes. The regional slope of the area is from northeast to southwest direction. There are number of drains in the area especially in Muktsar-Malout belts to take out rainwater/seepage water which gets accumulated in the low-lying areas.

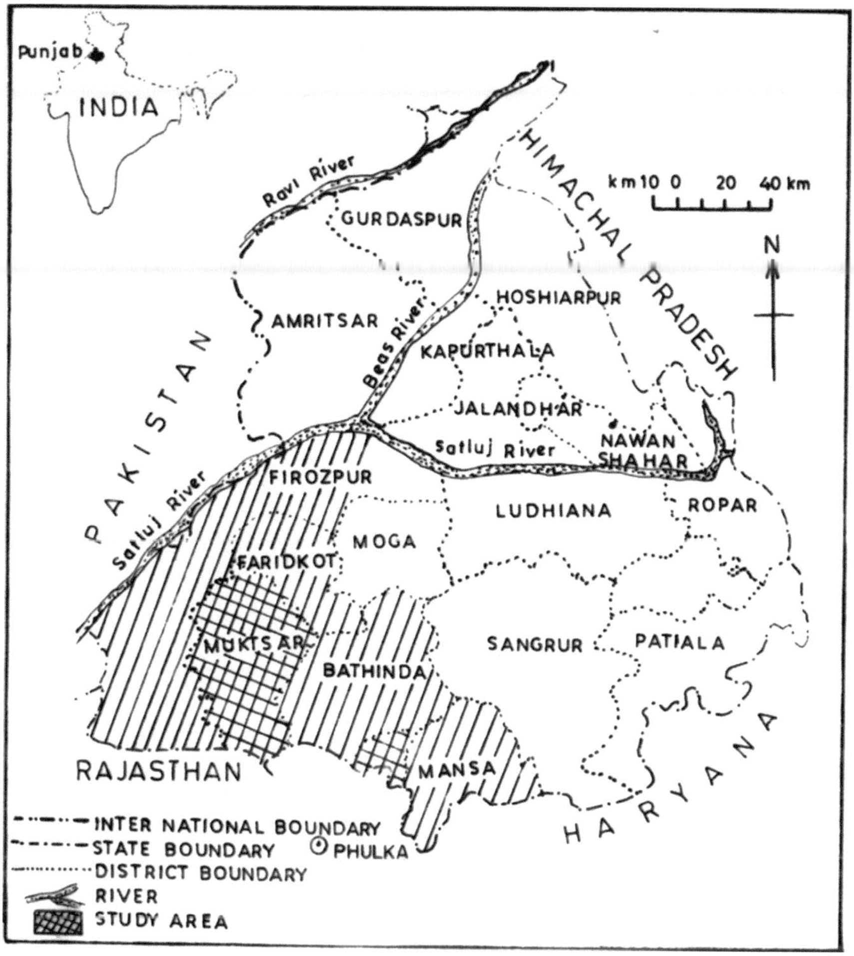

Fig. 1 The location of the study area and southwest districts of Punjab state (location of basement granite, Phulka lying in Haryana state is also shown)

3 Aquifer Geometry and Groundwater Conditions

The area comprises of Indo-Gangetic alluvium of Quaternary age. Groundwater mainly occurs in alluvium comprising sand, silt, kankar (calcium carbonate concretions), and gravel beds. Clay beds occur at all depths making the aquifer system multiple as revealed by the study of borehole lithologs. The regional aquifer system in the area along with the depth profile of basement granite sample is shown in Fig. 2. At shallow depths, groundwater occurs under unconfined conditions, whereas the aquifers occurring at more than 50 m depth hold groundwater under semi-confined/confined conditions. The results of studies carried out by Surinaidu

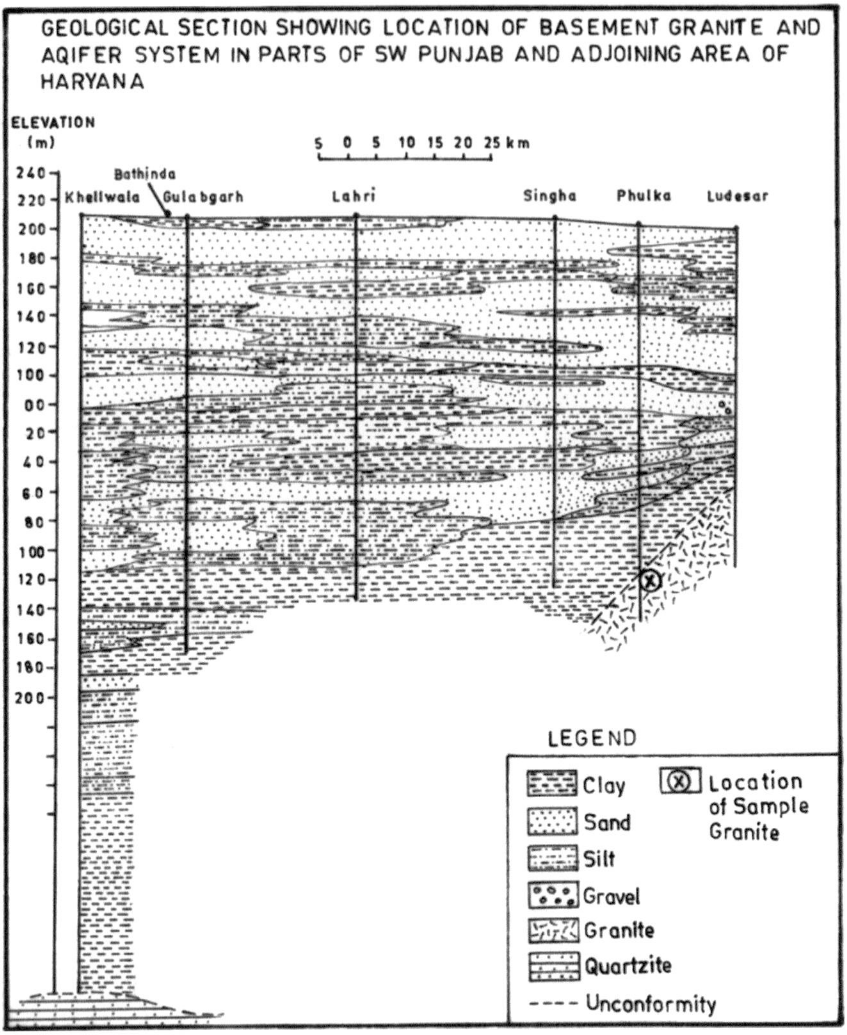

Fig. 2 Regional correlation of aquifer system in the southwest Punjab along with location of basement granite at Phulka (modified after UNDP 1979)

et al. (2010) based on resistivity survey indicate the presence of thick saline sediments up to 26 m depth. These saline sediments have influenced the groundwater quality in the area.

Average depth to groundwater level ranges 8–10 m below ground level in Talwandi Sabo block, whereas in Muktsar-Malout belt, shallow depths up to 2 m are predominating making parts of the areas as waterlogged. Waterlogging has led to the development of saline/alkaline soils in the area due to capillary rise of salts to the land surface. The transmissibility values range from 500 to 700 m^2/day, and aquifers have hydraulic conductivity ranging from 12 to 5 m/day. The groundwater flow is from NE to SW direction in general, and average hydraulic gradient is about 0.32 m/km. Water levels in the area have been rising for the last more than 100 years. In some areas in Talwandi Sabo block, water level has stabilized now. Groundwater is currently drawn out by number of handpumps having up to 15–30 m depth and shallow tubewells having up to 50 m depth in spite of the fact that chemical quality of groundwater is unfit or marginally fit for irrigation and drinking water purposes (Singh and Kishore 2010). However, along the canals, due to seepage of canal water, freshwater lenses have been formed having good quality of groundwater with electrical conductivity less than 2000 μS/cm at 25 °C. Such freshwater lenses are more common in Muktsar-Malout belt especially around Rajasthan canal and Sirhind feeder.

4 Methodology

Groundwater samples were collected from handpumps and tubewells in hard polythene bottles of one liter capacity. Water bottles were washed and rinsed before sampling. Handpumps and tubewells were pumped for 3–5 min to remove stagnant water. For trace element and uranium analysis, separate water samples were collected in 250 ml bottles. The samples were acidified with nitric acid to bring down pH and were kept under refrigerating conditions. GPS locations were recorded at the sampling sites. Uranium content in water samples was measured by laser fluorimeter (developed by Quantalase Enterprises Pvt. Ltd., Indore, India) at Guru Nanak Dev University, Amritsar. The laser fluorimeter has been successfully used for detection of trace quantities of uranium present in water by many workers (Sahoo et al. 2009; Kumar et al. 2011a, b; Singh et al. 2012; Bajwa et al. 2015). The instrument works on the principle of detection of fluorescence of uranium complexes in the sample. Under UV excitation, uranium salts emit green luminescence, which is measured by a sensitive PMT. Since the fluorescence yield is proportional to the intensity of excitation source and concentration of uranium in the sample, measurement of fluorescence gives information about concentration of uranium in the sample. The accuracy of laser fluorimetric technique for uranium analysis of water samples has been verified by using ICPMS technique by Shenoy et al. (2012). According to their study, the results of laser fluorimetric technique were compared well with ICPMS technique. They further opined that laser fluorimeter is reliable, simple, faster, and cost-effective.

A limited number (5) of samples of sediments/soils and calcrete (Kankar) were collected from top zone up to 50 cm in polythene bags. Calcium carbonate concretions were separated manually for analysis. The sediment and calcrete samples were analyzed for trace elements using EDRX technique at RCIL, Punjab University, Chandigarh. The basement granite sample collected earlier from Phulka drilling site was analyzed using ICP-AES facilities available in NGRI, Hyderabad.

5 Results and Discussion

5.1 Uranium Concentration in Groundwater

The uranium concentration in the groundwater drawn from handpumps (up to 30 m depth) and shallow tubewells (up to 50 m depth) ranges from 1.78 to 261 μg/L (Table 1). Uranium concentration shows spatial variations, and even within a short distance of 200 m, a variation is noticed. In majority of samples, uranium concentration exceeds the permissible limit of 15 μg/L prescribed by WHO (2004) and 60 μg/L prescribed by AERB, DAE (2004). The results of uranium concentration in groundwater in the area are shown in Table 1. The data reveal that in Muktsar district, U concentration ranges from 1.78 to 237 μg/L. Uranium hot spots, i.e., samples with concentration more than permissible limit (60 μg/L) for drinking water (AERB, DAE 2004), occur at villages Daula, KotBhai, Burj Mahima, Khara, Sangh, and Thandewal. In Talwandi Sabo area forming parts of Bhatinda district, U concentration ranges from 77 to 261 μg/L. Samples of Bhatinda district show

Table 1 Uranium concentration at different locations in the study area

Sr. No.	Location Muktsar district	GPS location	Type of well	Uranium concentration (μg/L)
1	Bherkhera	30° 15′ 15″ 74° 29′ 30″	HP	2.75
2	Alamwala	30° 15′ 00″ 74° 26′ 01″	HP	12.8
3	Midha	30°18′ 29″ 74° 20′ 06″	HP	1.78
4	Gatewali	30° 18′ 29″ 74° 20′ 06″	HP	174[a]
5	ButtarSaron	30° 21′ 25″ 74° 41′ 02″	HP	142[a]
6	Doda	30° 22′ 54.2″ 74° 38′ 12.4″	HP	12.8
7	Sangh	30° 27′ 14.5″ 74° 22′ 47.3″	HP	70.7[a]
8	Thandewal	30° 28′ 14.3″ 74° 34′ 51.6″	HP	128[a]
9	Warring	30° 30′ 44.2″ 74° 38′ 23.6″	HP	20.2
10	Khara	30° 31′ 58″ 74° 43′ 02″	HP	147[a]
11	Burj Mahima	30° 16′ 08.3″ 74° 55′ 07″	HP	73.6[a]
12	Burj Mahima	30° 15′ 22.5″ 74° 47′ 22″	TW	106[a]
13	KotBhai	30° 15′ 47.7″ 74° 42′ 19″	HP	182[a]
14	Damla	30° 13′ 09″ 74° 4′ 9.8″	HP	237[a]
Talwandi Sabo area (Bhatinda District)				
15	Malkana	29° 56′ 1.6″ 76° 02′ 18.8″	HP	31.2
16	Burj	30° 00′ 42.7″ 75° 10′ 4.1″	HP	77[a]
17	Shekpura	30° 00′ 11.4″ 75° 09′ 34.4″	HP	45.8
18	Shekpura	30° 00′ 08.2″ 75° 09′ 26″	HP	78.5[a]
19	Talwandi	29° 59′ 10.7″ 75° 07′ 17.2″	HP	107[a]
20	Talwandi Sabo	30° 00′ 12.1″ 75° 04′ 34.2″	HP	261[a]

[a]Concentration of uranium more than permissible limit (60 μg/L) AERB, DAE (2004)

comparatively higher U concentration than those of Muktsar district. Dilution of groundwater due to seepage from vast canal network could be the reason for such variations.

5.2 *Geochemistry of Calcrete (Calcium Carbonate Concretion)*

Samples of kankar (calcium carbonate) were collected from selected sites to know the chemical composition. They were analyzed using XRF techniques at CIL, Punjab University, Chandigarh. The results of XRF analysis of calcium carbonate concretions (calcrete) locally known as kankar from two sites, i.e., Burj Mahima and Daula, are shown in Tables 2 and 3. The results indicate higher concentration of calcium (as calcium oxides) which ranges from 332,383 to 34,091 μg/g. Heavy metals, i.e., Pb, Ni, Cu, Zn, show low concentration. Arsenic was also detected. Uranium concentration ranges from 1 to 3 μg/g. At Daula, higher concentration of uranium, thorium, strontium (Sr), and arsenic was observed. Thorium, strontium, and compton show an increase with increasing U concentration in calcretes.

Table 2 Results of XRF analysis of kankar (calcium carbonate concretions) sample collected from Burj Mahima

CaO	TiO_2	V	Cr	MnO	Fe_2O_3	Co
130.0 KCps	5.6 KCps	0.2 KCps	0.1 KCps	2.9 KCps	5.7 KCps	0.6 KCps
332,383 PPM	0.31%	55 PPM	88 PPM	0.12%	2.13%	39 PPM

Ni	Cu	Zn	Ga	As	Rb	Sr	Y
0.1 KCps	0.1 KCps	0.2 KCps	0.2 KCps	0.0 KCps	3.8 KCps	15.1 KCps	2.3 KCps
27 PPM	15 PPM	28 PPM	7 PPM	19 PPM	49 PPM	181 PPM	16 PPM

Zr	Nb	Mo	Sn	Sb	Cs	Ba	La
13.7 KCps	1.1 KCps	−0.6 KCps	0.2 KCps	0.2 KCps	−0.0 KCps	0.2 KCps	0.0 KCps
122 PPM	7 PPM	2 PPM	1 PPM	1 PPM	3 PPM	190 PPM	14 PPM

Ce	Pb	Th	U	Compton
0.0 KCps	0.5 KCps	0.2 KCps	−0.1 KCps	105.7 KCps
40 PPM	18 PPM	4 PPM	1 PPM	70.17%

Eval2 V2.2.426 Admin 29.07.2016 12:40:42
Sample: Burj Mahima
Measured on 30.10.2013 16:21:57
Sample measured by Admin
Measurement method: GEO-QUANT T

Table 3 Results of XRF analysis of kankar (calcium carbonate concretions) sample collected from Daula

CaO	Sc	TiO_2	V	Cr	MnO	Fe_2O_3
13.3 KCps	0.0 KCps	14.2 KCps	0.3 KCps	0.2 KCps	2.4 KCps	12.0 KCps
34,091 PPM	11 PPM	0.55%	61 PPM	107 PPM	0.06%	3.08%

Co	Ni	Cu	Zn	Ga	As	Rb	Sr	Y
2.0 KCps	0.2 KCps	0.2 KCps	0.5 KCps	0.4 KCps	0.1 KCps	9.9 KCps	20.4 KCps	5.3 KCps
107 PPM	29 PPM	15 PPM	56 PPM	12 PPM	26 PPM	95 PPM	174 PPM	24 PPM

Zr	Nb	Sn	Sb	Cs	Ba	La	Ce	Pb
33.8 KCps	2.5 KCps	0.3 KCps	0.3 KCps	0.0 KCps	0.6 KCps	0.0 KCps	0.1 KCps	0.9 KCps
214 PPM	11 PPM	3 PPM	3 PPM	11 PPM	429 PPM	27 PPM	76 PPM	21 PPM

Th	U	Compton
0.4 KCps	−0.0 KCps	149.7 KCps
9 PPM	3 PPM	85.08%

Eval2 V2.2.426 Admin 29.07.2016 12:42:40
Sample: Daula
Measured on 29.10.2013 16:45:17
Sample measured by Admin
Measurement method: GEO-QUANT T

5.3 Geochemistry of Sediment Samples

The results of XRF analysis of sediments from two sites, i.e., Burj Mahima (Table 4) and KotBhai (Table 5), indicate that U concentration in sediments varies from 2 to 3 µg/g. Strontium (Sr), thorium (Th), and compton increase with the increasing concentration of uranium in sediments. Arsenic concentration ranges from 19 to 22 µg/g. Iron oxides (Fe_2O_3), TiO_2, and MnO increase with increasing concentration of arsenic and uranium. Low concentrations of heavy metals, i.e., Pb, Cr, Mo, Ni, Co, and others were observed. Other elements do not show any significant correlation with U concentration in sediments.

5.4 Geochemistry of Basement Granite

A granite core sample collected earlier from Phulka site in adjoining area in Sirsa district of Haryanaw which was collected earlier during drilling carried out by CGWB under UNDP project was chemically analyzed at NGRI, Hyderabad, using ICP. The main objective was to know the uranium content in it as many workers (Kochhar et al. 2009) opined that basement granite is responsible for uranium concentration in groundwater observed in the area. The results (Table 6) indicate that the uranium content is 3.688 µg/g which is below the average concentration, i.e., 4 µg/g (Bowie and Plant 1983). Thorium (Th), rubidium (Rb), and barium concentrations are also normal. Heavy metals, i.e., Cu, Zn, Pb, and Cr, show low concentration.

Table 4 Results of XRF analysis of sediment sample collected from Burj Mahima

CaO	Sc	TiO_2	V	Cr	MnO	Fe_2O_3
8.1 KCps	0.0 KCps	12.2 KCps	0.3 KCps	0.1 KCps	2.2 KCps	14.9 KCps
27,975 PPM	10 PPM	0.63%	79 PPM	116 PPM	0.07%	3.86%

Co	Ni	Cu	Zn	Ga	As	Rb	Sr	Y
1.2 KCps	0.2 KCps	0.1 KCps	0.4 KCps	0.4 KCps	0.1 KCps	7.7 KCps	10.5 KCps	4.2 KCps
80 PPM	42 PPM	18 PPM	52 PPM	14 PPM	28 PPM	102 PPM	124 PPM	27 PPM

Zr	Nb	Mo	Sn	Sb	Cs	Ba	La	Ce
25.7 KCps	1.8 KCps	−0.7 KCps	0.2 KCps	0.3 KCps	0.0 KCps	0.5 KCps	0.0 KCps	0.1 KCps
229 PPM	12 PPM	1 PPM	3 PPM	3 PPM	12 PPM	454 PPM	27 PPM	94 PPM

Pb	Th	U	Compton
0.6 KCps	0.3 KCps	−0.0 KCps	108.3 KCps
22 PPM	11 PPM	3 PPM	83.49%

Eval2 V2.2.426 Admin 29.07.2016 12:41:17
Sample: Burj Mahima 1
Measured on 19.12.2013 16:02:58
Sample measured by Admin
Measurement method: GEO-QUANT T

Table 5 Results of XRF analysis of sediment sample collected from KotBhai

CaO	Sc	TiO_2	V	Cr	MnO	Fe_2O_3
44.1 KCps	−0.0 KCps	9.6 KCps	0.2 KCps	0.1 KCps	2.2 KCps	8.7 KCps
112,702 PPM	6 PPM	0.41%	55 PPM	94 PPM	0.06%	2.48%

Co	Ni	Cu	Zn	Ga	As	Rb	Sr	Y
1.1 KCps	0.1 KCps	0.1 KCps	0.4 KCps	0.3 KCps	0.1 KCps	7.4 KCps	28.8 KCps	4.0 KCps
60 PPM	22 PPM	15 PPM	48 PPM	10 PPM	22 PPM	76 PPM	266 PPM	20 PPM

Zr	Nb	Mo	Sn	Sb	Cs	Ba	La	Ce
27.4 KCps	1.7 KCps	−0.8 KCps	0.2 KCps	0.3 KCps	−0.0 KCps	1.1 KCps	0.0 KCps	0.1 KCps
182 PPM	8 PPM	1 PPM	1 PPM	2 PPM	4 PPM	748 PPM	24 PPM	64 PPM

Pb	Th	U	Compton
0.8 KCps	0.3 KCps	−0.1 KCps	136.7 KCps
21 PPM	6 PPM	2 PPM	77.36%

Eval2 V2.2.426 Admin 29.07.2016 12:43:25
Sample: KotBhai
Measured on 29.10.2013 16:07:36
Sample measured by Admin
Measurement method: GEO-QUANT T

Table 6 Trace element concentration in the basement granite sample

Element	Concentration (μg/g)
U	3.688
Th	30.313
Pb	34.01
Rb	238.2
Ba	245.8
Cu	1.88
Zn	22.95
Cr	5.9

6 Conclusions

Uranium concentration in the groundwater drawn from handpumps (up to 30 m depth) and shallow tubewells (up to 50 m depth) ranges from 1.78 to 261 μg/L. There is significant spatial variation in uranium concentration in groundwater. In majority of samples, uranium concentration exceeds the permissible limit of 15 μg/L prescribed by WHO (2004) and 60 μg/L prescribed by AERB, DAE (2004). The results of geochemical analysis of sediments/soils indicate that uranium concentration ranges from 1 to 3 μg/g. The sediments are rich in calcium, and they also contain arsenic and iron oxides. The results of basement granite sample collected from Phulka (29° 29′; 75° 08′) close to the study area during exploratory drilling carried out by Central Ground Water Board, Government of India, under UNDP project (1975–78) indicate that the average uranium concentration in granite is 3.688 μg/g and falling below the normal range of concentration 4 μg/g. Further, the granite core sample shows negligible (0.01 mR/h) radioactivity. In addition, granitic basement occurring at the depth of 242 m or more at other places could not have contributed uranium to groundwater occurring at shallow depth (up to 50 m). Therefore, the role of basement granite is ruled out as a cause of uranium contamination of groundwater. The role of fly ash in imparting uranium contamination in groundwater has been already ruled out by Alrakabi et al. (2012). Further, areas located away from thermal power plants such as Muktsar-Malout and Talwandi Sabo could not have been affected by fly ash generated by Bathinda thermal power plant. The studies carried out by the Punjab Agricultural University, Ludhiana, have also ruled out the role of phosphatic fertilizers in imparting uranium contamination (personal communication from Head, Department of Soils, Punjab Agricultural University, Ludhiana). In view of the above, the possible source of uranium could be mobilization of uranium from sediments/soils under favorable geochemical environment (high HCO_3, pH, and TDS). Similarly, mechanism of uranium mobilization has been proposed by Jurgens et al. (2009) in Central Valley, California, USA.

Acknowledgements KPS thanks UGC for sanctioning a major project. The authors also thank Director, NGRI, Hyderabad, for providing ICP facilities and helping in chemical analysis of granite core samples. The authors also thank Director RCIL and Shri Tejbir Bambra for providing facilities for XRF analysis. Thanks are also to Chairman, Centre of Advanced Study in Geology, Punjab University, Chandigarh, for providing facilities.

References

Agency for Toxic Substances and Diseases Registry (ATSDR) (1999) Toxilogical profile for uranium. US Department of Health and Human Services, US Public Health Services, Atlanta, GA, USA

Alrakabi M, Singh G, Bhalla A, Kumar S, Srivastva A, Rai B, Singh N, Shohi JS, Mehta D (2012) Study of uranium contamination of groundwater in Punjab State in India using X-ray fluorescence technique. J Radioanal Nucl Chem doi:10.1007/s10967-011-1585-x

Atomic Energy Regulatory Board, Department of Atomic Energy (AERB, DAE) (2004) Drinking water specifications in India. Mumbai, India

Bajwa BS, Kumar S, Singh S, Sahoo SK, Tripathi RM (2015) Uranium and other heavy toxic elements distribution in the drinking water samples of SW-Punjab, India. J Radiat Res Appl Sci. doi:10.1016/j.jrras.2015.01.002

Bowie SHU, Plant JA (1983) Natural radioactivity in the environment. In: Thornton I (ed) Applied environmental geochemistry. Academic Press, New York pp 481–494

Domingo JL (1995) Prevention by chelating agents of metal-induced development toxicity. Reprod Toxicol 9:105–113

Jurgens BC, Fram MSF, Belitz K, Burow KR, Landon MK (2009) Effects of groundwater development on uranium: Central Valley, California, USA. Ground Water. http://dx.doi.org/10.11111/j.1745-6584.2009.00635.x

Kochhar N, Gill GS, Naresh T, Dhadwal V, Balaram V (2009) Chemical quality of groundwater in relation to incidence of cancer in parts of SW Punjab, India. Asian J Water Environ Pollut 4:107–112

Kumar A, Rout S, Narayanan U, Mishra MK, Tripathi RM, Singh J (2011a) Geochemical modeling of uranium speciation in the subsurface aquatic environment of Punjab state in India. J Geol Min Res 3(5):137–146

Kumar A, Tripathi RM, Sabyasachi R, Mishra MK, Ravi PM, Ghosh AK (2014) Characterization of groundwater composition in Punjab state with special emphasis on uranium content specialization and mobility. Radiochim Acta 102(3):239–254

Kumar A, Usha N, Sawant PD, Tripathi RM, Sanu S, Mishra M (2011b) Risk assessment for uranium in subsurface water of Punjab state, India. Int J 17(2):38

Sahoo SK, Mahapatra S, Chakrabarty A, Sumesh C, Jha VN, Tripathi RM, Puranik VD (2009) Distribution of uranium in drinking water and associated age dependent radiation dose in India. Radiat Prot Dosiometry 136:108–113

Scott RC, Barker FB (1962) Data on Uranium and Radium in Groundwaters of United States. U.S. G.S Professional Paper no 426

Shenoy NS, Verma A, Kumar SA, Pandey S, Kumar SD, Reddy AVR (2012) A comparative analysis of uranium in portable waters using laser fluorimetry and ICPMS techniques. J Radioanal Nucl Chem 294(3):413–417

Singh KP (2010) Uranium in groundwater in Punjab state, India: status paper. Punjab Envis Cent 8:1–13

Singh KP, Kishore N (2010) Uranium in groundwater. Some observations from parts of Talwandi Sabo block, District Bathinda, Punjab. Proceedings Punjab Science Congress, February 7–9

Singh L, Kumar R, Kumar S, Bajwa BS, Singh S (2012) Health risk assessments due to uranium contamination of drinking water in Bathinda region, Punjab state, India. Radioprotection. doi:10.1051/radiopro/2012042

Surinaidu L, Mahesh J, Padalunaidu G, Rajesh R, Acharya PS, Sharma VS, GurunadhaRao VVS (2010) Geophysical Studies for Assessment of Groundwater potential in old Amritsar, Jallandhar, Ludhiana and Muktsar districts. In: Rao MS, Khobragade S, Kumar B, Singh RD (eds) Water Availability and Management in Punjab, National Institute of Hydrology, Roorkee, pp 173–180

UNDP (1979) Groundwater Studies in the Ghaggar River Basin, India, Project Report DP/UN/IND-74-009/1, New York

United States Environmental Protection Agency (USEPA) (1991) EPA Standard Default Exposure Factors, Washington, DC, USA

World Health Organization (WHO) (2004) Guidelines for drinking water quality, 3rd edn. Geneva, Switzerland

Conjunctive Use of Surface and Groundwater in Efficient Manner

D.K. Chadha

Abstract Conjunctive use of groundwater and surface water is being adapted in many countries to enhance the intensity of irrigation and host at often benefits like cope with the water logging and salinity problems in canal irrigated areas, blending of water reduce the contamination load etc. In India, a number of scientific studies have been conducted which conclusively proved its importance and benefits, but the implementation of this technology has been overlooked because of apathy at various levels. The implementation of conjunctive use program has to be considered now seriously if we really mean "each drop of water counts" for food security in views of the climate change scenario. The chapter presents details of the different aspects of the conjunctive use with the assertion that institutional arrangements and action plan be made for taking up the projects for which detailed hydrogeological studies have already been completed and found favorable for its application.

Keywords Aquifer · Conjunctive use · Water logged · Inland salinity Haryana

1 Introduction

The term conjunctive use generally refers to a coordinated and planned utilization of two or more sources of water viz groundwater and surface water. Thus, conjunctive use management can be defined as the management of multiple water resources in a coordinated operation to the end that the total yield of such a system over a period of years exceeds the sum of the yields of the separate components of the system resulting from an uncoordinated operation.

D.K. Chadha (✉)
Global Hydrogeological Solutions, G-66, Block G, Vikas Puri, India
e-mail: devinderchadha27@gmail.com

D. Saha et al. (eds.), *Clean and Sustainable Groundwater in India*,
Springer Hydrogeology, https://doi.org/10.1007/978-981-10-4552-3_8

The conjunctive use is extremely important particularly in canal command areas and is planned with the following broad objectives:

- To mitigate the effect of canal water supplies shortage.
- To improve the dependability of the existing water supply.
- To alleviate the problem of high water table and salinity.
- To facilitate the use of saline groundwater by blending with freshwater.

Projects using conjunctive use technology are yet to take up in India. However, they are occasionally referred in the hydrogeological reports of departments, discussed at scientific gatherings and publications (e.g., Bredehoeft and Young 1983; Sahuquillo 2002; Chadha 1991; Chadha 2013). To improve drought, water security and for long-term climate change adaptation, conjunctive use represents one of the most important tool in terms of field-level water management. Deploying the groundwater storage associated with aquifers to buffer the high flow variability and drought propensity of surface water (Foster et al. 2010a) is the key characteristic of conjunctive use and is thus capable (at varying levels of efficiency) of achieving:

1. Greater water supply security, using natural aquifer storage, provides the opportunity.
2. Higher net water supply, than we can get using only one source.
3. Timely irrigation-water delivery—since groundwater can be put in use rapidly to offset any deficit in canal water availability at critical times for crop growth.
4. Reduced environmental impact—by counteracting land water logging and salinization, and excessive river flow depletion or aquifer over exploitation.
5. Increasing water supply to tail-end distributaries where much reduced water volume reaches.

2 Types of Conjunctive Use

The conjunctive use can be defined in terms of

(i) Spontaneous use and
(ii) Planned conjunctive use.

2.1 Spontaneous Conjunctive Use

Spontaneous or impulsive (unplanned, unregulated and unmanaged) groundwater use many a time results in depletion to water levels in aquifer that make problems in deployment of cost-effective ground-level irrigation pumps and/or that induce saline groundwater encroachment. Whereas in many irrigation canal command areas water used is more than requirement and many a times excessive leakage cause water logging and loss to agricultural productivity.

2.2 *Planned Conjunctive Use*

Planned conjunctive use often followed in the canal command areas recognizes that there shall be an upper limit on how much groundwater can be withdrawn based on average rate of recharge in the area under consideration. It needs to find a balance of groundwater use along with availability for avoiding long-term water-level decline, as well as to counter rising water tables and reducing the water logging and soil salinization. Allocation of water in the area of conjunctive management needs to be on an "integrated basis," as surface water and groundwater in alluvial aquifers are generally hydraulically connected.

3 Conjunctive Use Plan

For successful adaptation of conjunctive use technology, number of strategies need to be adopted depending upon the running period of canal/distributary and the hydrogeological framework groundwater resource availability of aquifers. Some of the strategies are listed below:

Strategy 1: Allocating different parcels of land permanently for a particular use. For example, a parcel of land can be allocated for surface water and another parcel for irrigation by groundwater.
Strategy 2: Allocating surface and groundwater in time so that in particular season (say Kharif) only surface water is used and in the other season like Rabi, only groundwater is explored.
Strategy 3: This strategy is a combination of 1 and 2. Some part of the command will be permanently on surface water and some other permanently on groundwater. Beside some part will get surface water in one season and groundwater in the other.
Strategy 4: Both ground water and surface water can be used with variations within the year as well as from year to year depending on rainfall, surface and groundwater availability.

Foster et al. (2010a, b) has identified the cause which hamper promotion of rational and better planned conjunctive use in established command areas as:

- Dominance of socio-politically powerful farmers in areas dominated with surface water supply and who refuse to reduce canal intakes, thus available flow to reach marginal areas dwindles.
- Poor understanding of potential role of groundwater and conjunctive use technique among water resource specialist and irrigation engineers, and due to absence of domain specialists in controlling departments.

- Separate agencies for surface water and groundwater management, hinder realistic and adequate change is water use. Problems of rural electrification and adequate power supplies for pumping of groundwater in canal command areas.
- Faulty water pricing policy with a huge cost difference (as felt by users) between groundwater and surface water for irrigation due to highly subsidized canal water.

These hurdles must be overcome by prolong mass awareness regarding the risks of water logging and salinization. As well as introducing revised water allocation criteria incorporating incentives for balanced groundwater use and campaigning about collateral benefits of groundwater pumping.

4 Problem of Water Logging and Salinity

There is no denying of the fact that the water logging problem emerged mainly as a direct effect of seepage from applied irrigation as well as from unlined canal systems. But it can also be caused to some extent by excess soil moisture due to periodic flooding, overflow of run-off, and obstructed sub-surface drainage.

Adverse effects of canal irrigation system, i.e., water logging in Western Yamuna canal (Haryana) was first received attention in 1850, followed by Nira Irrigation Project in the Deccan in 1884. By 1907, water logging problem was noticed in the canal of Punjab (CGWB 1999). In Punjab, the water logged area was estimated to be 109 million km^2 in 1958. In Haryana as per 1966 estimate, area affected by water logging was 0.0065 million km^2. In Uttar Pradesh, the estimated area was of the order of 0.0081 million km^2. In Rajasthan, it was 0.0035 million km^2 in 1968.

The water logging status in Rajasthan, Punjab, Haryana, and Uttar Pradesh is given in Table 1. The comparative statement of 1972, 1976, and 1998 reveals the change under water logged areas. In 1998, the area prone to water logging in the states was 65,860 km^2 and water logged area as 28,220 km^2.

Table 1 Status of water logging and groundwater saline areas in some states

Sl. No	District	Nagraj et al. (1972) km^2	Nat. Comm. for Irr. (1976) km^2	Water logged areas (0–2 m) (Nov. 1998) km^2	Areas prone to water logging (2–3 m) (Nov. 1998) km^2	Areas underlain by saline groundwater (Nov. 1998) km^2
1	Haryana	–	6200	5478	5455	11,438
2	Punjab	14,270	10,900	4875	4500	3058
3	Rajasthan	NR	3480	2779	1680	141,036
4	Uttar Pradesh	6900	8100	20,566	59,680	–
	Total	21,170	22,480	28,220	65,860	144,094

In canal command areas, law is silent on the compensation to the farmer who loses their land because of water logging. There is a need to consider compensating the poor farmers for such an eventuality.

5 Conjunctive Use Studies in India—Overview

In order to draw the attention of the scientists and engineers to check the problems of water logging and salinity and to promote the concept of conjunctive use, Central Ground Water Board organized a seminar on "Conjunctive Use of Surface Water and Ground Water Resources" in 1986. The key recommendation was to use the groundwater stored in the aquifers to enhance the intensity of irrigation in India as is being done in other countries and reduce the areas under water logging.

The status report on "water logging and its remedy by conjunctive use" emphasized details of the conjunctive use mechanism and the status of different studies carried out in six river basins (Chadha and Sinha Ray 2000). The studies brought out the benefits in increasing the intensity of irrigation and the cost benefit ratio of the respective projects.

For enabling greater water supply security in Indian conditions, Foster and Steenbergen in 2011 strongly emphasized spontaneous conjunctive use of water by extraction from shallow aquifers in canal-commands which is driven by the capacity for groundwater to buffer the variability of surface water availability.

Considering seriousness of the problem and the importance of the conjunctive use, Government of India sanctioned six detailed studies in irrigation commands so that it can bring out up-to-date analysis and recommendations for application of conjunctive use to solve water logging problem and also increase in the intensity of irrigation. An overview of six completed studies is given in Table 2.

Based on the encouraging results, four additional studies were added under Phase II, covering Nagarjuna Sagar Project, Andhra Pradesh; IGNP—phase II Kosi Canal Command Area, Bihar and Gandak Command Area and Rajasthan, and were awarded.

6 Hydrogeological and Engineering Feasibility

Pandit (1983) listed the following hydrogeological and engineering parameters while implementing the conjunctive use studies in Haryana.

Location of augmentation tube wells along the carrier channels is governed by the following hydrogeological and engineering considerations:

i. Adequate potential aquifers zone should be present within a depth of 150 m which should be capable of yielding mln.mum discharge of 50–70 ha m per year.

Table 2 Overview of the results of conjunctive use studies in different irrigation canal command areas

	Canal name	Total area under study km^2	Water logged area km^2	Conjunctive use planning
1	IGNP-1	5530	1150	The water logged areas will reduce to 785 km^2. The present groundwater development is cropping intensity increase to 120% and groundwater development to 18%. Cost benefit ratio 1.03
2	Hirakud	1570.18	Pre-monsoon —174 Post-monsoon —1494	The present irrigation intensity is 170% but will be increased to 200% by using surface water 90% and groundwater 10%. Cost benefit ratio is 1.66
3	Sharda Sahayak	8978	4874	To control water logging, it is proposed to reduce surface water utilization from the present 1481.8 to 1015.9 MCM and the groundwater utilization to be increased from 906.32 to 1356.23 MCM; cropping intensity from 153 to 200%; cost benefit ratio 1.56
4	Tungabhadra	6354	Pre-monsoon —98.75 Post-monsoon —229.00	Cropping intensity will be increased from 61.43 to 116%. Cost benefit ratio 3.45
5	Ghataprabha	10,370	July-143, Aug-344	Cropping intensity will increase to 140% and cost benefit ratio varies 1.25–5.34
6	Mahi Kadana	3717	Pre-monsoon —3370 Post-monsoon —160.80	It is proposed for an irrigation intensity of 185% till 2015. The present utilization of surface water and groundwater in the proportion of 65:35 is optimal situation of conjunctive use

ii. The aquifer zones present below the base of the unconfined aquifer, as far as possible, should be exploited. It is done so that the water table is not adversely affected.
iii. The carrier channels should preferably be lined so that the water due to seepage losses is not circulated.
iv. There should be adequate storage in the aquifer, and the annual recharge is enough to sustain the draft from the contemplated augmentation tube wells for obvious reasons.
v. Depth to water table should not be very deep, and it should be between 7 and 12 m.

vi. In areas where saline groundwater underlies the freshwater, the cone of depression should remain well above the saline–fresh interface to check ingression of saline into freshwater by upconing.

The implementation of conjunctive use by diverting the groundwater to the canal system created some management problems in case of Narwana branch along which under 100 deep tube wells were installed for conjunctive use, resulted in lowering of the water table of the shallow aquifer. The management problem created by lowering of water table in Narwana Branch area was overcome by resorting to following policy decision:

i. Water supply is made available to the area served by the affected shallow tube well by running of nearby augmentation tube wells at the option of the shareholders at the normal rates of direct irrigation tube wells.
ii. Alternatively, it is compensated in respect of meeting the cost of lowering their pumping sets @ Rs. 300.00 per meter depth.

In case for the conjunctive use brackish groundwater is to be used, the significant factors for mixing saline groundwater with canal water are as follows:

1. To mix the brackish groundwater with the canal water so that the quality of mix water is within the permissible limit for drinking and irrigation use. Gupta (1984) carried out experimental studies and concluded that the various factors which are significant for such a mixing are as follows:

   a. Pre-sowing irrigation will have to be given preferably by canal water.
   b. Subsequent watering can be given by mixed water.
   c. Tube wells be run for the period the water course is running.
   d. First outlet will have to be 20–60 m away from the point of mixing.
   e. Power availability constraints will lead to utilization of tube wells for only thousand to 1500 hr per day.

2. To develop the saline groundwater at the canal nakka points to mix it with canal water for its use in agriculture.

The following important points to implement the conjunctive use projects efficiently are as follows (Chadha and Sinha Ray 2000):

i. Identification of land by proper definition, collect information on the extent, characteristics, genesis, and classification of the salt-affected soils in different parts of the country.
ii. Study salt and water dynamics in irrigated agriculture, conduct detailed hydrological surveys, and provide suitable criteria for surface and sub-surface drainage for controlling the salt and water balance in the soil.
iii. To create an institutional arrangement or an expert group.
iv. Study the factors governing the chemical composition of surface and underground waters, evolve methods to check deterioration and pollution of these waters, and utilize waters of different qualities for agricultural purpose.

v. Find physiological mechanism associated with salt tolerance of crop and evolve tolerant crop varieties for saline and sodic conditions. Develop technologies for the reclamation and utilization of salt-affected lands in the country.
vi. To implement pilot projects in different canal command areas under World Bank or UNDP programs.
vii. Training and research in the field of salt-affected soils and related subjects to central and state organizations.
viii. Evaluation.

7 Conclusion

The conjunctive use mechanism is being used by a number of countries for achieving enhanced irrigation returns besides solving the problem of water logging and inland salinity. It is therefore recommended that conjunctive use of the project areas where studies have been completed should be considered for implementation with support from international institutions if we really mean "each drop of water counts" for maintaining food security under climate change scenario. Although the conjunctive use of groundwater from deeper aquifers with surface canal water is successful in encouraging, the conjunctive use of saline groundwater which covers large areas in north India with surface canal water can be carried forward with field experiments and institutional arrangements.

Because of the failure in implementing conjunctive use projects, it is now pointed out that Indian hydrogeologists responsible for dealing with the problems of water logging, salinity, and water scarcity have to advocate in favor of conjunctive management of surface and groundwater and facilitate its implementation by enabling institutional modifications at policy level.

References

Bredehoeft JD, Young RA (1983) Conjunctive use of groundwater and surface water for irrigated agriculture: risk aversion. Water Resour Res J 19(5):1111–1121

Central Ground Water Board (1999) Bhujal News, Special issue on conjunctive—use studies, vol 14(1&2), pp 1–113

Chadha DK (1991) Conjunctive use of groundwater and surface water resources in Kurushetra area, Haryana State, India—a case history. XXIII I.A.H. Congress "Aquifer exploitation"—Canary Islands, Spain

Chadha DK (2013) Conjunctive use—coping with water logging and salinity. Water Energy Int 43–46

Chadha DK. and Sinha Ray SP (2000) Water Logging and its Remedy by Conjunctive Use, Central Ground Water Board, pp 1–141

Foster S, Steenbergen F (2011) Conjunctive groundwater use: a 'lost opportunity' for water management in the developing world. Hydrogeol J 19(5):959–962

Foster S, Hirata R, Misra S, Garduno H (2010a) Urban ground water use policy: balancing the benefits and risks in developing nations. GW-MME strategic overview series 3. World Bank, Washington, D.C. www.worldbank.org/gwmate.CitedDec2010

Foster S, van Steenbergen F, Zuleta J, Garduno H (2010b) Conjunctive use of ground water and surface water: from spontaneous coping strategy to adaptive resource management. GW-MATE strategic overview series 2. World Bank, Washington, DC

Gupta MC (1984) Possibilities for mixing of brackish and saline groundwater with fresh water in canals and water courses. Document on studies for the use of saline water in the command areas of irrigation projects Haryana, India, pp 47–54

Pandit RN (1983) Conjunctive use of ground water in surface irrigation systems-performance in Haryana. In: Proceedings of the seminar on Ground water development—a perspective for the year 2000 A.D., pp 281–292. Water Resources

Sahuquillo A (2002) Conjunctive use of surface water and groundwater. UNESCO Encyclopedia

Silveira L (ed) Life-support systems (Chap. 2.9). EOLSS Publishers

Efficient Conjunctive Use of Surface and Groundwater Can Prevent Seasonal Death of Non-Glacial Linked Rivers in Groundwater Stressed Areas

Shashank Shekhar, Suman Kumar, Rajiv Sinha, Sanjeev Gupta, Alexander Densmore, S.P. Rai, Manoranjan Kumar, Ajit Singh, Wout Van Dijk, Sunil Joshi, Philipa Mason and Dewashish Kumar

Abstract The surface and groundwater system is an interdependent system. In the case of alluvial plains, they are best manifested by the relationship between a river and groundwater system. This relationship is vibrant and dynamic in rain-fed rivers or the non-glacial linked rivers. It has been typically observed that many a case such rivers originate from lesser elevation and in downstream stretches; during summers, their flow is maintained by groundwater contribution. In groundwater stressed regions, often these rivers get disconnected from aquifer. Thus there could be different levels of stream-aquifer interaction during monsoon and non monsoon seasons. In this context, the article examines effect of long-term groundwater abstraction on such river flows in a conceptual framework. It also proposes efficient conjunctive use of surface and groundwater to prevent death of such rivers.

Keywords Surface water · Ground water · Conjunctive use · Aquifer Stream–aquifer interaction

---

S. Shekhar (✉) · S. Kumar · M. Kumar
Department of Geology, University of Delhi, New Delhi, India
e-mail: shashankshekhar01@gmail.com

R. Sinha · S. Joshi
Indian Institute of Technology, Kanpur, Kanpur, India

S. Gupta · A. Singh · P. Mason
Imperial College, London, England

A. Densmore · W. Van Dijk
Durham University, Durham, UK

S.P. Rai
NIH, Roorkee, Roorkee, India

D. Kumar
National Geophysical Research Institute, Hyderabad, India

D. Saha et al. (eds.), *Clean and Sustainable Groundwater in India*,
Springer Hydrogeology, https://doi.org/10.1007/978-981-10-4552-3_9

1 Introduction

The growing population and improvement in the standard of living has led to increase in water uses world over. The water resource in many parts of India is under stress (Chatterjee et al. 2009; Rodell et al. 2009; Shekhar 2006a; Tiwari et al. 2009; Mukherjee et al. 2015). The demand–supply imbalance requires a scientifically planned strategy. The two major components of water, the surface and groundwater systems, remain in dynamic equilibrium (Shekhar and Prasad 2009). Hence, one of much talked about such strategy is efficient conjunctive use of surface and groundwater. The strategy involves rational and symbiotic development of both surface and groundwater resources to supply a given urban or irrigation canal command areas (Foster and Steenbergen 2011). This involves detailed estimation of the water demand and allocation of the surface and groundwater sources optimally. The underlying principle is based on restricting uses to environmentally sustainable limits. The conjunctive use strategy is particularly popular in canal command areas. This is so because these areas have assured surface water sources as an alternate to the existing groundwater sources. An optimal and prudent utilization of canal water depends on canal allowances for different regions and efficiency of distributaries networks in different parts of the canal command areas. Further, the conjunctive use strategy links it to the groundwater potential of the region. It is here that efficiency of this strategy is required for sustainable water management.

The overexploitation of groundwater resources can certainly have adverse effect on stream–aquifer interaction. The present article examines changes in stream–aquifer interaction linked to water uses in a conceptual framework. This becomes important as approximately 16% of assessment units in India (total 6607) are overexploited with regard to groundwater resources (CGWB 2014a), see Fig. 1.

2 Decline in Groundwater Levels in India

India experiences strong monsoon season, and monsoon rainfall is the main source of recharge. The deeper water level in a year is observed just before the onset of monsoon (pre-monsoon), while the shallowest water level is observed in the month of August. The groundwater level of November is considered to be very important for water recourses utilization planning, because the month is beginning of main extraction of groundwater for irrigation. Irrigation consumes about 90% of total groundwater extraction. If we observe water level difference between decadal mean of pre-monsoon water level (2003–12) and pre-monsoon 2013, it is obvious that a good part of India has shown decline in water levels (Fig. 2). This is mainly attributed to heavy abstraction of groundwater and demising recharge in many

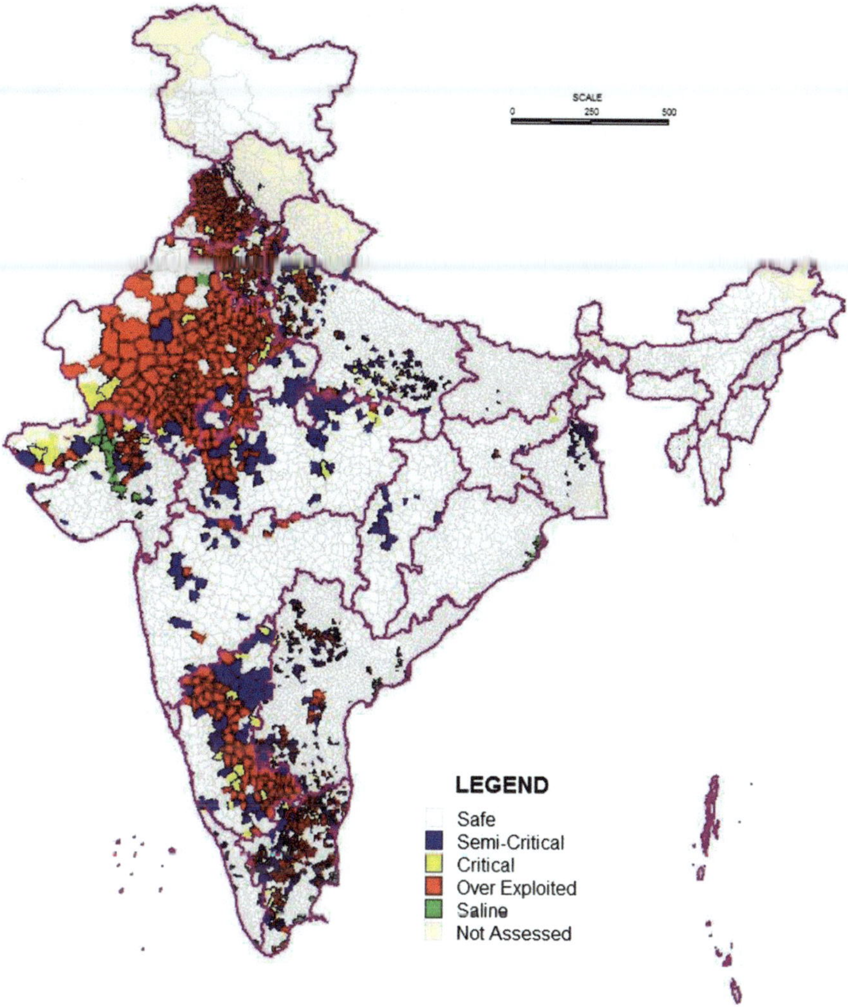

Fig. 1 Categorization of groundwater assessment unit in India (CGWB 2014b)

pockets. A significant part of India has depth to water level in the range of 10–20 m below ground level (mbgl), and some regions have still deeper water levels as observed in pre-monsoon 2013 (Fig. 3). In such regions, the stream–aquifer interaction for non-glacial fed rivers changes gradually. This gradual change is explained further below.

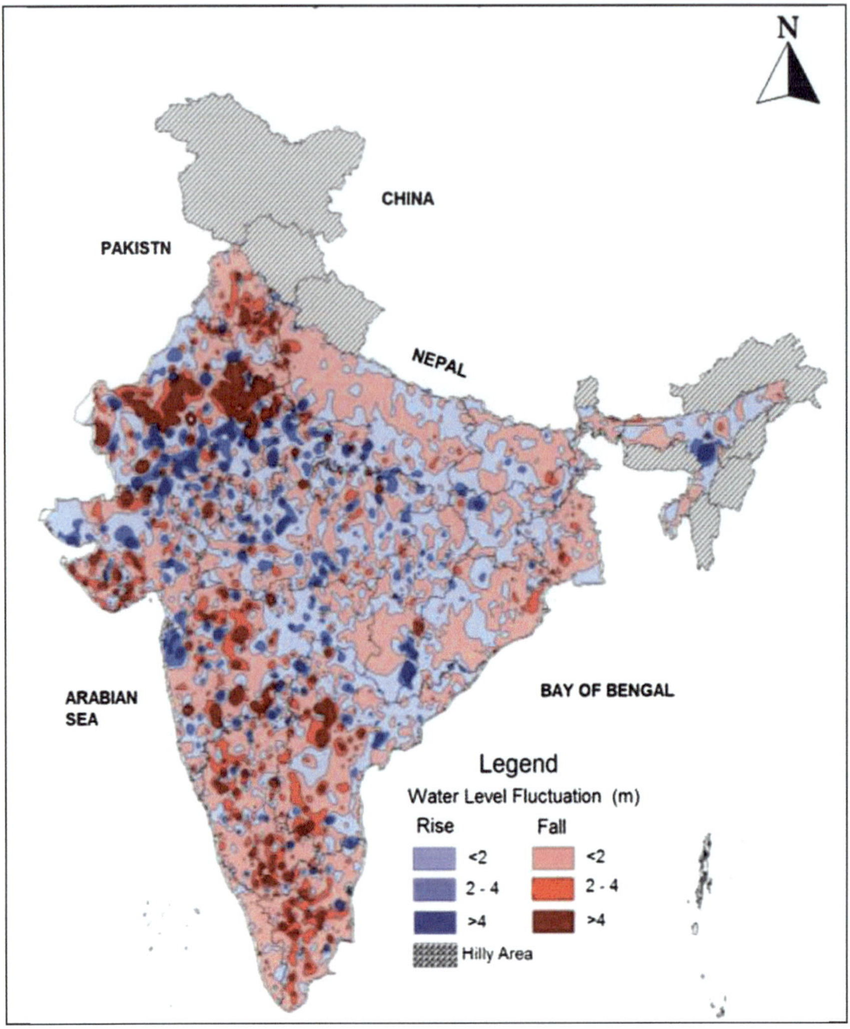

Fig. 2 Decadal water level fluctuation—decadal mean May (2003–2012) versus pre-monsoon 2013 (CGWB 2014b)

3 Stream–Aquifer Interaction in Conceptual Framework from Non-Glacial Rivers

The stream–aquifer interaction for non-glacial rivers in areas with declining water levels is explained with help of a schematic model (Fig. 4). The model demonstrates changes in stream–aquifer dynamics over the years with changes in groundwater level. It shows how the changes in water level over the basin have affected vertical connectivity in the river basin and the stream–aquifer interaction

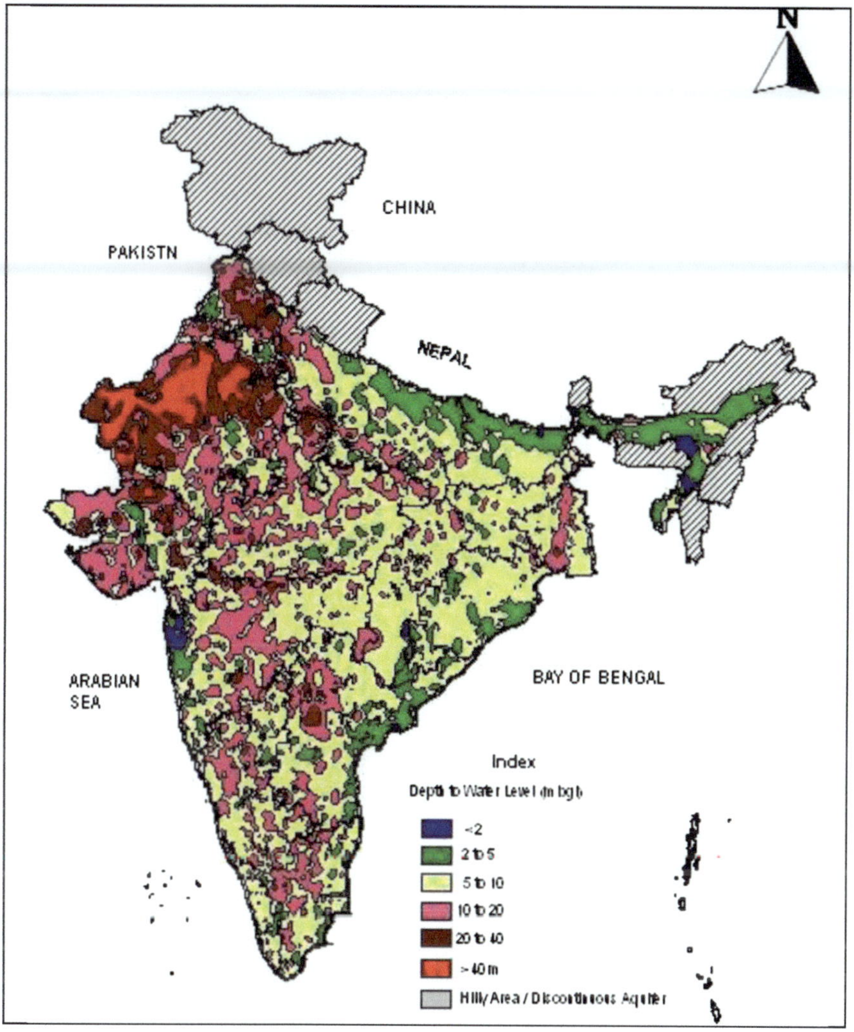

Fig. 3 Depth to water level map pre-monsoon 2013 (CGWB 2014b)

dynamics (Fig. 4). It is clearly observed that the river changes its nature from an effluent stream to influent stream as the water level declines. On account of this, the base flow in the river ceases there by, leading to very limited flow during non-monsoon months. The overall water system dynamics in the basin evolves to a stage that there is continuous decline in groundwater level and river flow. If the water level decline continues, the flow in such rivers during lean seasons becomes dead. An efficient conjunctive use of surface and groundwater resources in such areas would prevent seasonal death of rain-fed or non-glacial linked rivers.

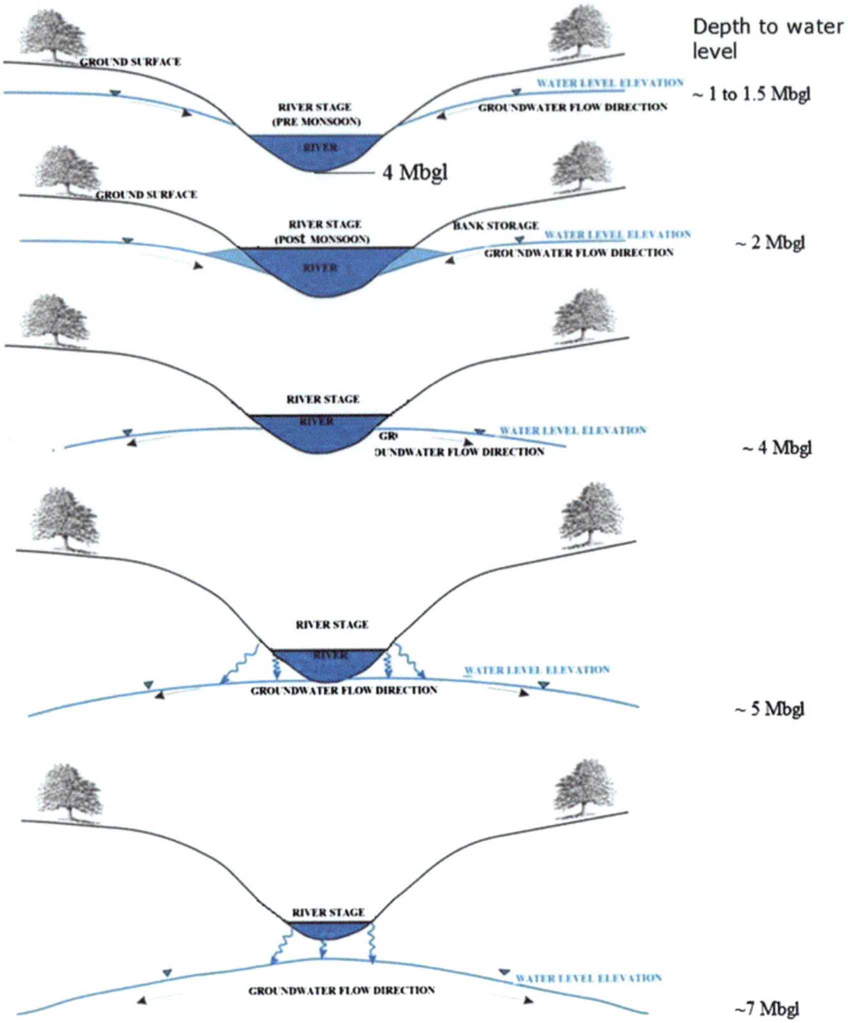

Fig. 4 Schematic model showing changes in vertical connectivity and stream–aquifer interaction in the Ghaggar basin vis-à-vis changes in groundwater levels (Shekhar 2006b)

4 Discussion and Conclusions

Some of the groundwater stressed areas in India such as Punjab and Haryana have excellent network of functional canal irrigation system. The network of canals supplies irrigation water for agricultural activity. When such assured canal water is available, farmers use minimal amount of groundwater. While at the same time when there is non-availability of assured canal water, farmers resort to groundwater pumping for irrigation purposes. This is an area where conjunctive use intervention is required. The conjunctive use strategy optimally allocates water demand to surface and groundwater sources in environmentally sustainable way.

These groundwater stressed areas also have non-glacial linked rivers such as Ghaggar and Chautang. It has been observed that some of these non-glacial linked rivers historically used to have sufficient flow in them during non-monsoon season because of effluent seepage from aquifer. However, in recent past these rivers are dead in some stretches during lean season. This phenomenon can be explained with the help of schematic model above (Fig. 4). Seasonal death of such rivers could have been easily prevented by efficient conjunctive use of surface and groundwater. A visit to one such basin (Ghaggar River) revealed overreliance on groundwater abstraction for agricultural uses. This could be on account of skewed canal supply or canal water distribution-related management issues. However, if prudent and efficient conjunctive use policy would have adopted in the basin, it would have restricted sharp decline in water levels and disconnection of stream–aquifer system. This in turn would have prevented death of the river during non-monsoon periods. Thus, it is essential that sustainable agricultural groundwater abstraction should be estimated in conjunction with the canal allowance. It would have been ideal to work out water balance of the basin considering to sustainable groundwater yield. Such estimate needed to have rider that stream-aquifer connectivity should be retained.

We conclude that the aim should be for maximization of agricultural output with limitation that water resources exploitation remains within environmentally sustainable limits. When we try for this, the best approach is to maintain the inflow and outflow directions of the dynamic equilibrium between surface and groundwater sources. This will certainly not alter base flow or propose over irrigation by canal system so as to induce soil salinity.

Acknowledgements The financial support through MoES, India, and NERC, UK funded project entitled "The structure and dynamics of groundwater system in north western India under past, present and future climates" is duly acknowledged. The authors are thankful to Chairman, CGWB & Dipankar Saha Member (SAM) for their support.

References

CGWB (2014a) Dynamic Groundwater Resources of India (31 March 2011), Central Ground Water Board, Ministry of Water Resources. River Development & Ganga Rejuvenation, Faridabad, India

CGWB (2014b) Ground Water Year Book, 2013–2014. Central Ground Water Board, Ministry of Water Resources, Faridabad, India

Foster S, Steenbergen FV (2011) Conjunctive groundwater use: a 'lost opportunity' for water management in the developing world? Hydrogeol J 19:959–962

Gupta BK, Chatterjee R, Mohiddin SK, Singh PN, Shekhar S, Purohit R (2009) Dynamic groundwater resources of National Capital Territory, Delhi: assessment, development and management options. Environ Earth Sci J 59:669–686

Mukherjee A, Gupta A, Ray RK, Tewari D (2015 On line) Aquifer response to recharge—discharge phenomenon: inference from well hydrographs for genetic classification. J Appl Water Sci. doi:10.1007/s13201-015-0293

Rodell M, Velicogna I, Famiglietti JS (2009) Satellite-based estimates of groundwater depletion in India. Nature 460:999–1003

Saha D, Harvey CF, Taylor RG, Ahmed KM, Bhanja SN (2015) Groundwater systems of the Indian Sub-Continent. Abhijit Mukherjee. J Hydrol Reg Stud. doi:10.1016/j.ejrh.2015.03.005

Shekhar S (2006a) An approximate projection of availability of the fresh groundwater resources in South West district of NCT Delhi, India—a case study. Hydrogeol J 14:1330–1338

Shekhar (2006b) Environmental flows and river rejuvenation. J Geol Soc Ind 88(6): 813–814

Shekhar S, Prasad RK (2009) The groundwater in Yamuna flood plain of Delhi (India) and the management options. Hydrogeol J 17:1557–1560

Tiwari VM, Wahr J, Swenson S (2009) Dwindling groundwater resources in northern India, from satellite gravity observations. Geophys Res Lett 36(18):1–5

Specific Yield of Unconfined Aquifers in Revisiting Efficiency of Groundwater Usage in Agricultural Systems

Himanshu Kulkarni, Uma Aslekar and Dhaval Joshi

Abstract Groundwater has been the backbone of water for agriculture in India. Millions of farmers depend on groundwater for sustaining their agricultural livelihood. Any discussion on groundwater would be incomplete without taking into consideration the resource unit i.e. aquifers. Aquifer storage capacity is not necessarily reflected in discussions on water use efficiency. Efficiency estimation in agricultural systems has centred on a crop-per-drop approach. We take up three traditional approaches to efficiency in the context of two case studies from western Maharashtra which are part of shallow unconfined Deccan basalt aquifer systems, viz. Pabal and Randullabad. We compare the efficiencies of these two aquifers and propose a more comprehensive methodology to address efficiency in groundwater-based agricultural systems taking into consideration the characteristics of aquifers, mainly the specific yield. Such an efficiency index is useful in gauging specific impacts from initiatives that include supply and demand-side interventions.

Keywords Groundwater · Water use efficiency · Aquifers · Deccan basalts Specific yield · Maharashtra

1 Introduction

India's groundwater story has many similarities with that from other parts of the world. However, India's groundwater usage is quite unique, in more ways than one. While India has emerged as the largest groundwater extractor in the world, this precious natural resource still remains a blind spot in the national discourse on water management and governance. With about 30 million irrigation wells increasing at a rate of 0.8 million wells per year, India has also emerged as the leading groundwater anarchy across the globe (Shah 2010). Currently, about 85%

H. Kulkarni (✉) · U. Aslekar · D. Joshi
Advanced Center for Water Resources Development and Management (ACWADAM), Plot 4, Lenyadri Society, Sus Road, Pashan, Pune 411021, India
e-mail: acwadam@vsnl.net

D. Saha et al. (eds.), *Clean and Sustainable Groundwater in India*,
Springer Hydrogeology, https://doi.org/10.1007/978-981-10-4552-3_10

rural water supply comes from groundwater (DDWS 2006) while more than 60% irrigation in agriculture is through groundwater (Ministry of Agriculture 2014). The output from groundwater-irrigated areas is 1.5–3 times more than that due to surface water irrigation (Mukherji and Shah 2005). The use and exploitation of groundwater has not been limited to rural areas and agriculture but has rapidly transgressed into other sectors such as urban water supply and industrial use. According to a study conducted by Centre for Science and Environment (Narain 2012), about 48% of urban water supply is dependent on groundwater, through formal and informal water supply systems. Hence, not only has groundwater become the mainstay of rural water supply for both domestic and agricultural purposes, but also it has significantly contributed to filling out the gaps in urban public water supply as well as in the irrigation command areas based on surface water projects.

The above statistics on groundwater usage imply that on an average, at least 1 billion Indians use groundwater every day with some 700 million Indians using groundwater every day in rural India. What is even more interesting is the fact that at least 420 million Indians use groundwater in agriculture during at least one season of the year. Groundwater forms a reliable buffer against both weather vagaries and climate change. Groundwater is less vulnerable to climate fluctuations than surface water and is therefore an important buffer to prevent migration of farmers and stabilize agricultural activities in times of drought (Moench 2002). In an agrarian country like India, this becomes very critical since most of the population dependent on this resource has either been small landholders, farmers with limited entitlements or farmers from drought prone regions. Small and marginal farmers make up 78% of the total; each of these farmers operates less than 2 ha land but together access 32% of the total land; however, they own 45% of wells and tube wells and account for almost 40% of the groundwater-irrigated area (Mukherji and Shah 2005). The growth of groundwater in India has been fuelled by millions of farmers gaining access to groundwater across different hydrogeological and agroecological typologies in India.

The impacts of such a large dependency on groundwater resources have meant that groundwater exploitation is evident in many regions of the country. Moreover, groundwater contamination is also evident in some isolated regions. Many villages in India have witnessed drinking water schemes in villages slipback to the pre-scheme conditions, while agriculture production drops as groundwater levels fall. As per the Planning Commission report (2007), about 60% of the blocks in the country either face issues of availability and quality or in some cases, both. Quality issues related to geogenic pollution such as fluoride and arsenic have emerged in some regions of the country while issues related to salinity have sprung up in others (CGWB 2011). Groundwater competition has been a cause and an effect of the growing groundwater dependency in India, often resulting in potential conflict around groundwater resources (Kulkarni et al. 2015).

The iniquitous development of groundwater across the country has led to extreme variation in the impact on agriculture. Even at local scales, one comes across a disaggregated situation where some pockets of a region are under severe exploitation and while neighbouring ones have not witnessed any significant groundwater resources development. Larger socio-economic, and in some cases, political inequity are only magnifying the challenges of groundwater competition and conflict.

Most responses to the groundwater problem have centred around augmenting the resource through soil and water conservation approaches in general and artificial recharge of groundwater in particular. Many promising programmes on watershed management are inclusive of specific interventions on groundwater recharge with good examples of Managed Aquifer Recharge also evident from many regions (Dillon et al. 2009). However, management of water demand, especially in agriculture has remained a major challenge, particularly given the highly decentralised (called 'atomistic' by Shah 2010) nature of groundwater access in the country. Crop-water budgeting coupled with micro-irrigation techniques has progressively found their way into practices and policies on groundwater management (Das and Burke 2013). More crop per drop is increasingly becoming the mantra behind improved efficiencies of water in agriculture, particularly in areas that depend upon groundwater resources. This paper puts forth some preliminary ideas regarding revisiting the aspect of water efficiencies given India's groundwater dependency and attempts to locate the concept of participatory groundwater management as a means of achieving aquifer-level water application efficiency in agriculture in India.

2 Irrigated Agriculture and Water Use Efficiency in India

Irrigation is the largest consumer of freshwater resource across the world accounting for 70% withdrawals and about 90% consumptive use. Globally, out of about 301 million ha irrigated land, about 38% is dependent on groundwater (Siebert et al. 2010). Nearly 38 million ha of land in India is irrigated by groundwater resources, out of the total 62 million ha area equipped for irrigation. Food security, therefore, will be increasingly affected by the status of groundwater in India's aquifers. Moreover, with increasing allocations to non-agricultural sectors such as industry and urban centres, competition and conflict around water resources in general, and groundwater in particular, will only increase (Joy et al. 2008).

The region of Nepal, Bangladesh, Pakistan and China together indicates groundwater extraction of about 300 km^3 of groundwater annually, which comes to about half of the world's total use (Shah et al. 2003). Ever since the advent of green revolution in India, which coincided with the advent of pumping and drilling technology, the emphasis on groundwater for irrigation increased manifold. Today, more than 60% of net irrigated area, in the country, is dependent on groundwater (Vijay Shankar et al. 2011). As a result, cropping pattern changes, mainly the shift from rain-fed farming to irrigated farming became possible. The decentralised and fugitive nature of groundwater as a resource allowed access to millions of farmers in the country to tap into this invisible yet reliable resource.

Traditionally, groundwater-based irrigation systems have relied on dug wells and springs as sources of water. Whether they were the *rahats* in northern India or the *mhots* in western and Peninsular India, these versions of the Persian Wheel method of extraction of groundwater from dug wells had been the mainstay of groundwater irrigation in the subcontinent for centuries. While the Persian Wheel and its adaptations have progressively given way to modern wells (tube wells, bore wells) and pumps, a few of these still exist in districts like Kolar in Karnataka and Dungarpur in Rajasthan. Figure 1 illustrates the comparison between the drawdown in a well fitted with a Persian Wheel and one fitted with a modern submersible pump. Both wells tap a hard-rock aquifer with similar aquifer properties and raise pertinent questions about water use efficiency. Modern pumping systems are able to provide more water over a shorter period of time as compared to that extracted through the Persian Wheel. Persian Wheels on the other hand work on the principle of 'steady state drawdown' and extract a limited amount of water at zero drawdowns bringing out a different dimension regarding the efficiency of well use.

Efficiency of water use has to be assessed in at least three different ways—irrigation, productive and economic efficiencies—according to Vaidyanathan and Sivasubramaniyan (2004). They describe each of these as follows:

- Irrigation efficiency—ratio of consumptive use of irrigation to gross irrigation supplies.
- Productive efficiency—quantum of production per unit of water utilised.
- Economic efficiency—comparison between outcomes of current allocation including use of available water (and associated inputs) and their 'optimum' allocation and use.

Most popularly though, agriculturists have preferred to look at a 'crop-per-drop' approach, which focuses on understanding the total biomass produced against water supplied. Irrigation engineers and scientists have used the term water use efficiency to describe how effectively water is delivered to the crops and to indicate the amount of water wasted in the process (Sharma et al. 2015). Agricultural water productivity is the net benefit from the crops to the amount of water used to benefit

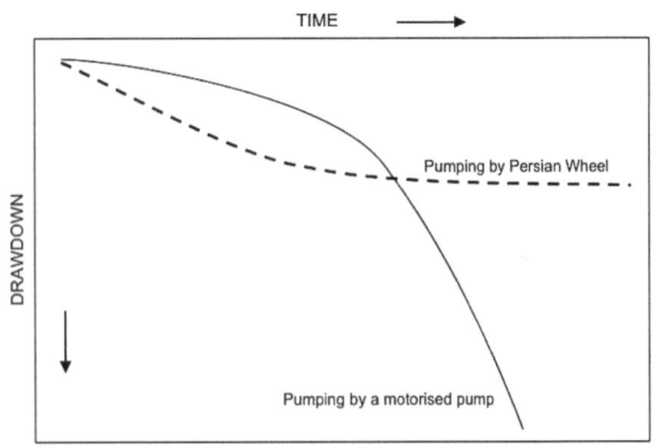

Fig. 1 Time-drawdown plots for a well fitted with Persian Wheel and motorised pump respectively

those crops (Molden et al. 2007). In broadest sense, it reflects the objectives of producing more food, income, livelihood and ecological benefits at less social and environment cost per unit of water consumed (Sharma et al. 2015).

Having said that there is very little said about the environment cost when it comes to groundwater-dependent irrigation systems. Both the above approaches that of the agriculturists and irrigation engineers is limited in its sense since it does not take into consideration the net effect of agricultural pumping on the resource unit of groundwater, i.e. the aquifers providing groundwater. The exclusion of groundwater from the estimates of major river flows in India is a classical example, highlighted through the work of Ranade (2005). Aquifers are used for meeting multiple needs such as water for drinking, domestic, livestock and agriculture. The exploitative consumption of groundwater by agriculture may tend to have various social costs in terms of increasing vulnerability of drinking water scarcity, livestock development etc. Thus, while discussing water use efficiency or water productivity as is often quoted, it is necessary to take into consideration the natural conditions in which groundwater is present.

3 Participatory Groundwater Management in the Heterogeneous Basalt Aquifers of West-Central India

In order to understand the role of groundwater for agricultural sustenance and its effective management, it is necessary to take into consideration the context of the resource, which in case of groundwater is an aquifer. Aquifers are geologic formations that have two essential properties. Firstly, they must possess enough storativity for groundwater to accumulate within aquifer openings, and secondly, they must have sufficient transmissivity for groundwater to flow from one part of the aquifer to another. Storativity and transmissivity not only vary across aquifers, but can also vary within a single aquifer, attributing its heterogeneity so that different wells yield water in different quantities.

Groundwater management and governance in India must take into consideration the concept of the common pool nature of aquifers (Kulkarni and Shankar 2009). Aquifer diversity in India is a function of six broad hydrogeological settings that define the groundwater typology of the country (Kulkarni et al. 2015). The Deccan Volcanic Province of west-central India, constituted largely by basalt rocks is one of these settings. The Deccan basalts (also referred to as the Deccan Traps) occupy an area of more than 500,000 km^2 in west-central India and average 600 m in thickness. The lavas that formed the Deccan basalts erupted in the Tertiary period some 65 million years ago.

The Deccan basalts are formed from many individual lava flows, varying from a few metres up to 100 m in thickness (Saha and Agrawal 2006). A typical sequence within a single flow comprises of a vesicular-amygdaloidal upper layer and a denser compact lower layer. An alternating sequence of vesicular-amygdaloidal and compact types of basalt gives rise to two aquifer systems (Kulkarni et al. 2000). Groundwater occurs within the shallow weathered and fractured zones of these two types of basalts. Shallow aquifers exist over most of the area covered by the Deccan basaltic lavas. The shallow aquifer is normally considered to be unconfined, extending to depths of 15–20 m only. Aquifer transmissivities are usually low, in the range of 20–200 m^2/day (Deolankar 1980; Kulkarni et al. 2000). Higher transmissivities reflect the presence of well-developed sheet joints. The specific yield, or storativity, of the shallow aquifer is also variable, laterally and with depth. Specific yield varies between 2 and 12% (Lawrence and Ansari 1980; Macdonald et al. 1995).

The basaltic aquifer is important in many regions of western and central India as it is the only source of freshwater. The shallow weathered aquifer has been exploited for many centuries for domestic water supply and for minor irrigation by shallow, large-diameter wells. These wells, which are typically 4–10 m in diameter, are ideally suited for exploiting low permeability aquifers since they provide considerable well storage. Deeper confined aquifers are known to exist within the basaltic lava sequence; however, they are considered to be generally of limited extent (Versey and Singh 1982).

The heterogeneous nature of a basalt aquifer has many implications in agriculture. First, well-yields are quite disparate even for wells tapping a single aquifer. This implies that some wells yield water throughout the season, while others are seasonal. Managing such an aquifer remains a big challenge, whether it is supply augmentation by recharge or demand management that includes equitable distribution and efficient usage of groundwater. However, results from ACWADAM's programme on participatory groundwater management in the Deccan basalts of Maharashtra have proved promising on integrating supply and demand-side interventions, particularly with regard to the management of shallow unconfined basalt aquifers.

This paper uses two case studies in developing an 'efficiency-based' approach to compare the impacts of a three-year initiative on participatory groundwater management in the case of the shallow unconfined Randullabad aquifer with a business-as-usual approach of groundwater development and usage in the shallow unconfined Pabal aquifer. Both these areas are situated within the agro-climatic zone of western Maharashtra on the upland plateau formed by a stack of basalt lavas. The region falls in the rain shadow zone of the Western Ghats, typical of great variation in rainfall magnitude and pattern. Pabal village is located in Pune district while Randullabad lies within Satara district. A comparison between of the basic features of the two aquifers is provided in Table 1 as a first step in developing the rationale towards improved understanding of aquifer-level efficiencies.

Pabal village is relative highly populated centre. There are 860 households in the village compromising a population of 3857 (Census of India 2011). The average rainfall in Pabal is 400–600 mm. It comprises of vesicular-amygdaloidal basalts, which form the primary unconfined aquifer systems in the area. All the drinking and agricultural wells are located in this aquifer. Based on the study (Macdonald et al. 1995), it was found out that the Pabal aquifer is in the

Table 1 Salient features of Pabal and Randullabad aquifers

Salient characteristics	Pabal aquifer	Randullabad aquifer
Geological features	Vesicular-amygdaloidal basalt (VAB) + upper fractured portions of the underlying compact basalt (CB); VAB shows well-developed sub-horizontal sheet joints while CB is sub-vertically jointed	VAB with horizontal sheet joints underlain by a columnar basalt (columnar—largely vertical joints)
Aquifer thickness	15 m	18–20 m
Aquifer scale (spatial)	200 ha	268 ha
Transmissivity (range) in m^2/day	10–500 m^2/day	10–100 m^2/day
Specific yield (range)—as fraction from pumping tests	0.008–0.09	0.01–0.05
Specific yield—as fraction—by water balance method	0.05	0.025

overexploited[1] category i.e. groundwater discharge rate exceeds recharge rate. Flood irrigation is the common practice in the village with major crops such as potato and bajra grown in Kharif season while onion, wheat and maize grown in Rabi season. Dug wells form the primary source of groundwater in Pabal.

Randullabad village includes 395 households comprising a population of 1857 (Census of India 2011). The village receives moderate rainfall zone, with average annual rainfall of 600–700 mm. The entire village is dependent on groundwater for fulfilling its drinking water, domestic water and irrigation water demands. Randullabad is one of the more progressive villages in western Maharashtra and has a history of water management practices going back at least 30–40 years. The village passed a resolution in its Gram Sabha, banning 'drilling of bore wells' in the village, in the year 2000. There are more than 160 dug wells, with virtually no bore wells. More than 60% agricultural land of the village is irrigated. Farmers have developed strong market links and are getting good returns from agriculture. Based on the study, it was found out that the unconfined aquifer system of Randullabad comprises of amygdaloidal basalts and highly weathered compact basalt. In terms of groundwater development, it falls under safe category.

The Participatory Groundwater Management (henceforth PGWM) programme was taken up in Randullabad to engage the community in a process of hydrological monitoring, mapping of aquifers, characterising them and finally, arriving at a set of groundwater management protocols. Engaging the communities in the process of monitoring and decision-making on aquifer management is a critical aspect of

[1]Groundwater overexploitation maybe defined as a state in which, for some years, the average abstraction rates from the aquifer is greater than or closer to average recharge rate (Custodio 2002).

PGWM programme. PGWM involves building capacities around understanding of groundwater, setting up a participatory hydrological monitoring network in the village (for monitoring of well water levels, rainfall measurement and surface run off measurement) and a continuing dialogue with the village community on data, results of analysis and institutional decisions.

Various measures were adopted as part of the decision support system within PGWM. These measures included sharing of wells, using drip and sprinkler irrigation sets for water application to crops and changes in cropping pattern. The community had already, in the year 2000, realised the importance of sustaining their unconfined aquifer system as a preference over developing the confined aquifers in the village and had banned drilling of bore wells in the village, through a Gram Sabha resolution. This decision is in place till date, reflecting the high level of collective decision-making and ownership regarding the aquifer(s) in the village.

4 Understanding Ground Water 'Efficiency': Business-as-Usual Versus PGWM

In order to understand groundwater efficiency in the case of a typical aquifer setting within the Deccan basalt, we take up these two cases that represent similar basaltic aquifer systems from the State of Maharashtra in India. Pabal is based in Pune district while Randullabad is based in Satara district (Fig. 2).

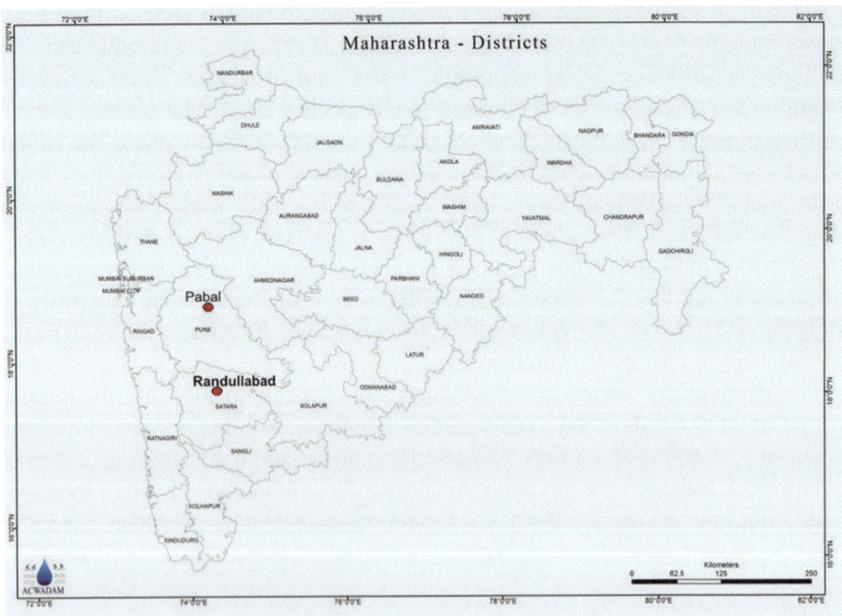

Fig. 2 Location of Pabal and Randullabad in Maharashtra

The Pabal aquifer was studied earlier too, through an intensive mapping and monitoring as it witnessed changes through various stages of development and subsequently exploitation (Kulkarni 1987; Macdonald et al. 1995). The Randullabad aquifer was part of the PGWM initiative that ACWADAM, undertook between 2010 and 2013. These two cases had many similarities, including aquifer conditions, cropping pattern and almost entire reliance on large-diameter dug wells. In both the cases, wheat is the crop for which a significant proportion of irrigation is provided. While other crops are quite similar, wheat can by assumed as the crop for which irrigation from the aquifer causes significant drawdowns. It is also a winter (Rabi) crop when groundwater provides the only source of crop-water (irrigation) without any other sources (rainfall). The major difference in the two cases was the PGWM interventions in Randullabad versus a business-as-usual approach in Pabal. Table 2 summarises the features of both these aquifers on the basis of key parameters measured for this study.

In order to compare the situations in the case of the two aquifers, it is useful to examine different kinds of productivity parameters. Productivity of groundwater use in agriculture can be formulated in three different ways:

1. Crop per drop: the total production in terms of applied water (kg/litre)
2. Crop per unit area: total production in terms of unit area (kg/hectare)
3. Water per unit area: total water applied in terms of unit area (m^3/hectare)

Based on these three productivity indices approaches, we illustrate how these productivity 'indices' can be compared for the two aquifers (Table 3).

The productivity of the Randullabad aquifer is more when the 'crop-per-drop' and 'crop-per-unit-area' approaches are used. On the other hand, the efficiency of water used per unit area in Pabal is greater than that in Randullabad. However, when we bring in the context of aquifer drawdown in terms of groundwater

Table 2 Key parameters measured in the case of Pabal and Randullabad aquifers

Parameters	Pabal aquifer	Randullabad aquifer
Aquifers, their status, irrigation, management practice etc.	Overexploited aquifer, flood irrigation, some watershed development, no demand-side management	Aquifer in balance, sprinkler irrigation, systematic watershed development including groundwater recharge and PGWM
Groundwater abstraction for irrigation (in m^3 over irrigated area)	320,000 m^3	504,680 m^3
Net irrigated area (ha)	400 ha	500 ha
Aquifer specific yield	0.05	0.025
Annual aquifer drawdown (m)	3.5	5.7
Cropping	Potato, onion, wheat, maize	Potato, onion, wheat, sorghum
Indicative yield (quintals/ha) for wheat	<10 q/ha	15 q/ha

Table 3 Comparison of different productivity indices for Pabal and Randullabad aquifers

Productivity index	Pabal	Randullabad	Comparison of productivities
Crop per drop (kg/m^3)	1.25 kg/m^3	1.48 kg/m^3	$P_P < P_R$
Crop per unit area (kg/ha)	10 quintal/ha	15 quintal/ha	$P_P < P_R$
Water applied per unit area (m^3/ha)	800 m^3/ha	1009.25 m^3/ ha	There is less water applied in Pabal than in Randullabad (per unit area of land)

P_P Productivity (Pabal); P_R Productivity (Randullabad)

abstraction for agriculture, we found that the efficiency of Pabal system is better than the efficiency of Randullabad system. Although by traditional approach of 'crop per drop' which does not take into consideration the net effect on the aquifer system, we found that the net productivity in terms of water applied is more in case of Randullabad while it is less in the case of Pabal system.

A classical crop-per-drop approach considers water (or groundwater) an input variable just like other variables such as land-size, fertilizers, and pesticides. Agricultural productivity (production per unit of land) is measured against the water input (volume of water) as one measure of efficiency. However, in this case, one ends up ignoring the net effect of the production or productivity on the resource (aquifer) due to the extraction of water for the input.

Consistent abstraction rates that are more than or close to recharge rates for an aquifer tend to lead to overexploited systems over time (Custodio 2002). The context of aquifer becomes crucial while talking about groundwater-based agricultural systems, especially the factors governing efficiency of water usage from the aquifer. The storativity of an aquifer and its capacity to transmit water i.e. transmissivity are two inherent properties of aquifers, and therefore, how much water a well will yield will depend upon the aquifer characteristics. The first impact of groundwater extraction from an aquifer is the 'drawdown' in the aquifer. Hence, determining the 'crop per unit area per unit drawdown' for an aquifer is crucial in understanding the status of groundwater resources in an aquifer system or even a micro-watershed.

Modern agricultural approaches tend to increase the efficiencies of water applied by changing the method of water application in the form of micro-irrigation systems such as drips and sprinklers. Although these systems ensure efficiencies of water application as compared to traditional methods of flood irrigation, by significantly reducing the conveyance and evaporation losses, they do not necessarily give a complete picture unless the resource unit on which they are dependent, i.e. the aquifer is taken into consideration. Hence, this paper makes a strong proposition for including the specific yield of an aquifer (as an indicator of the unit for groundwater storage capacity of the aquifer) to estimate a compounded measure of efficiency of water usage in agriculture. The following section, including Table 4, discusses the scope of using aquifer drawdown and specific yield as parameters to determine groundwater use efficiency in agricultural systems.

Table 4 Comparing productivity indices within the context aquifers

	Pabal	Randullabad	
Water per unit area per unit drawdown in aquifer (litre/ha/m aquifer drawdown)	228.5	177	$E_P > E_R$
Water per unit area per unit specific yield of aquifer (m^3/ha/specific yield)	16,000	40,374	$E_P < E_R$
Yield per m^3 per m drawdown in aquifer (kg/m^3/m aquifer drawdown)	0.36	0.25	$E_P > E_R$
Yield per m^3 per unit specific yield of aquifer (kg/m^3/unit specific yield)	25	59.2	$E_P < E_R$

As shown in the above table, in Randullabad, the total yield was 0.25 kg/m^3/m of aquifer drawdown against 0.36 kg/m^3/m of aquifer drawdown in the case of Pabal aquifer. In other words, in Randullabad, there was lesser yield with the same amount of water and per unit drop in aquifer level than in Pabal. Thus, for every meter drawdown in the aquifer, agricultural efficiency seems better in Pabal than in Randullabad. This could be because of the higher transmissivity of the Pabal aquifer. While transmissivity is a reasonable indicator of the yield factor in the aquifer, it does not completely justify a measure of the aquifer storage, and therefore, the sustainability aspect of the aquifer. This aspect of sustainability must be reflected in the efficiency measure. Therefore, when the specific yield of the aquifer is considered, the total yield per unit of water per unit specific yield of aquifer is 59.2 kg/m^3/unit specific yield for Randullabad while it is 25 kg/m^3/unit specific yield for Pabal. Thus, specific yield of the aquifer system provides a completely opposite picture when it comes to understanding efficiency in terms of groundwater use. What it implies in this case is that effective recharge measures coupled with demand-side management under PGWM suggest a greater degree of efficiency in the case of Randullabad aquifer as compared to that for Pabal aquifer.

Ostrom (1990) in her seminal work on 'commons' described common pool resources in terms of resource stock and resource units. Resource stock is generally replenished on an annual (or even over longer time-cycles) basis while the resource unit is the property of that resource system which cannot be altered. In case of groundwater systems, the resource stock forms the available groundwater every year which can be abstracted by for various purposes while the resource unit are the aquifer systems which store and transmit these resource stocks i.e. groundwater. Any efficient management system needs to take into consideration the nature of the resource units, its characteristics and behaviour in order to arrive at any management plans for the resource. Unless the resource unit is understood clearly and defined in a manner that can be quantified, the efforts around management of resource by communities, governments or private entities would remain futile.

India's aquifers have helped millions of farmers gain access to groundwater resources for achieving food security. The aquifers are facing multiple challenges today. Supply augmentation, through systematic recharge and demand management such as the PGWM model hold the potential to address challenges regarding groundwater sustainability at local scales. Improving the efficiency of groundwater usage is of prime importance as part of a consolidated aquifer management effort in India. It is necessary to take into consideration the aspect of aquifer storage capacity represented by specific yield of unconfined aquifers while rethinking efficiency for groundwater-based agricultural systems.

5 Conclusion

Popular measures of efficiency such as crop per drop are not enough to fully describe the efficient use of groundwater in agriculture. Drawdowns in an unconfined aquifer are not only a consequence of the amount of water pumped from irrigation wells, but also a function of the specific yield of the aquifer. Hence, while an approach of crop-per-drop or land irrigated by one unit of water indicates productivity of cropping systems and water used, aquifer characteristics like specific yield are necessary to understand water-related efficiency at scales of aquifers. Such estimates of efficiency can be used to compare aquifers managed through different initiatives. Further research is required to understand how efficiency estimation using this approach can be applied to groundwater management in different hydrogeological settings prevalent in different regions of India.

Acknowledgements Many of the results from the Pabal study were obtained during the research project that formed a collaboration between Savitribai Phule Pune University (erstwhile University of Pune) and British Geological Survey of which Himanshu Kulkarni was part. The Randullabad case study was part of the PGWM programme undertaken by ACWADAM with support from Arghyam Trust. All the institutions mentioned here and the associated team members are gratefully acknowledged. The concept and salient findings included in the paper were presented at Bhoojal Manthan organised by CGWB, MoWR, Government of India in August 2015. The authors are grateful for CGWB's invitation to present the findings at Bhoojal Manthan.

References

CENSUS OF INDIA (2011) Chandramouli, C., and Registrar General

Central Ground Water Board (2011) Dynamic Ground Water Resources of India (as on March 2009). CGWB Publication

Custodio E (2002) Aquifer overexploitation: what does it mean? Hydrogeol J 10(2):254–277

Das SVG, Burke J (2013) Smallholders and sustainable wells: a retrospect: participatory groundwater management in Andhra Pradesh (India). Food and Agriculture Organisation of the United Nations, Rome

Deolankar SB (1980) The Deccan basalts of Maharashtra, India—their potential as aquifers. Ground Water 18(5):434–437

Department of Drinking Water and Sanitation (2006) Summary of Nation-wide Statistics from Rajiv Gandhi Drinking Water Mission. New Delhi: Department of Drinking Water Supply, Ministry of Water Resources, Government of India

Dillon P, Kumar A, Kookana R, Leijs R, Reed D, Parsons S, Ingleton G (2009) Managed aquifer recharge-risks to groundwater dependent ecosystems—a review. Land and Water Australia, Canberra

Joy KJ, Gujja B, Paranjape S, Goud V, Vispute S (2008) Water conflicts in India: a million revolts in the making

Kulkarni HC (1987) A study of the Deccan basaltic unconfined groundwater system from the Pabal area of Shirurtaluka, Pune district, Maharashtra state. Unpublished Ph.D. thesis, University of Pune, p 285

Kulkarni H, Shankar PV (2009) Groundwater: towards an aquifer management framework. Econ Polit Wkly 13–17

Kulkarni H, Deolankar SB, Lalwani A, Joseph B, Pawar S (2000) Hydrogeological framework of the Deccan basalt groundwater systems, west-central India. Hydrogeol J 8(4):368–378

Kulkarni H, Shah M, Shankar PV (2015) Shaping the contours of groundwater governance in India. J Hydrol Reg Stud 4:172–192

Lawrence AR, Ansari SU (1980) The shallow aquifer of the Betwa basin, India. Indo-British Betwa Groundwater Project Report, WD/OS/80/2

Macdonald DMJ, Kulkarni HC, Lawrence AR, Deolankar SB, Barker JA, Lalwani AB (1995) Sustainable groundwater development of hard-rock aquifers: the conflict between irrigation and drinking water supplies from the Deccan basalts of India

Ministry of Agriculture, Government of India (2014) Web-based land use statistics information system. Various Land Use Statistics Reports, Directorate of Economics and Statistics, Department of Agriculture and Cooperation. www.lus.dacnet.nic.in/dt_lus.aspx

Moench M (2002) Water and the potential for social instability: livelihoods, migration and the building of society. Nat Resour Forum (vol 26(3): 195–204). Blackwell Publishers Ltd

Molden D, Oweis TY, Pasquale S, Kijne JW, Hanjra MA, Bindraban PS, Hachum A (2007) Pathways for increasing agricultural water productivity

Mukherji A, Shah T (2005) Groundwater socio-ecology and governance: a review of institutions and policies in selected countries. Hydrogeol J 13(1):328–345

Narain S (2012) Excreta matters. State of India's Environment, A citizens' report, Centre for Science and Environment (CSE), New Delhi

Ostrom E (1990) Governing the commons: the evolution of institutions for collective action. Cambridge University Press, Cambridge

Planning Commission Report (2007) Report of the expert group on 'Groundwater Management and Ownership.' Government of India, New Delhi

Ranade R (2005). 'Out of Sight, Out of Mind'. absence of groundwater in water allocation of Narmada basin. Econ Polit Wkly 2172–2175

Saha D, Agrawal AK (2006) Determination of specific yield using water balance approach—a case study of Torla Odha water shed in Decan Trap Province, Maharashtra State, India. Hydrogeol J 14:625–635

Shah T (2010) Taming the anarchy: groundwater governance in South Asia, Routledge

Shah T, Roy AD, Qureshi AS, Wang J (2003) Sustaining Asia's groundwater boom: an overview of issues and evidence. Nat Resour Forum (Vol. 27, No. 2, pp. 130–141). Blackwell Publishing Ltd

Sharma B, Molden D, Cook S (2015) Water use efficiency in agriculture: measurement, current situation and trends. Managing water and fertilizer for sustainable agricultural intensification, p 39

Siebert S, Burke J, Faures JM, Frenken K, Hoogeveen J, Döll P, Portmann FT (2010) Groundwater use for irrigation–a global inventory. Hydrol Earth Syst Sci 14(10):1863–1880

Vaidyanathan A, Sivasubramaniyan K (2004) Efficiency of water use in agriculture. Econ Polit Wkly 2989–2996

Versey HR, Singh BK (1982) Groundwater in Deccan basalts of the Betwa basin, India. J Hydrol 58(3):279–306

Vijay Shankar PS, Kulkarni H, Krishnan S (2011) India's groundwater challenge and the way forward. Econ Polit Wkly 46(2):37–45

Toward a Managed Aquifer Recharge Strategy for Gujarat, India: An Economist's Dialogue with Hydro-geologists

Tushaar Shah

Abstract Gujarat state in Western India exemplifies all challenges of an agrarian economy founded on groundwater over-exploitation sustained over decades by perverse energy subsidies. Major consequences are secular decline in groundwater levels, deterioration of groundwater quality, rising energy cost of pumping, soaring carbon footprint of agriculture and growing financial burden of energy subsidies. In 2009, Government of Gujarat asked the present author, an economist, to chair a Taskforce of senior hydro-geologists and civil engineers to develop and recommend a managed aquifer recharge (MAR) strategy for the state. This chapter summarizes the recommended strategy and its underlying logic. It also describes the imperfect fusion of socioeconomic and hydro-geological perspectives that occurred in course of the working of the Taskforce and highlights the need for transdisciplinary perspectives on groundwater governance.

Keywords Managed aquifer recharge · Energy–irrigation nexus
Groundwater depletion · Groundwater economics

1 Gujarat's Groundwater Governance Challenge

Besides its age-old dynamism in maritime trade and commerce, during recent decades, Gujarat has become increasingly known for its crisis of groundwater. During the Colonial era, while the British concentrated public irrigation investments in the Indo-Gangetic plains in the north and on Godavari, Krishna, Cauvery basins in the south, much of the Bombay Presidency, of which Gujarat was a part, remained left out. Here, supplemental irrigation using bullocks and leather buckets to lift water from shallow open wells remained a widespread tradition (Shah 2009a). Use of Persian wheels was common too. At the turn of the twentieth century,

T. Shah (✉)
International Water Management Institute, IWMI Tata Water Policy Program,
"Jal Tarang", Near Smruti Apartment, Anand 388001, Gujarat, India
e-mail: t.shah@cgiar.org

D. Saha et al. (eds.), *Clean and Sustainable Groundwater in India*,
Springer Hydrogeology, https://doi.org/10.1007/978-981-10-4552-3_11

Gujarati farmers were among the first to import Lister–Petter engines from England and press them into irrigation service. Around 1910, many owners of diesel engines in central Gujarat began laying buried cement pipelines to sell well irrigation service to other farmers. Shah (1993) interviewed an old farmer who supplied, as far back as in 1935, irrigation water to farmers in four neighboring hamlets using his own buried pipelines. So productive was well irrigation considered that the Colonial government imposed a groundwater cess which generated much controversy in Bombay newspapers (Hardiman 1998). Localized, informal groundwater markets using buried cement pipeline networks existed in large parts of central, south and north Gujarat by 1950 (Shah 1993). With the availability of electricity and electric pumps, these markets became central to Gujarat's agriculture. With rapid expansion in well irrigation, groundwater depletion emerged as a major challenge especially in dry north Gujarat, Saurashtra and Kutch. With declining groundwater levels, bullock-bailers became history by the early 1980s; demand for electric pumps soared and a complex political economy emerged around the nexus between electricity supply and groundwater irrigation. The Electricity Board found the transaction costs of metering far-flung tube wells scattered over a vast countryside prohibitive, and farmers complained of the tyranny of meter readers and their extortionary demands from them. In 1987, yielding to a long and strident farmer agitation, Gujarat Electricity Board changed from metered pricing of agricultural electricity consumption to flat tariff based on connected horsepower of irrigation pumps (Shah 1993). This had the multi-pronged effect of increasing the demand for electric pumps, raising the operating factor[1] of electric tube wells and enhancing the depth and breadth of local water markets (ibid). At the same time, the new formula also deepened further Gujarat's groundwater depletion crisis.

Situated in the western-most part of India, the state of Gujarat with a population of over 60 million is enclosed on three sides by a 1600-km Arabian Sea coastline (Fig. 1). Geographically, the eastern mainland of Gujarat is humid in the south and semi-arid in the north. In the west, the Saurashtra peninsula is divided from the mainland by the Gulf of Kutch, which is separated from the mainland Gujarat to the east and Pakistan to the north by the low-lying desert called the Rann of Kutch. The state can be conveniently divided into four hydro-geological zones: (a) humid south Gujarat with alluvial aquifers; (b) semi-arid central and eastern Gujarat with alluvial aquifers in some parts and hard-rock aquifers in others; (c) arid north Gujarat and Kutch with mostly alluvial aquifers and (d) semi-arid Saurashtra with mostly hard-rock aquifers.

The state displays great variability in its agro-meteorological and climatic conditions. Most of Gujarat, except some southern districts, represents one of the most water-stressed regions of India. The hydrology of the state is characterized by wide spatial variations in rainfall (377 mm in Kutch to 1875 mm in the Dangs), and high year-to-year fluctuations in rainfall with the coefficient of variation ranging from

[1]Operating factor is the number of hours of actual operation of a tube well/year as a % of hours of possible operation/year.

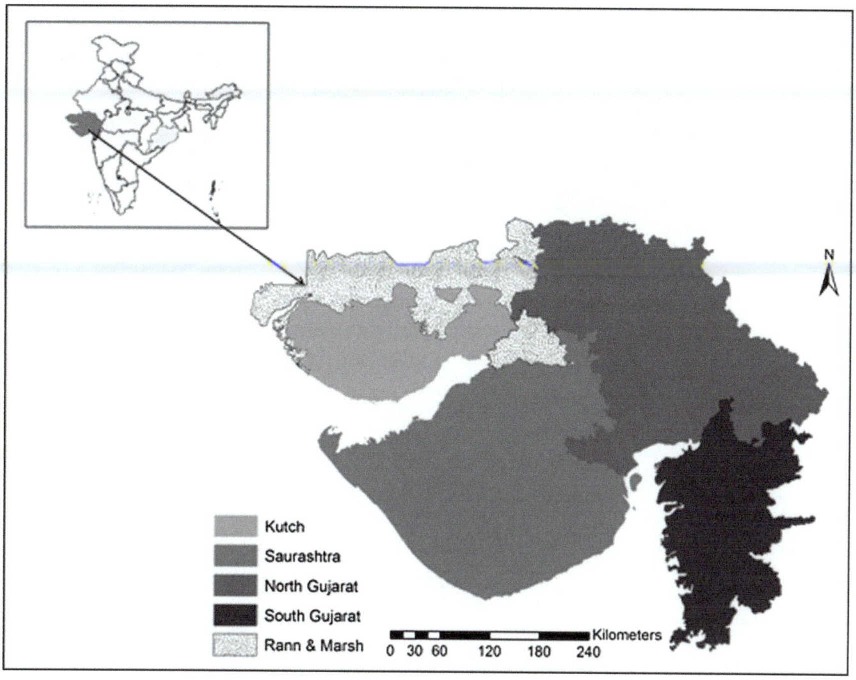

Fig. 1 Map of Gujarat and its four major regions

25% in Dangs to 80% in Kutch (IRMA-UNICEF 2001). The state has unconsolidated, semi-consolidated and fissured geological formations (Fig. 2).

The state has a history of frequent droughts and famines. A famine Gujarat suffered in 1900 was so severe that a third of the population and most of the livestock perished (Yagnik and Sheth 2005). Today, Gujarat is one of India's most rapidly growing states in industrial and economic terms. Yet, some 57% of the state's population is rural and more than 45% depends on agriculture for its livelihood. Even as the share of agriculture in the state's domestic product has declined to about 15%, the number of workers employed in agriculture has not declined as rapidly. As a result, the performance of the agrarian economy affects large sections of the population. Through the 1980s and 1990s, Gujarat's agricultural economy recorded a negative trend rate of growth, which has been a matter of great concern for the state's planners (Bagchi et al. 2005).

After 2000, thanks to a clutch of imaginative policies by the state government (Shah et al. 2009), Gujarat's agricultural economy has experienced a dramatic turnaround, with annual agricultural growth rates approaching an unprecedented 10%. This has eased somewhat the concerns for agrarian livelihoods. However, the state's groundwater situation has continued to remain precarious. At the turn of the millennium, groundwater depletion had assumed the proportions of a crisis in semi-arid alluvial north Gujarat and Kutch, and in hard-rock areas of Saurashtra.

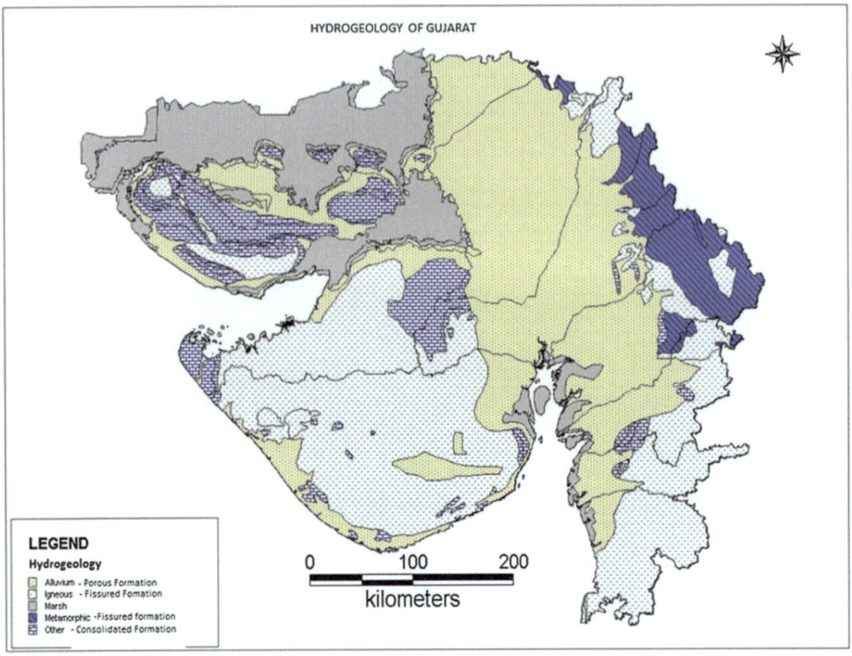

Fig. 2 Hydrogeology of Gujarat

India's Central Ground Water Board (CGWB) assessed that out of Gujarat's 184 talukas,[2] 31 were already withdrawing every year more groundwater than the long-term annual recharge creating a negative groundwater balance. Twelve more talukas were on the border-line, drafting 90% of the estimated 'safe yield'. Of the remaining, 69 talukas were drafting 65% of the 'safe yield' but experiencing rapid development of the resource. Figure 3 shows the differing levels of groundwater development in the districts of Gujarat.

High fluoride concentrations and salinity levels in groundwater, the main source of drinking water supply, were detected in 16 of Gujarat's 25 districts in the early 2000s. In parts of north Gujarat, groundwater tables had dropped so low that 'tube well companies' of farmers replaced numerous individual-owned tube wells to share the large investment—and the risk of failure—associated with installing deep tube wells, fitted with 90 horsepower to 120 horsepower submersible pumps and buried pipeline networks for water distribution.

Gujarat has been by far the most proactive and advanced among Indian states to take cognizance of the groundwater depletion challenge. Recognizing that perverse subsidies to agricultural electricity supply is a major cause for groundwater depletion, Government first tried to correct these perverse incentives by metering

[2]A taluka is a sub-district administrative unit covering roughly 100–120 villages.

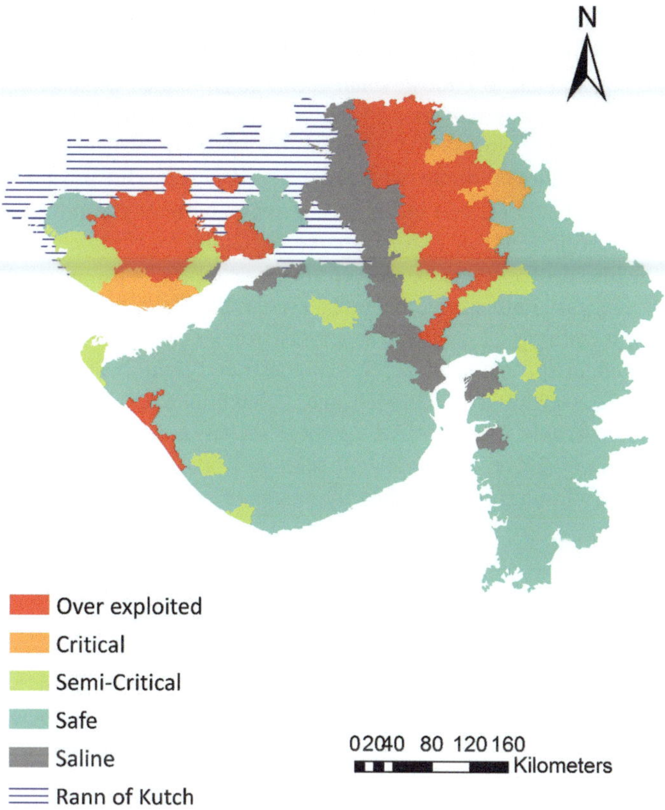

Fig. 3 Over-exploited, critical, semi-critical and safe blocks of Gujarat

tube wells. However, when farmer opposition to this proposal proved too strident, it implemented *Jyotigram Yojana* for rewiring the countryside separating tube wells from other rural electricity consumers and imposed a strict ration on farm power supply (Shah and Verma 2008; Shah et al. 2008). The government also actively supported farmer movement for decentralized water harvesting and groundwater recharge (Shah 2000, 2008). All these helped somewhat to alleviate groundwater depletion in some parts of the state, notably the hard-rock districts of Saurashtra where recharge efforts by local individuals and communities developed into a veritable mass movement that arguably helped to turn around the region's agriculture (Shah 2010). Thanks partly to this success, there was growing recognition of the need for more systematic and organized attempts to enhance groundwater recharge and improve groundwater regime in the state. One option was to let local communities take charge of recharge works as they did in Saurashtra. However, there were several reasons why it was thought an integrated MAR strategy might be useful. First, farmers' outlook to recharge tends to be opportunistic; they prefer sites that can benefit their wells over the best available recharge sites. Second, farmer communities are not allowed to construct recharge structures in reserve forest lands where only government departments would be allowed to operate. Third, communities are constrained to choose small recharge structures consistent with their resources and skills; these may not be the best. Fourth, communities are unlikely ever to undertake structures that involve transport of water beyond village boundaries even within a single watershed or sub-basin. But most importantly, despite government support and incentives, farmers in many groundwater-depleted districts showed little enthusiasm for recharge. By 2008, for example, 482 MCM of the total 807 MCM of local recharge capacity was created only in six Saurashtra districts (Bhavnagar, Jamnagar, Junagadh, Porbandar, Rajkot, Surendranager and Amreli). The remaining 22 districts accounted for less than half of the recharge capacity created in the entire state. Six of the most groundwater-depleted north Gujarat districts—Banaskantha, Mehsana, Gandhinagar, Kutch, Patan and Sabarkantha—had only 106.5 MCM of local recharge capacity (CGWB 2013). Clearly, community-driven decentralized recharge work boomed in hard-rock areas but remained very limited in alluvial aquifer areas of north Gujarat where depletion is severe and energy use in pumping is high. Arguably, in north Gujarat's porous and permeable alluvial aquifers, small recharge structures create flat recharge mounds that soon dissipate, with little impact on water levels in wells, whereas in hard-rock Saurashtra, wells surrounding recharge structures remain full to the brim at least for some time after a major rainfall event. This may explain vibrant local recharge efforts in Saurashtra and lukewarm response in north Gujarat. The upshot is that left wholly to community initiative, groundwater recharge efforts were likely to remain confined to aquifers with specific features, to small structures within village boundaries, to non-forest lands. A state-level MAR strategy was expected to complement community effort, undertake basin-scale projects, including in forest lands, if justified.

To this end, in late 2008, Government of Gujarat constituted a Taskforce to develop a managed aquifer recharge (MAR) strategy for the state. The mandate of the Taskforce was not to develop detailed, aquifer-wise MAR plan but to explore potential gains from making large-scale MAR key plank of its water policy. The Taskforce took MAR to mean a process by which recharge to the groundwater reservoir is augmented at a rate exceeding that under natural conditions of replenishment. Any man-made structure, scheme or facility that adds water to an aquifer was considered to be part of an MAR system.[3]

The remainder of this chapter outlines the work of this Taskforce, the dialectics that took place especially between the economist chairman and hydro-geologists/ irrigation engineers as prominent members. These brought different perspectives to the work of the Taskforce and the strategy it recommended. Section 2 briefly

[3]Dillon et al. (2009: 2) define MAR as "purposeful recharge of water to aquifers for subsequent recovery or environmental benefit. Aquifers, permeable geological strata that contain water, are replenished naturally through rain soaking through soil and rock to the aquifer below or by infiltration from streams." According to them, aquifer recharge can be unintentional, unmanaged or managed.

outlines the terms of reference and the methodology evolved to compute the broad contours and scale of the MAR intervention required and possible. Section 3 presents the socioeconomic perspective. Section 4 discusses three alternative strategic plans that emerged from the Taskforce's work. Section 5 concludes the chapter.

2 Taskforce for Managed Aquifer Recharge

The Taskforce was dominated by high-level hydro-geologists and irrigation engineers with over 200 years of collective experience, mostly in the government. The Government of Gujarat invited the present author, an economist-cum-social scientist, to chair the Taskforce and oversee its work. The terms of reference (ToR) written by a hydro-geologist were as follows: (i) review of the conditions of the surface and groundwater resources of the state, (ii) study the hydrogeology of the state to identify suitable aquifers for recharge, (iii) study the results of the experiments carried out for artificial recharge by state government and central government as well as NGOs, (iv) suggest construction of different types of recharge schemes adopting different techniques to harness maximum runoff of the rainwater in dried aquifer system and to improve groundwater quality, (v) assign priority to areas with no surface water irrigation facility. The ToR also provided for the Taskforce to consult other experts/experienced persons/organizations and obtaining their guidance as needed. Suggestions for financing plan and preliminary assessment of energy saving and potential for carbon credit too were elicited. The Taskforce met 9 times over an 18-month period.[4] Gujarat's 185 river basins (17 for Gujarat region, 71 for Saurashtra region and 97 for Kutch region) were regrouped into 11 regional basins (see Fig. 4 and Table 1).

The methodology adopted was similar to the one used in the National Groundwater Recharge Master Plan (NGRMP) that the Central Groundwater Board (CGWB) had developed in 2004 for the country as a whole.[5] The key steps of CGWB methodology are summarized in Fig. 5. The NGRMP viewed dewatered aquifers as very attractive and technically feasible 'warehouses' for storing water subject to appropriate lithological conditions, porosity and permeability of aquifer material. In order to deal with the diversified geological, climatological and topographic features of the Indian landmass, the CGWB classified Hydro-geological Units (HGUs) into porous and hard-rock formations. Porous HGUs were further classified into unconsolidated and semi-consolidated HGUs. Hard-rock HGUs too were sub-classified into igneous, volcanic and carbonate rock formations.

[4]The work done by the Taskforce is described in the Taskforce report which can be accessed from p.reghu@cgiar.org.

[5]http://cgwb.gov.in/documents/MASTER%20PLAN%20Final-2002.pdf.

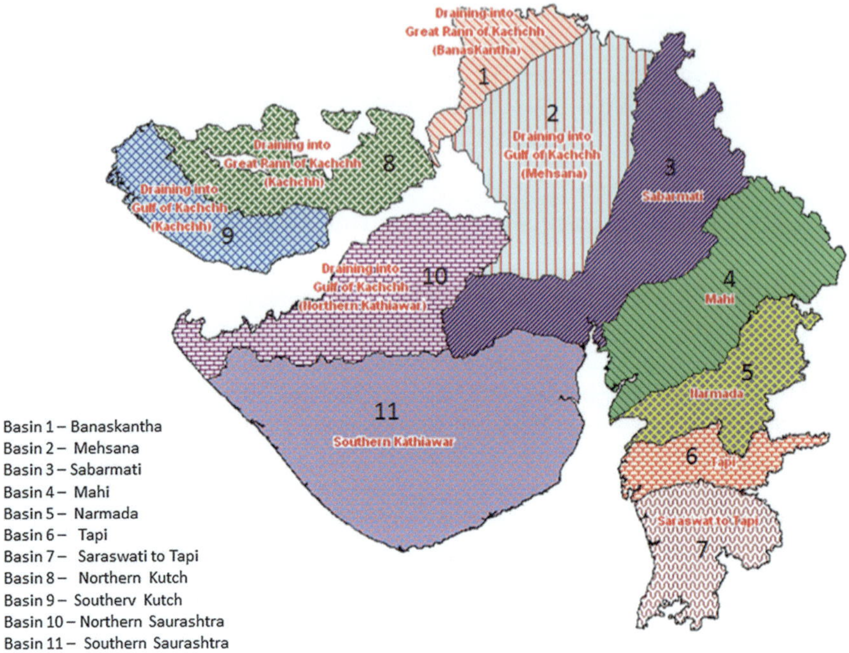

Fig. 4 Eleven major river basins of Gujarat

Table 1 Gujarat's 11 major river basins

	Summary name	Region	Basin geography
1	Banaskantha	North	Luni and other rivers draining into Great Rann of Kutch (including parts of Banaskantha and Patan)
2	Mehsana		Rivers draining into Rann of Kutch (including parts of Banaskantha, Patan, Mehsana, Gandhinagar, Ahmedabad and Surendranagar)
3	Sabarmati		Sabarmati
4	Mahi		Mahi
5	Narmada	South	Narmada
6	Tapi		Tapi
7	Saraswati to Tapi		Saraswati to Tapi
8	Northern Kutch	Kutch	Rivers draining into Great Rann of Kutch
9	Southern Kutch		Rivers draining into Gulf of Kutch
10	Northern Saurashtra	Saurashtra	Rivers draining into Gulf of Kutch
11	Southern Saurashtra		Southern Kathiawar rivers

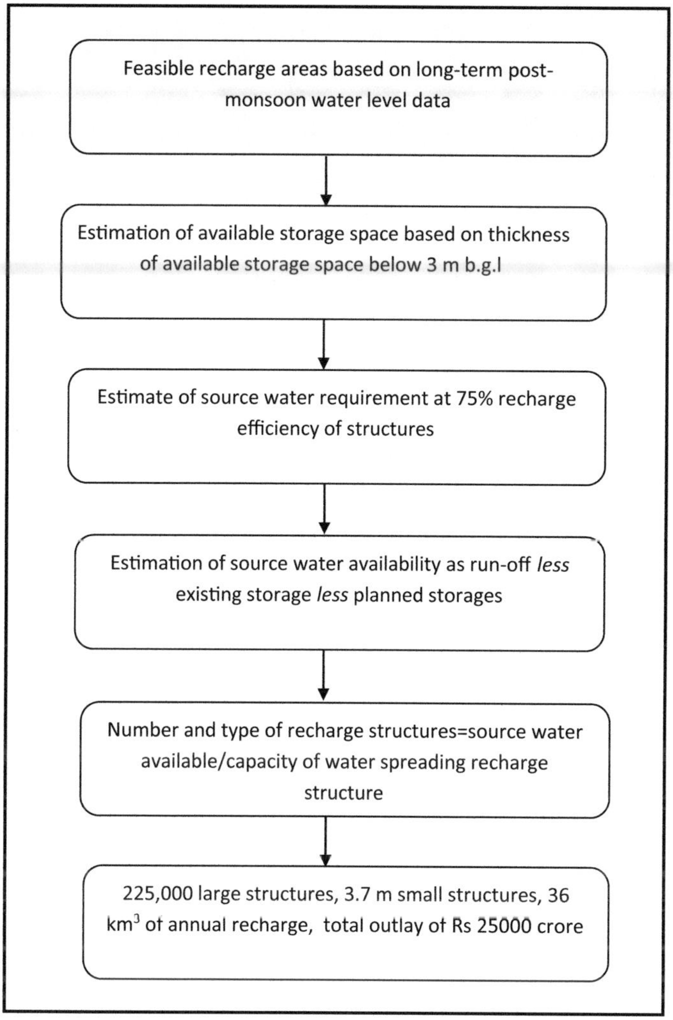

Fig. 5 Methodology used by CGWB to develop India's groundwater recharge master plan

Area under irrigarion by surface and ground water in 11 major river basins in India is produced in Fig. 6.

For effective MAR, the NGRMP specified two preconditions: (1) availability of non-committed surplus monsoon runoff in space and time; (2) suitable hydro-geological environment and sites for cost-effective MAR. For (1), it focussed on the rainfall pattern in space and time in different HGUs. For (2), preferred aquifers were those with high vertical hydraulic conductivity and low horizontal hydraulic conductivity which absorb large quantities of water and do not release them quickly.

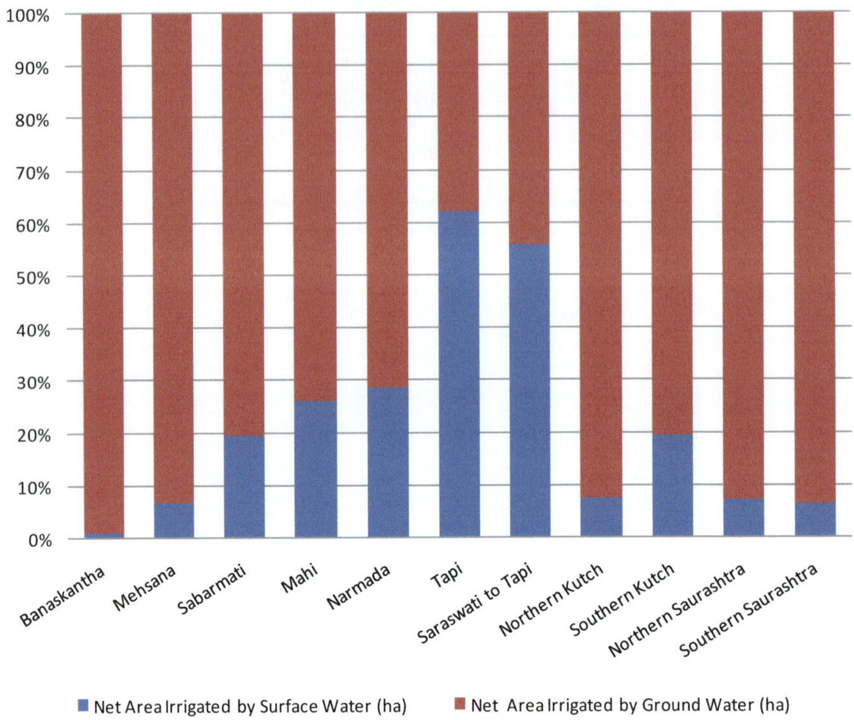

Fig. 6 Contribution of surface and groundwater to net irrigated area in 11 major basins

Using groundwater level data from its the National Hydrograph Network Stations, depth to water table map was prepared for post-monsoon period based on decadal average of the depth to groundwater with contour intervals of 3 m. Similarly, a long-term decadal post-monsoon water level trend map was prepared to identify ranges of water level rise and fall on a long-term basis. These two maps were combined to identify variations in the depth to water table along with their decadal trends. HGUs were then classified into 4 categories with water levels at (1) 3–6 m b.g.l. and a declining trend of more than 10 cm per year, (2) 6–9 m b.g.l. and declining at more than 10 cm per year, (3) more than 9 m b.g.l. and (4) 3 m or less b.g.l. Category 4 was ruled out for MAR. In order to estimate available storage space, the thickness of available unsaturated zone below 3 m b.g.l. was estimated for different range of water level depth at 3 m intervals which were then averaged to arrive at the thickness and volume of the unsaturated zone in each HGU. This volume was then multiplied by the average specific yield for the respective HGU to compute the water required for MAR to saturate the aquifer to 3 m b.g.l.

The NGRMP first estimated that saturating the vadose zone to 3 m b.g.l. for the country as a whole can create a sub-surface storage potential of 59.06 M ha m of which 43.65 M ha m would be retrievable. However, in view of the large variability in monsoon runoff and sub-surface storage potential across HGUs, it estimated the *feasible* groundwater storage as 21.42 M ha m. Most of these new storages would be in HGUs with high monsoon runoff as well as high sub-surface storage potential. Based on experience gained in pilots and experimental projects by CGWB, an average recharge efficiency of 75% was applied to recharge structures, giving a multiple of 1.33 to be applied to derive source water requirement.

The surface water available in different sub-basins was estimated based on data provided by state Governments on total runoff availability, runoff committed to existing surface storages, provision for future planning and uncommitted surplus water available. Using these, in a 1996 exercise, the CGWB had estimated non-committed monsoon runoff availability in 20 major river basins to be 87.19 million hectare meters (M ha m). The availability so estimated was for the entire sub-basin and not for the areas identified for MAR. Therefore, the actual available water for MAR was then apportioned pro rata from the total water availability in the basin.

Based on these computations for each HGU, the number, type and design of new recharge structures such as percolation tanks, check dams, recharge shafts needed in a particular block/watershed/HGU were recommended bearing in mind extant hydro-geological situation, existing density of structures, land availability and such other factors. Data and experience derived from numerous pilots implemented by CGWB and other agencies were used to develop these recommendations.

The Gujarat Taskforce used this same methodology. However, an important difference was the depth below ground level (b.g.l.) to which the vadose zone is to be saturated through MAR. The NGRMP assumed 3 m b.g.l. would be appropriate. But the hydro-geologists on the Gujarat Taskforce argued that 8 m b.g.l. would put to rest any threat of waterlogging and secondary salinization.

Accordingly, much of the work undertaken by the hydro-geologists on the Taskforce involved the following steps:

(a) Identification and prioritization of areas in severe need of MAR;
(b) Estimation of storage space and the quantity of water needed to arrest the declining water levels, to begin with;
(c) Mapping and measuring areas according to their post-monsoon depth to the water table;
(d) Estimation of sub-surface storage space and the quantity of water needed for MAR to saturate the unsaturated zone up to 8 m b.g.l.;
(e) Quantification of surface water requirement for MAR in each basin/region; and
(f) Estimation of surplus average annual runoff availability as source water; these were taken from an elaborate exercise that Tahal Consultants had undertaken in a 1997 report[6] for Government of Gujarat; in this, surface water runoff was estimated for different basins by taking into account different dependability ratios of average long-term rainfall in different basins and Gujarat's allocations from trans-boundary rivers.

---

[6]'Water Resource Planning for the state of Gujarat', May 1997.

Table 2 Values of specific yield and coefficient of replenishment assumed

	Formation	Specific yield	Coefficient of replenishment
1	Alluvial deposits	0.10	0.12
2	Metamorphic formations	0.015	0.018
3	Igneous formations	0.015	0.018
4	Semi-consolidated formations	0.020	0.024

(g) Identifying areas of poor chemical quality of groundwater and analyzing the scope of improving water quality by suitable recharge measures.

In order to compute the water requirements for MAR, the Taskforce needed estimates of specific yield as well as the 'Coefficient of Replenishment' which is the volume of water in cubic meters that can be put into storage in a1 m^2 column of aquifer to a height equal to the thickness of the aquifer, when the water table is raised by 1 m. The unsaturated zone comprises a wide variety of formations with different hydraulic characteristics. The coefficient of replenishment of various formations in the vadose zone was derived by taking into account the relation of porosity with specific yield. The estimates used for four distinct formations encountered in Gujarat are presented in Table 2.

The volume of vadose zone was then multiplied by coefficient of replenishment on an area-specific basis to compute the volume of water required to arrest the declining water levels and to saturate the aquifer to 8 m b.g.l. The volume of water needed to saturate unsaturated portions of vadose zone was computed as:

$$W_n = \sum (\mathrm{GWL}_i - 8) * A_\mathrm{i} * \mathrm{CoR}_i \text{ for areas with GWL} > 8 \text{ m}$$

where

W_n volume of water needed for MAR to bring groundwater level to 8 m b.g.l. in *n*th basin
$\mathrm{GWL_i}$ depth to static water level in *i*th segment of the basin
A_i area in m^2 of *i*th segment of the basin
$\mathrm{CoR_i}$ Coefficient of Replenishment applicable to *i*th segment of the basin

The Taskforce then developed a Basic MAR strategy for Gujarat which met two requirements; first, that MAR should be planned only on a scale permitted by the availability of average annual non-committed surplus water available within that basin after reserving the water required for all existing, under-construction and planned future schemes for surface storage; and second, that the basins identified for MAR should have suitable hydro-geological environment and sites with sufficient space to create sub-surface reservoirs by accommodating the volumes of water being recharged.

CGWB's observation well data from 1998 to 2007 were used to identify 95019 km^2 out of 196,000 km^2 of Gujarat's geographical area that was found appropriate for MAR where post-monsoon water level was below 8 m and groundwater quality was suitable for MAR. Of these 43,728 km^2 were alluvium, 31,130 km^2 were igneous rocks, 15,269 km^2 were sedimentary formations, and 5775 km^2 had metamorphic formations. North Gujarat with 39,677 km^2, Saurashtra with 34,416 km^2 and Kutch with 19,024 km^2 were considered suitable for MAR, while south Gujarat with barely 2784 km^2 was found to have had little need for MAR.

Table 3 sets out the broad contours of the Basic MAR strategy of the Taskforce which recommended construction of 7700 percolation tanks, 9900 check dams and 31,250 injection wells besides modification of 32,000 dug wells for MAR at an estimated total capital cost of US$437 million.[7] If implemented, the Basic MAR strategy would contribute additional average aquifer recharge of the order of 2700 Mm^3/year in the state as a whole, 17% of estimated current annual recharge. Finally, even as its work was in progress, the Taskforce was told by the government to exclude allocating any runoff from Narmada, Tapi and Tapi to Saraswati (basins 5, 6 and 7) basins to MAR because this was to be reserved for the proposed Kalpasar project[8] to create a freshwater reservoir in the Gulf of Khambhat. This meant that 3/4th of the non-committed surplus water in the state was not to be allocated for MAR even during the 15–20 years that the planning and construction of the Kalpasar reservoir are expected to take before it begins to get filled up.

Like the National Groundwater Recharge Plan that the CGWB had developed earlier (Shah 2006), the Basic MAR strategy maximized MAR investments in basins in the South which least needed MAR and provided little resources in North, Saurashtra and Kutch where MAR is most needed. Even after investing US$437 million, the Basic MAR strategy would not enable the state to even arrest groundwater depletion in over-exploited basins, leave alone saturating the vadose zone to 8 m b.g.l.

As columns [7] and [8] show, even after investing nearly half a billion US dollars on a MAR program, the most that can be achieved under the Basic MAR strategy recommended by the Taskforce is to arrest annual groundwater depletion. That too can happen only if groundwater draft is reduced, through proactive demand management, to close the annual depletion henceforth so that MAR would be devoted entirely to meeting the accumulated past deficit. Column [8] shows that in Banaskantha and Mehsana, even after proactive demand management, it would take 40–50 years to meet the accumulated groundwater deficit through MAR. As shown in columns 9 and 10 in Table 3, in the absence of proactive demand management, raising water level to 8 m b.g.l. can be never achieved in all those basins where non-committed surplus runoff available for MAR is equal to or smaller than annual groundwater deficit. Raising water level to 3 m b.g.l., as proposed under the National Groundwater Recharge Plan, is naturally out of the question.

---

[7]US$ = Rs. 50 at the time of the Taskforce report.

[8]http://en.wikipedia.org/wiki/Kalpasar_Project. The Kalpasar project has been on the drawing board for several decades and may take at least 20–25 years to get fully commissioned.

Table 3 Broad contours of the basic MAR strategy

	Basin	Volume of water (Mm^3) needed for MAR to				Non-committed surplus runoff available for MAR	With demand management, years of MAR needed to		Without demand management, years of MAR needed to	
		Arrest annual groundwater depletion	Raise water level to 8 m b.g.l.	Raise water level to 3 m b.g.l.	To achieve 2 and 3		Arrest annual decline in water level	Raise water level to 8 m b.g.l.	Arrest annual decline in water level	Raise water level to 8 m fb.g.l.
	[1]	[2]	[3]	[4]	[5]	[6]	[7]	[8]	[9]	[10]
1	Banaskantha	476	3271	5204	3747	69	7	47	Never	Never
2	Mehsana	480	6803	13,637	7283	168	3	40	Never	Never
3	Sabarmati	67	3533	10,092	3600	431	0	8	0	10
4	Mahi	29	2613	6401	2642	320	0	8	0	10
5	Narmada	11	779	1686	790	721	0	1	0	1
6	Tapi	0	7.5	43	7.5	525	0	0	0	0
7	Saraswati to Tapi	123	5.4	19	128.5	6980	0	0	0	0
8	Northern Kutch	118	3127	3338	3245	103	1	30	Never	Never
9	Southern Kutch	267	3271	6296	3538	191	1	17	Never	Never
10	Northern Saurashtra	40	494	1840	534	573	0	1	0	3
11	Southern Saurashtra	0	2895	9790	2895	1057	0	3	0	3

Finally, removing the non-committed surplus waters of Narmada, Tapi and Saraswati-to-Tapi from MAR allocation drastically reduced the scope of Basic MAR strategy even further, which now cannot even arrest the persistent depletion (assuming that further groundwater abstraction over current levels is brought to a complete halt) leave alone saturating the vadose zone up to 8 m b.g.l., which would not be feasible in foreseeable future in Banaskantha, Mehsana, Kutch and much of Saurashtra. Given that the completion of the Kalpasar project is likely to take 15–20 years, not using non-committed surplus waters of Narmada, Tapi and Saraswati-to-Tapi basins for MAR in the interim is a wasted opportunity.

3 The Economist's Perspective and the Reality of Gujarat's Irrigation Economy

As an economist on the Taskforce, I believed that our approach to the Basic MAR strategy was largely supply driven and overlooked at least six basic notions of economics and business planning:

(a) ***Demand-side Perspective***: The core premise that the Basic MAR strategy followed was: 'recharge where there is surplus water and room available in the aquifers'. This did not pay heed to the fact that areas so identified for MAR are in no real need of MAR and the areas really in need are excluded from its preview.

(b) ***Criticality of Opportunity cost***: The Basic MAR strategy took a narrow view of the costs and benefits of alternative courses of action. In particular, it did not factor into its analysis the full opportunity cost to Gujarat's economy of not closing the accumulated groundwater deficit in regions with high levels of groundwater depletion.

(c) ***Irrelevance of Sunk Cost***: Relatedly, the Basic MAR strategy took the strong view that no matter what, past investments in surface water storages must be prioritized in the allocation of runoff even when aquifers have become the principal source of irrigation and domestic water supply. Economic reasoning would suggest that these investments were incurred in the past when Gujarat's use of groundwater storage was minimal compared to the present. Past investments in surface storages should now be treated as 'sunk costs' at least to the extent that their relevance had declined relative to aquifer storage. In today's Gujarat, aquifer storage has become more critical to manage than surface storages. Standard economics proposes that one should not let sunk costs influence one's decisions, because doing so would not be rationally assessing a decision exclusively on its own merits. For instance, one does not avoid a buying laptop today simply because one is in possession of a working machine purchased years ago.

(d) ***The Incremental Principle***: In economics, bygones are forever bygones. Rational decision making requires that we choose from alternative courses of action whose incremental gains exceed its incremental cost. If reallocating 1 BCM of runoff to MAR today creates more benefits to the society as a whole compared to diverting it to surface storages, MAR allocation should be the chosen course of action.

(e) ***The Proportional Principle***: Allocate resources among alternatives in proportion to their payoff. In its broadest sense, this requires that the time, effort and resources we allocate to alternative courses of action should be determined by the potential payoff these offer. In business economics, this translates into the ABC principle of inventory management which argues that material with high opportunity costs should be more carefully managed than the rest.

(f) ***Partial versus Total solutions***: Even when one is deciding about a particular sector of the economy—water in the present case—one must factor in costs and benefits attendant to a decision in all other sectors of the economy (e.g. energy) and not just the water sector. In another context, when we are planning for the water economy of Gujarat, we cannot plan only for what the government will do; we must factor in how people, communities, industries, civil society and other players can be enjoined to contribute to that effort. The Taskforce's mandate was to develop a MAR strategy for Gujarat as a whole, not only for the government; it was to benefit the economy as a whole, not just the water sector.

Many surface irrigation systems in Gujarat were constructed decades ago when the groundwater economy was much smaller than today. However, today, most basins of Gujarat—except in south—have experienced runaway increase in demand for groundwater for irrigation and other uses. This *demand-side perspective* needs to be worked into evolving a MAR strategy. Of the total of 3.39 million ha of net irrigated area in Gujarat, 2.76 million is served by groundwater wells, while only 0.63 million ha are served by canal irrigation. In keeping with the '*proportionality principle*', managing groundwater storage is more important for sustaining Gujarat agriculture than prioritizing surface storages.

Groundwater storage is also more efficient in meeting the needs of the state compared to surface water storage. As Table 4 shows, based on data compiled by agencies of the Government of Gujarat, groundwater is only 30% of Gujarat's total water resources but it meets 82% of the state's domestic water needs, 65% of industrial water demand and serves 79% of the state's gross irrigated area. To irrigate 1 ha, Gujarat needs over 13,500 m^3 of surface storage. While some of this contributes to soil moisture and groundwater recharge, a good deal of which is lost to non-beneficial evaporation. In contrast, one irrigating ha with tube wells uses only 3500 m^3 of groundwater storage. A million m^3 removed from surface storage and dispatched to aquifers reduces canal irrigated area by 72.2 ha but increases groundwater irrigation by 285 ha ('*the incremental principle*'). Under this situation, it is hard to understand why existing, under-construction and planned future surface water structures (which represent partly '*sunk costs*' except where they serve as efficient MAR structures) should have a higher priority over MAR. Given the

Table 4 Role of surface and groundwater resources in Gujarat's water economy

S. No.	Details	Surface water	Groundwater	Total
1	Total rainfall precipitation in a normal year (Mm^3)			1,96,000
2	Water resources (Mm^3/year)	37,400	15,800	53,200
3	Water resources utilized (Mm^3/year)	12,000[a]	11,500	23,500
4	% of domestic water requirements met	18	82	100
5	% of industrial water requirements met	35	65	100
4	Gross area irrigated (m ha)	0.87	3.26	4.13
5	Storage volume/ha (m^3)	13,777	3531	5690

[a]Actual surface water storage in major, medium and minor irrigation projects (excluding Sardar Sarovar Project) plus recharge structures (check dams, percolation tanks, etc.) during year 2003–04

current structure of Gujarat's irrigation economy, losing a million m^3 of aquifer storage imposes a much higher *opportunity cost* on Gujarat economy and society compared to a million m^3 of surface storage.

4 Energy and Carbon Footprint of the Groundwater Economy

The Basic MAR strategy recognized, but did not factor into its workings, the sizeable energy and carbon footprint of groundwater irrigation economy of Gujarat. In this sense, it adopted a *partial* and not a *general*, cross-sectoral perspective. By 2008, Gujarat's farmers had pressed into service 780,000 electric pumps and 480,000 diesel pumps on deep tube wells, shallow tube wells and dug wells for irrigation (GoG 2009). In addition, some 400,000 applications for new tube well connections were pending with electricity companies. Farmers used around 12 billion units (kWh) of power—costing US$1.2 billion to generate at power stations (at US$0.1/kWh)—to pump groundwater. Electricity that Gujarat farmers use in pumping groundwater is nearly 4 times what the Sardar Sarovar Project, Gujarat's biggest hydroelectric project generates in a year.[9] Given that 80% of Gujarat's groundwater irrigated area is served by electricity and 20% by diesel pumps,[10] the average electricity used per hectare of area irrigated by electric pumps is 5400 kWh/ha. If electricity to farmers was priced commercially, its use on such

[9]http://www.infraline.com/details/ssnnl-awarded-for-sustainable-energy-162234.htm, accessed on July 9, 2013.

[10]Numbers are hard to come by to support this assumption. However, diesel pumps are used mostly for lifting water from surface canals and ponds and from shallow wells which are declining. Moreover, subsidized electricity costs a fraction of diesel cost involved in pumping. So I believe that the 80–20 division is close to the reality.

scale would be prohibitively costly. But farmers can use such expensive groundwater for irrigation because they pay less than 1/5th of its true cost to the society.

A critical issue is the regional variations in electricity use/ha of tube well irrigation. South and central Gujarat, constituting approximately 1/4th of the state's land area, are water surplus and have high and more reliable rainfall and more rainy days. Here much irrigation is through surface canals and groundwater is close to the surface; as a result, the energy needed to irrigate per hectare here is around 1500–2000 kWh/ha. In contrast, in Saurashtra, Kutch and north Gujarat regions, the rainfall is lower, less reliable, and is received on fewer rainy days compared to south Gujarat. There is little surface irrigation, and with decades of over-exploitation, groundwater levels are deep. Here, the energy needed per ha of tube well irrigation can go as high as 10,000–12,000 kWh/ha. The real benefit of aggressive MAR in these basins would be in terms of reducing electricity use in agriculture which Basic MAR strategy all but ignored (*Partial versus general perspective*).

In 2008, of the estimated 11,500 Mm^3 of groundwater that Gujarat farmers lifted, an estimated 9000 Mm^3 was lifted using 780,000 electric pumps. If these use 12 billion kWh of power at generating stations and 9 billion kWh at wellhead, by implication the average dynamic head from which a representative tube well in Gujarat lifts water is over 115 m,[11] assuming 30% combined pump and motor efficiency. Since the dynamic pumping head—which drives energy use/m^3 lifted—declines rapidly in response to rise in static water level, aggressive MAR program can have major impact of reducing energy use in pumping. Saturating unsaturated vadose zone up to 8 m b.g.l. would reduce the dynamic head of pumping to around 30 m on average. Every meter decline in the dynamic pumping head from which a representative tube well lifts groundwater saves Gujarat 0.072 billion kWh of electricity. Lifting 9000 Mm^3 of groundwater for irrigation to a dynamic head of 30 m instead of 115 would need only around 3.30 billion kWh of farm power at power stations. Stabilizing groundwater levels state-wide to 8 m b.g.l. would thus save Gujarat 8.4 billion kWh of electricity at generating stations valued at US$840 million/year. This 'energy dividend' would be much larger under a more ambitious MAR strategy of raising water levels to 3 m b.g.l. as was proposed under the National Recharge Plan of the CGWB earlier.

Indirect support to this claim is offered by column 5 in Table 5 which presents basic data on farm power supply in different basins. In Tapi and Saraswati-to-Tapi, where post-monsoon groundwater levels are close to 8 m b.g.l. (so that there is no MAR investments needed there), the connected load per tube well is less than 5 hp. If a comprehensive MAR strategy raised post-monsoon groundwater levels in other basins to 8 m b.g.l., as in Tapi and Saraswati-to-Tapi, the aggregate connected load for tube wells in the state as a whole can be expected to decline from 6.03 GW

---

[11]Using the formula $E = KVC/e$ where E = energy needed for lift (kWh); K = constant 0.0026; V = volume of water in m^3; C = dynamic head in m and e = electro-mechanical efficiency assumed to be 0.30.

Table 5 Electricity use in groundwater pumping

Sr. No		Number of tubewell connections	Total connected load (horsepower)	Average load (horsepower)/connection	Average electricity consumption (kWh)/connection at the user end[a]	Estimated total farm power dispatched at generating stations (million kWh)[b]
1	2	3	4	5	6	7
1	Banaskantha	58,983	610,896	10.4	11,186	880
2	Mehsana	75,525	2,145,805	28.4	30,685	3090
3	Sabarmati	119,403	1,329,491	11.1	12,025	1920
4	Mahi	38,137	406,884	10.7	11,522	590
5	Narmada	17,544	172,409	9.8	10,613	250
6	Tapi	25,010	123,106	4.9	5316	170
7	Saraswati to Tapi	51,170	233,873	4.6	4936	330
8	Northern Kutch	12,434	239,837	19.3	20,832	350
9	Southern Kutch	14,784	288,587	19.5	21,081	410
10	Northern Saurashtra	91,631	662,553	7.2	7809	960
11	Southern Saurashtra	259,841	1,880,902	7.2	7818	2710
Total		764,461	8,094,344	10.6	11,435	11,650

[a]Assuming average pumping hours of 1500/year per farm power connection
[b]Transmission and distribution losses for farm power supply assumed at 25%

(8.09 million horsepower) to 2.85 GW (3.82 million horsepower), resulting in power savings equivalent to 2.2 Sardar Sarovar Dams, Gujarat's largest hydro-electric project with a generation capacity of 1450 MW.[12] At a conservative US $1 million investment required to create a MW of coal- or gas-based thermal power generation capacity retired from irrigation to other uses, this can mean equivalent of US$3.18 billion of investment saved in power generation.

Closely linked with the 'energy dividend' is the associated carbon credit since 3/4th of Gujarat's power comes from coal- or gas-fired thermal plants. We know that a kWh of electricity emits 1.34 lb of CO_2, 0.0111 lb of methane and 0.0192 lb of nitrous oxide, all greenhouse gases (Shah 2009b). Savings of 8.4 billion kWh of farm power in groundwater pumping would mean reduced GHG emissions of 5.08 billion kg of CO_2, 4.26 million kg of methane and 7.3 million kg of nitrous oxide. International prices of carbon credits have crashed during recent times. However, at a notional rate of US$10/mt of C, the CO_2 emission reduction through an aggressive MAR strategy in Gujarat would produce carbon credits worth US $19 million/year.

5 Proactive MAR Strategy I and II

Strictly from the economic point of view, thus using more of available runoff to meet the accumulated groundwater deficit and minimizing the energy costs of pumping would make sound sense, given that the cost of electricity use in groundwater irrigation is a sizeable portion of the state's entire GDP from agriculture (irrigated and rainfed) and even larger share of GDP from area irrigated by electric tube wells. In many areas of north Gujarat, where groundwater is pumped from great depths using subsidized power, losses suffered by electricity companies per ha irrigated by tube wells exceeds farmers' net earnings from irrigated agriculture (Narula et al. 2011[13]). The case for MAR becomes even stronger if carbon credit is valued. Even in financial terms, it makes strong economic sense to allocate runoff to MAR than to gravity irrigation. If Government of Gujarat was a commercial enterprise, it would happily sacrifice US$2.5 million it collects from farmers as canal irrigation tax and deploy the water so released to raise groundwater level through MAR and reduce the deadweight of power subsidy. Every Mm^3 of water released for surface irrigation in 2008 earned an Irrigation Service Fee of less than US$1000; diverted to MAR, it could save over US$10,000 in electricity saved. If carbon credit can be encashed, diverting surface water from gravity flow irrigation to MAR would be even more attractive in financial terms.

[12]https://en.wikipedia.org/wiki/Sardar_Sarovar_Dam visited on June 20, 2013.

[13]"Addressing the Water Crisis in Gujarat, India", March 2011, http://water.columbia.edu/files/2011/11/Gujarat-WP.pdf.

Tables 6 and 7 compare two alternative versions of Proactive MAR strategies with the Basic MAR strategy of the Taskforce outlined earlier. Proactive MAR strategy I assumes that 80% of all utilizable surface runoff in *every* basin will be sequestered for MAR until such time as the accumulated groundwater deficit is met and water levels in all basins are raised to 8 m b.g.l. The remainder 20% left in surface storages would be reserved for urban water supply, rural drinking and industrial water requirements with no water released for gravity flow irrigation except where doing so is the best approach to MAR. This implies that meeting accumulated groundwater deficit in each basin would have priority over all 'existing, under-construction and planned' surface storage structures. Proactive MAR strategy II assumes even bolder posture and takes 80% of the runoff in *all* basins of the state, pools the surplus runoff so obtained and reallocates it for MAR to basins in proportion to their MAR requirement to raise water level to 8 m b.g.l. This would imply large-scale inter-basin transfer (IBT) of surface water from southern region—the only surplus region—to the rest of Gujarat for the express purpose of MAR.

Table 6 Water available for MAR under Proactive MAR strategy I and II

Region	Water required in Mm^3 for			Water available in Mm^3		
1	2	3	4	5	6	7
	Arresting annual depletion	Getting WL to 8 m b.g.l.	Total	Basic MAR strategy	Proactive MAR strategy I	Proactive MAR strategy II
				Non-committed surplus	80% runoff in basins devoted to MAR	80% of runoff in the state with MAR driven IBT
Northern	1052	127,865	128,917	456	5586	13,162
Southern	134	11,093	11,227	7734	20,152	1197
Kutch	385	82,673	83,058	230	530	8376
Saurashtra	40	71,379	71,417	1390	3646	7179
Total			294,621	9810	29,913	29,913

Table 7 Years of MAR needed to arrest annual depletion and to raise WL to 8 m b.g.l. under alternative MAR strategies

	Basic MAR strategy		Proactive MAR strategy I		Proactive MAR strategy II	
	Arresting annual depletion	Getting WL to 8 m b.g.l.	Arresting annual depletion	Getting WL to 8 m b.g.l.	Arresting annual depletion	Getting WL to 8 m b.g.l.
Northern	Never	Never	1 year	23 years	0	10
Southern	0 year	1.5 years	0	0	0	0
Kutch	Never	Never	1	157 years	0	10
Saurashtra	1 year	Never	0	120 years	0	10

Relaxing the condition of reserving runoff for 'existing, under-construction and planned future surface' storage structures creates scope for larger, more aggressive MAR program in north Gujarat, Kutch and Saurashtra that badly need it as column 6 in Table 6 shows. Inter-basin transfer (IBT) of water from South to North, Saurashtra and Kutch increase greatly the scale of MAR activity possible in the latter regions. As Table 7 shows, even committing 80% of surplus runoff generated within each basin is insufficient to cover the accumulated groundwater deficit in Kutch and Saurashtra in less than a century. To meaningfully pursue the goal of raising groundwater levels, state-wide, inter-basin transfer (IBT) from South to North, Kutch and Saurashtra is essential, provided annual groundwater abstraction is not allowed to grow beyond current level. In sum, if Gujarat is serious about resolving its groundwater crisis, it must:

(a) reconsider the notion that only non-committed surplus water is available for recharge; (b) accept to use some surface storage in each basin for direct groundwater recharge rather than gravity irrigation; and (c) use IBT for direct groundwater recharge as well as providing surface irrigation *in lieu* of groundwater pumping.

While all Taskforce members, bar the chairman, ruled out (a), (b) and (c) as neither practical nor desirable, the reality is that Gujarat's political leaders, administrators and economic planners have always accorded the highest priority to the 'energy dividend' of using surplus runoff in South to raise groundwater levels in the North, Saurashtra and Kutch. Jayanarayan Vyas, a Chartered Accountant and business planner who served as a former Minister of Water Resources in Gujarat government, argued that the entire capital cost of the ambitious Sardar Sarovar Project on river Narmada could be paid off by 'energy savings' from rising groundwater levels in the North. To quote him verbatim:

> It is estimated that around 50% (1350 MW) [of electricity used in groundwater irrigation] will be saved as a result of the Sardar Sarovar Project by way of surface water irrigation and recharge of groundwater aquifers. Added to this is the production of 1450 MW of power from the project. In order to have around 2800 MW at the delivery point, a power project of 6000 MW (at 65% PLF and 20-25% transmission and distribution losses) is required. The present cost of setting up such a power project at an estimated rate of Rs. 55 million per MW would be Rs. 330 billion ($\sim$US$8 billion). This factor alone would more than compensate for the estimated cost of the [Sardar Sarovar] project. (Vyas 2001: 50)

As Fig. 7 shows, the Sardar Sarovar Project was designed to spread surplus surface waters over north Gujarat, Kutch and Saurashtra to use surface irrigation to crowd out groundwater pumping and to accelerate groundwater recharge. Moreover, at the turn of the millennium, Government of Gujarat also invested US $150 million in 334-km-long *Sujalam Sufalam* spreading canal (Fig. 8) whose objective was to transfer surplus flood waters from Kadana reservoir in south-eastern Gujarat to fill up 21 rivers and streams, and several large and medium-sized reservoirs in north Gujarat, while recharging the aquifers *en route*. Now, the chief minister has announced a new ambitious US$1.7 billion scheme to lift to a head of 20 m a million acre feet (1.23 BCM) of Narmada water every year to fill up all dams in groundwater-depleted Saurashtra through a network of pipelines to both recharge parched aquifers and to ease pressure on groundwater

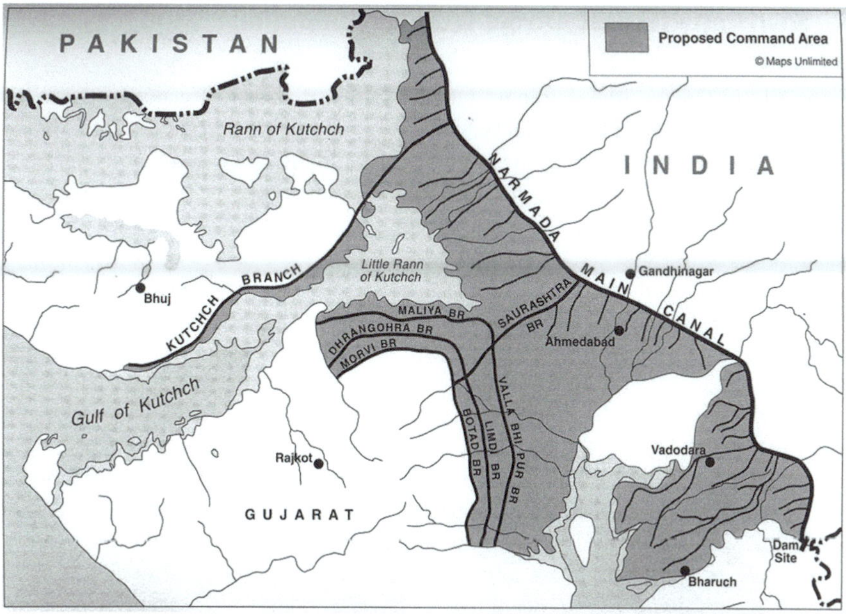

Fig. 7 Command area of Sardar Sarovar irrigation system

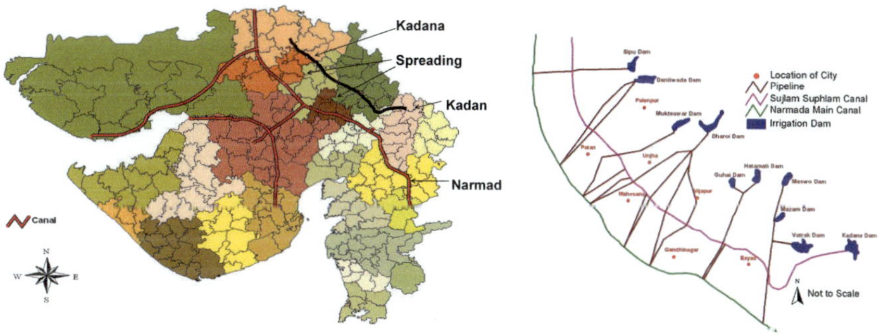

Fig. 8 Layout of Sujalam Sufalam recharge canal

(Khanna 2013). All these suggest that politicians and administrators took a more integrated view of surface-groundwater-energy co-management, while the dominant thinking in our Taskforce was limited and water-centric, focussed as it was on confining MAR to average annual non-committed surplus runoff after providing for 'existing, under-construction and planned future' surface structures.

6 Distributed MAR Direct from Precipitation

Especially in Gujarat, the dominant understanding of runoff as water resource has met with contestation from farmers and rural communities. The Taskforce began with the premise that the runoff from within the administrative boundaries of basins as well as from upper-riparian districts comprises the water resource that requires husbanding and managing. It is for runoff that MAR competes with 'existing, under-construction and planned future' surface storages (Fig. 9). However, since circa 1990, Saurashtra and Kutch regions have witnessed a veritable revolution in decentralized groundwater recharge directly from precipitation (Shah 2000). According to Thomas (1990), for India as a whole, it is estimated that 17.5% of rainfall is lost to evaporation soon after precipitation and another 12.5% contributes to natural groundwater recharge. Thanks to high temperatures and wind speed, direct non-beneficial evaporation in Gujarat may be somewhat higher than 17.5%. When farmers and local communities in Saurashtra and Kutch impound rainwater into hundreds of thousands of open wells, farm-bunds, check dams, percolation ponds and such structures, they accelerate and augment natural rainfall recharge at the expense of evaporation, ET and runoff. There is large body of evidence about the beneficial hydro-economic impacts of this popular mass movement for groundwater recharge in Saurashtra and Kutch (Shah 2000; Verma and Krishnan 2012; Shingi and Asopa 2002; Gandhi and Namboodiri 2009). The Taskforce examined all this evidence. Yet, the Taskforce report recommended MAR effort on only a negligible scale in these regions, thanks to its preoccupation with 'average annual non-committed surplus runoff'.

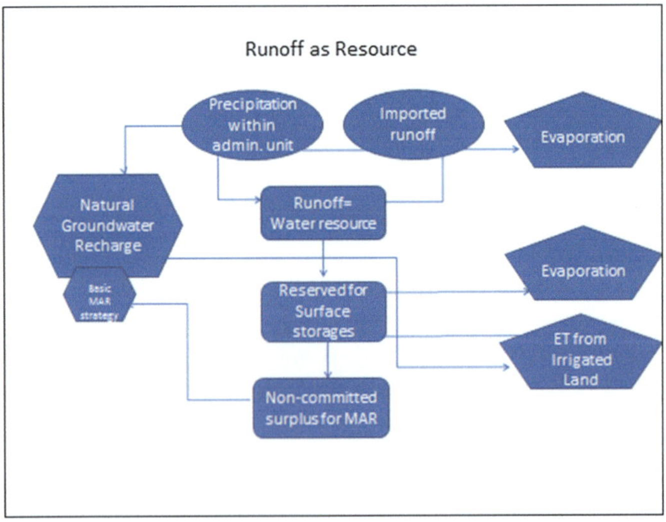

Fig. 9 MAR with runoff as resource

Arid and semi-arid districts of Saurashtra, Kutch and north Gujarat receive far lower average rainfall compared to the South. Since runoff-to-precipitation ratio declines rapidly with decline in average precipitation, the former basins appear unfit for large-scale MAR under the CGWB methodology. However, even these semi-arid districts have an excellent monsoon once every 3–5 years, with sizeable portion of its runoff wasted as floods for the lack of surface storages or MAR structures. Thus, a strategy of decentralized MAR designed for better than normal monsoon years can use such excess precipitation locally to meet several years' accumulated groundwater deficit. Table 8 shows that, during the 20-year period between 1988 and 2007, even Banaskantha and Mehsana, the most critical groundwater basins on Gujarat, had, respectively, 9 and 14 years with over 500 mm of precipitation, and 5 and 9 years, respectively, with more than 700 mm. The Taskforce emphasized that the accumulated groundwater deficits in North, Saurashtra and Kutch are too large relative to the 'non-committed surplus runoff' available for MAR intervention of required scale. However, Fig. 10 shows that if one treats precipitation as a resource, the picture changes entirely. If for the 20-year period during 1998–2007, the 11 basins had created sufficient decentralized MAR capacities to capture excess rainfall into aquifer storage during good years, no more than 10% of the aggregate rainfall during the 20-year period would be needed to bring the water level to 3 m b.g.l. (Fig. 11).

Thus, although the Taskforce was fully aware of the massive social movement for distributed aquifer recharge in progress and its benefits in Saurashtra and Kutch, its thinking did not fully incorporate different values that farmers and communities placed on different modes of water storage especially in closed or closing river basins. Table 9 presents a socioeconomic assessment of the preference that farming communities reveal for four dominant modes of water storage along several dimensions (Shah 2009b). Economic value of storage to farmers comes from space, time and form utilities that different storage modes offer. With 1.1 million private irrigation wells dispersed around the countryside, Gujarat is fully plumbed to use aquifer storage to the full. As a result, spatially distributed aquifer storage holds most attraction for farmers because it provides them maximum space, time and form utility. Aquifer storage loses the least to non-beneficial evaporation, offers the best protection against mid monsoon drought (Tuinhof et al. 2003). By assigning higher priority to 'existing, under-construction and planned future' large surface storages over decentralized MAR, the Taskforce recommendation of Basic MAR strategy rejected the valuation of different modes of storage set out in Table 9. The most likely reason could be that, in Indian situation, large surface storages offer the only significant source of water that the government has complete control over. These are also emerging as the increasingly important source of urban and industrial water supply. In Gujarat's specific context, water from large dams that the Irrigation Department supplies to municipalities and industries provides over 95% of its revenue, while irrigation that uses up the bulk of the dam water returns trivial amounts as water fees. The trade-off thus is between preserving and augmenting government-controlled water versus clearing the accumulated state-wide groundwater deficit through distributed MAR.

Table 8 Accumulated groundwater deficit relative to accumulated precipitation

	Gujarat's basins	Total rainfall-1988–2007 (m)	Volume of rainfall received within Gujarat portion of the basin (Mm^3)	Total surface water resources (Mm^3 and % of precipitation)	Volume of water needed for decentralized MAR to raise water level to		No. of years with >500 mm of rainfall	No. of years with >700 mm rainfall
					8 m b. g.l.	3 m b. g.l.		
1	Banaskantha	10	83,000	84 (2.02)[a]	3747	5204	9	5
2	Mehsana	13	233,000	562 (4.82)	7283	13,637	14	9
3	Sabarmati	17	585,000	2018 (6.90)	3600	10,092	17	12
4	Mahi	18	238,000	4320 (36.3)	2642	6402	18	15
5	Narmada	20	208,000	12,665 (121.8)	790	1686	19	17
6	Tapi	25	130,000	4880 (75.1)	7.5	43	20	19
7	Saraswati to Tapi	26	253,000	7645 (60.43)	128.5	19	20	20
8	Northern Kutch	8	68,000	306.4 (9.0)	3245	3338	7	2
9	Southern Kutch	8	70,000	364 (10.4)	3538	6296	7	3
10	Northern Saurashtra	11	145,000	1397 (19.3)	534	1840	11	6
11	Southern Saurashtra	13	464,000	3160 (13.62)	2895	9790	15	8
Total			2,477,000	37,401.4 (30.2)	28,410	60,939		

[a]Numbers in parentheses show total water resource of each basin over the 20-year period as a fraction of rainfall precipitation within these basins during these 20 years

[b]Parts of the districts

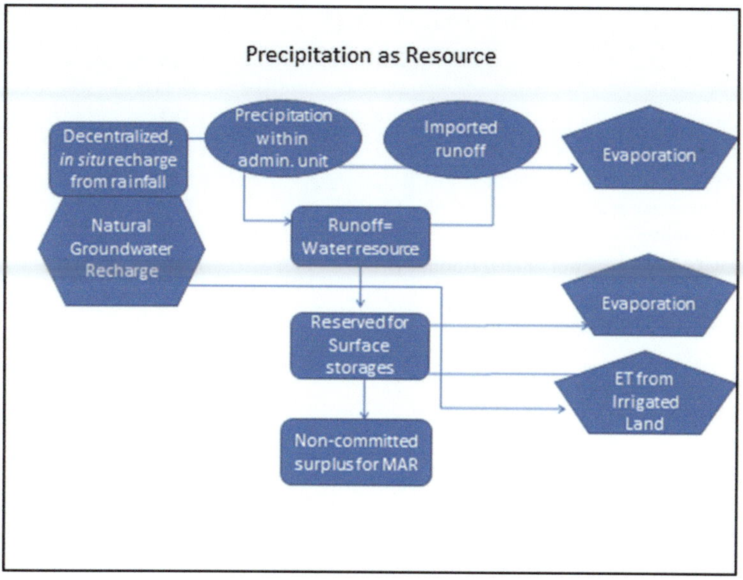

Fig. 10 MAR with precipitation as resource

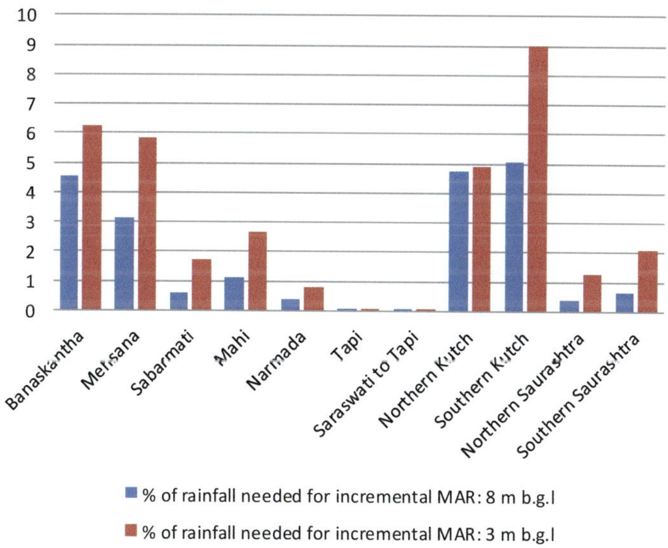

Fig. 11 Percent of accumulated precipitation needed for decentralized recharge to fill the accumulated groundwater deficit

Table 9 Pros and cons of different modes of storage

		Small surface storages	Large surface reservoirs	Aquifer storage (BAU)	Aquifer storage with distributed MAR
1	Makes water available where needed (space utility)	↑↑↑	↑↑	↑↑↑↑	↑↑↑↑↑
2	Makes water available when needed (time utility)	↑	↑↑	↑↑↑↑	↑↑↑↑↑
3	Level of water control offered (form utility)	↑	↑↑	↑↑↑↑	↑↑↑↑↑
4	Non-beneficial evaporation from storage	↓↓↓↓	↓↓	↓	↓
5	Non-beneficial evaporation from transport	↓↓	↓↓↓	↓	↓
6	Protection against mid-monsoonal dry spell (2–8 weeks)	↑↑	↑↑↑	↑↑↑↑↑	↑↑↑↑↑
7	Protection against a single annual drought	↑	↑	↑↑↑	↑↑↑↑↑
8	Protection against two successive annual droughts	↑	↑	↑↑	↑↑↑↑
9	Ease of storage recovery during a good monsoon	↑↑↑↑↑	↑↑↑↑	↑↑	↑↑↑
10	Social capital cost of water storage, and transport and retrieval structures	↓↓	↓↓↓↓↓	↓↓	↓↓↓
11	Operation and maintenance of social costs of storage, transport and retrieval structures	↓	↓↓	↓↓↓↓↓	↓↓↓
12	Carbon footprint of agricultural water use	↓	↓↓	↓↓↓↓↓	↓↓↓

7 Concluding Remarks

The analytics presented here have several methodological limitations. Actual MAR planning at aquifer or basin level would need to tackle these infirmities. The hydro-geological methodology used to plan MAR projects needs to take account of flows and fluxes as well as variability of water demand and supply in space and time. It needs to explicitly consider surface–groundwater interactions and distinguish between consumptive and non-consumptive water uses. Aquifer-level MAR plans would necessitate substantial modeling, scenario building and scenario evaluation component. Equally, the broad-brush economic approach to exploring alternative MAR scenarios also has its limitations. Simple economics with linear arithmetic computations of storage space and runoff for recharge used here needs to factor in the nonlinear rainfall–runoff relationship, soil moisture movement and hydro-geological principles. The mandate of the Taskforce, however, was not to develop detailed basin-/aquifer-wise MAR plans but to advise the government on whether large-scale MAR should be adopted as the center-pin of state's water policy and why. The main argument of this chapter is that the recommendation made would vary depending on whether it is based exclusively on hydro-geological thinking or on a fusion of thinking derived from hydrology and social sciences. The methodological infirmities of the Taskforce work as well as the economic analysis in this chapter do not significantly undermine this argument. We realize, however, that the actual implementation of a MAR strategy would require more detailed and rigorous analyses and modeling; and that to succeed, its implementation will need to be adaptive, inter-sectoral and multi-scalar in nature.

The Basic MAR strategy developed for Gujarat by the Taskforce using the CGWB methodology made several important contributions in identifying basins with sub-surface space in dewatered vadose zone, in computing the volume of space available using differential values of the coefficient of replenishment, estimating volumes of water needed to arrest annual groundwater depletion and to stabilize groundwater level at 8 m b.g.l. and computing the average annual runoff available in different basins. However, the average annual availability of non-committed surplus runoff after meeting the requirements of existing and planned surface storages emerged as a binding limitation on the scale and locale of MAR program. The Basic MAR strategy, which was the main recommendation of the Taskforce, argued for MAR in basins where it was least needed and plead non-availability of average annual non-committed surplus runoff for MAR where it is badly in need. The institutional format the Taskforce recommended, however, included farmers, communities and NGOs as stakeholders who could contribute to MAR besides government departments (Table 10).

Superimposing economic reasoning on hydro-geological calculus would have led to a more ambitious MAR based on greater allocation of runoff to MAR even with inter-basin transfer. The energy and carbon dividend of reducing the pumping head in 0.87 million irrigation tube wells (which number has since increased to 1.1 million) is so large that it justifies ambitious MAR interventions even at the cost of a portion of surface storage. Proactive MAR strategies I and II were presented to the Taskforce but were voted out. The strategy for distributed MAR direct from precipitation was presented too and was given space in the Taskforce report primarily because primary diversion of water by farmers and communities for distributed recharge has become far too widespread and dominant in rural Gujarat to ignore. This also helped the Taskforce to accept an institutional format for implementing MAR in which besides government, some role was assigned to individuals and communities besides government managers as key players as Table 10 shows.

Basic MAR strategy paid little heed to some simple but important economic ideas. Had all or most members of the Taskforce worked these concepts into their thinking, the Taskforce would likely have recommended a different, more ambitious MAR strategy. Table 11 suggests how each of these basic concepts could have altered the thinking of the Taskforce. We saw that despite hydro-geologists' preference for limiting MAR to average annual non-committed surplus runoff, Gujarat's administrative and political leaders have launched ambitious inter-basin water transfers such as *Sujalam Sufalam* recharge canal and SAUNI canal network for

Table 10 Role of different players in MAR

	Aquifers affected	Key players	Numbers of actors who can contribute	Recharge volumes/structure	Location of structures
Small structures for recharging wells, farm ponds and roof-water harvesting structures	Dynamic groundwater in hard-rock areas	Individual farmers and urban citizens	Millions	100–5000 m^3	Private farmlands and homes
Check dams, percolation tanks, sub-surface dykes, etc.	Dynamic groundwater in hard-rock areas	Communities using a common aquifer system	Tens of thousands	100,000–5,000,000 m^3	Common-property or government land
Large structures on government land for recharge to confined aquifers; improved conjunctive management of surface and groundwater	Confined aquifers; large alluvial aquifers especially in arid and semi-arid areas	Public agencies with hydro-geology expertise; canal system managers	Few	0.1–1 km^3 or more	Government wastelands or forest lands; command areas of canal irrigation systems

Table 11 How would have Taskforce thinking changed with inclusion of economic concepts?

	Economic concept	Its impact on Taskforce's work
1	Demand-side perspective	The Taskforce would have questioned the wisdom of recommending MAR where there was not need and not recommending it where it is needed most
2	Opportunity cost	The Taskforce would have generated support for some runoff allocation to MAR over surface storage in view of higher social opportunity cost of providing a m^3 of groundwater versus m^3 of surface water
3	Sunk cost	The Taskforce would have been less insistent on giving the first charge on all runoff to existing, under-construction and planned future storages in preference to MAR
4	Incremental principle	The Taskforce would have been more open to investing in MAR given that groundwater storage, although smaller than surface storage currently meets 3/4th of Gujarat's water needs. It would have recognized that a Mm^3 increase in groundwater availability means a great deal more than a Mm^3 of surface storage
5	Proportional principle	The Taskforce would have recognized that the effort and resources committed to MAR need to be proportional to the significance of aquifer storage to Gujarat's water needs of today and tomorrow
6	Partial versus total perspective	The Taskforce would have recognized that, besides increasing groundwater availability, the benefits of successful MAR would cut across several sectors be energy, agriculture, water and sanitation, public health (through reduced fluoride and nitrate concentration in groundwater), drought-resilience The Taskforce would have recognized that Gujarat's MAR strategy is not only about what the government departments can do but also about what people, communities, NGOs and other players can do with supportive policies and incentives from the government

Saurashtra. Intuitively, these leaders are viewing MAR as an integral response to multiple challenges such as groundwater depletion, quality deterioration, rising energy costs, and carbon footprint of pumping and of sustaining small-holder irrigation.

The key takeaway in the exercise was the criticality of collaboration between social scientists and water scientists in a genuine effort to understand what each discipline has to offer and how it can enrich the collective thinking. It is unlikely that a group of social scientists would have gone through the rigorous process of estimating available storage volumes in dewatered vadose zones of each basin like the Gujarat Taskforce did. Likewise, economists and social scientists can also make important contribution to water strategy formulation as we saw in much of this chapter. The ideal is to achieve a synthesis of these different perspectives that is as close to reality as practical.

References

Bagchi AM, Das P, Chattopadhyay SK (2005) Growth and structural changes in the economy of Gujarat, 1970–2000. Eco and Polit Weekly 40(28):3039–3047

CGWB (2002) Master plan for artificial recharge to groundwater in India. Central Ground Water Board, Government of India, New Delhi. http://cgwb.gov.in/documents/MASTER%20PLAN%20Final-2002.pdf

CGWB (2013) Master plan for artificial recharge to groundwater in India. Central Ground Water Board, Government of India, New Delhi. http://cgwb.gov.in/documents/MasterPlan-2013.pdf

Dillon P, Pavelic P, Page D, Beringen H, Ward J (2009) Managed aquifer recharge: an introduction. National Water Commission, Canberra, Australia, 65 p. Waterlines Report Series 13. http://www.nwc.gov.au/__data/assets/pdf_file/0011/10442/Waterlines_MAR_completeREPLACE.pdf

Gandhi V, Namboodiri NV (2009) Groundwater irrigation: gains, costs and risks. Indian Institute of Management, Ahmedabad, W.P. No. 2009-03-08

Government of Gujarat (GoG) (2009) First approximation report to increase groundwater resources through artificial recharge in Gujarat State. Narmada, Water Resources, Water Supply and Kalpsar Department, Gandhinagar

Hardiman D (1998) Well irrigation in Gujarat: systems of use, hierarchies of control. Eco Polit Wkly 33(25):1533–1544

IRMA (Institute of Rural Management, Anand)—UNICEF (United Nations Children's Fund) (2001) White paper on water in Gujarat. IRMA, Anand, India

Khanna R (2013) Work on SAUNI to start by October end. Times of India, Ahmedabad, 13th Oct 2013, Ahmedabad edition, p 9

Narula K, Fishman R, Modi V, Polycarpou L (2011) Addressing the water crisis in Gujarat, India. White Paper. Columbia Water Center, Earth Institute, Columbia University, New York

Shah T (1993) Groundwater markets and irrigation development: political economy and practical policy. Oxford University Press, Bombay

Shah T (2000) Mobilizing social energy against environmental challenge: understanding the groundwater recharge movement in western India. Nat Resour Forum 24(3):197–209

Shah T (2006) Note for the planning commission. Government of India on Groundwater Regulation: International Experience

Shah T (2008) India's master plan for groundwater recharge: an assessment and some suggestions for revision. Econ Polit Wkly 43(51):41–49

Shah T (2009a) Taming the anarchy: groundwater governance in South Asia. Resources for the Future Press, Washington D.C.

Shah T (2009b) Climate change and groundwater: India's opportunities for mitigation and adaptation. Environ Res Lett J 4(03):5005 (13 pp)

Shah T (2010) Farmer organisation for lift irrigation: irrigation companies and tubewell cooperatives of Gujarat, 1993. ODI Irrigation Management Network, Paper No. 26

Shah T, Verma S (2008) Co-management of electricity and groundwater: an assessment of Gujarat's Jyotirgram scheme. Econ Polit Wkly 43(7):59–66

Shah T, Bhatt S, Shah RK, Talati J (2008) Groundwater governance through electricity supply management: assessing an innovative intervention in Gujarat, western India. Agr Water Manage 95(11):1233–1242

Shah T, Gulati A, Pullabhotla H, Shreedhar G, Jain RC (2009) The secret of Gujarat's agrarian miracle after 2000. Econ Polit Wkly 44(52):45–55

Shingi PM, Asopa VN (2002) Independent evaluation of check dams in Gujarat: strategies and impacts, executive summary, vol 1. Indian Institute of Management, Ahmedabad

Thomas KC (1990) Water resources of India and its development prospects. In: Dhawan BD (ed) Big dams: claims and counterclaims. Commonwealth Publishers, New Delhi

Tuinhof A, Heederik JP, de Vries J (2003) Management of aquifer recharge and subsurface storage: a promising option to cope with increasing storage needs. In: Tuinhof A, Heederik JP (eds) Management of aquifer recharge and subsurface storage: making better use of our largest reservoir. Netherlands National Committee and International Association of Hydrogeologists, pp 3–18

Verma S, Krishnan S (2012) Assessing the agrarian impact of decentralized water harvesting at the basin scale: a discussion on methodology. IWMI-Tata Water Policy Research Highlight, 47, 9 p

Vyas J (2001) Water and energy for development in Gujarat with special focus on the Sardar Sarovar project. Int J Water Res Dev 17(1):37–54

Yagnik A, Sheth S (2005) The shaping of modern Gujarat: plurality, Hindutva, and beyond. Penguin Books, New Delhi

Co-solving Groundwater Depletion and Seasonal Flooding Through an Innovative Managed Aquifer Recharge Approach: Converting Pilot to a Regional Solution in the Ram Ganga Sub-basin

Prasun K. Gangopadhyay, Bharat R. Sharma and Paul Pavelic

Abstract Climate induced extreme events such as floods and droughts are often disastrous in incidences and affects Indian economy often. Low per capita surface water storage (225 m^3/capita[1]), few sites for additional storages facilities and depleting groundwater aquifers reduce the resilience of the communities to alleviate the day to day short age and larger seasonal shocks. India has a long history of storing and recharging runoff waters through community participation. Ongoing such programs are focused on drought-prone or socio-economically weak areas and exclude the flood prone zones. The present study aims to improve the groundwater resources availability through diverting flows from rivers or canals at times when these flows pose flood risk and recharging the groundwater through suitable artificial recharge structures. This method addresses the issue of groundwater depletion as well as reducing the flood risks. A geo-hydrological analysis in spatial platform using data available in public domain and detailed ground survey, a site was identified in Jiwai Jadid village of Milk Block of Rampur district of Uttar Pradesh, India. A community owned pond was retrofitted with recharge wells and associated infrastructure to draw excess monsoon water from a nearby flood-prone river. The preliminary results show a positive impact on groundwater table and water quality. However, to achieve the full benefit of the method it is required to implement it in larger scale. Ongoing government programs that are focused on livelihood improvement and natural resources management are the best options to scale up such effect in regional scale.

Keywords Flood · Ground water · Ramganga sub-basin · Irrigation
Aquifer · UTFI (Underground taming of flood water for irrigation)

P.K. Gangopadhyay (✉) · B.R. Sharma
International Water Management Institute (IWMI), 2nd Floor, CG Block C,
NASC Complex DPS Marg, Pusa, Opp. Todapur, New Delhi 110012, India
e-mail: P.Gangopadhyay@cgiar.org

P. Pavelic
International Water Management Institute (IWMI), c/o National Agriculture and Forestry Research Institute (NAFRI), Ban Nongviengkham, Xaythany District, Vientiane, Lao PDR

D. Saha et al. (eds.), *Clean and Sustainable Groundwater in India*,
Springer Hydrogeology, https://doi.org/10.1007/978-981-10-4552-3_12

1 Introduction

Floods and droughts are of recurrent occurrence in South Asia. In India alone, floods cause on average the annual economic loss of USD 7.5 billion (UNISDR 2015). More than 12% (4 Mkm^2) of total geographical area (32.9 Mkm^2) is flood prone and on average 7.5 mha area of land is affected by floods each year (NDMA 2016). Particularly during the monsoon season, large volumes of run off from the Himalayan range often cause great damage downstream. Rapid urbanization, growing developmental and economic activities in flood plains are the main causes of recurrent floods, which is accentuate by climate charge. As a result of high sediment loads from upper catchment and reduced carrying capacity of rivers over time, new areas are being flooded which were not considered as flood prone earlier. As an example, after the 2013 flood event in northern India, which caused 5000 deaths and massive destruction of public and private infrastructures; India was hit by two consecutive years of weak monsoons spread across 10 states. A study conducted by the Associated Chambers of Commerce and Industry of India has reported that the cost of drought could be as high as $100 billion for the year of 2016 (ASSOCHAM 2016). Besides, year-round agriculture production is heavily dependent on groundwater. Ground water pumped if exceeds the annual recharge which is primarily from rainfall, causes groundwater levels to fall. This has negative implications not only on agriculture, but also on water supply for domestic and industrial purpose, particularity, where they are aquifer depended.

To mitigate flood and drought events, common structural interventions are storage and diversion of water. Non-structural measures include awareness campaign, flood forecasting, relief and resettlement measures and the process continues regularly in an annual cycle. Larger dams are beneficial for storing water in monsoon and divert during the dry season, however, the practice is often costly and controversial primarily from ecological and socio-economic point-of view. Underground Taming of Floods for Irrigation (UTFI) involves diverting high water flows from rivers or canals at times when these flows pose flood risk, and recharging the aquifer through relevant techniques. This method, preferably in a cluster, is useful to reduce the flood risks locally and in the downstream areas, as well as improve the groundwater conditions (Pavelic et al. 2012, 2015).

India's long tradition of storing and recharging runoff waters through social mobilization for domestic supplies and agricultural use has extended across large parts of the country. This has probably started at the dawn of civilization with small human settlements on the banks of rivers. As the population increased, settlements developed into towns and cities and agriculture expanded. Techniques were developed to augment water availability by collecting and storing rainwater, tapping deeper aquifer and also gathering water from snow and glacier melt (CGWB 2007). In recent decade's over-exploitation of groundwater led to significant water level decline both in hard rock and alluvial area threatening sustainability of this resource. To tackle the issue, artificial recharge is one of the interventions, which aims at supplementing the natural replenishment of ground water storage by changing the natural condition artificially and building suitable structures. Compared to other recharge methods, retrofitting village ponds with recharge wells and shafts are preferred as they are cheap, simple in design and easy to manage through community participation (CGWB 2007). However, the areas with less permeable top layers like clay, alternative techniques are required and similar way to divert the water to aquifer through recharge wells are being practiced widely across the Ganga River Basin (Kaledhonkar et al. 2003; Kumari et al. 2014). Although the technical and social components of UTFI are not necessarily new to India—the mode of operation and areas targeted are designed with a new perspective.

After independence, poverty alleviation was the major agenda for Government of India and multiple community development programs were initiated to build the capabilities of the poor and marginal section of the society. Reactions of the society along with natural resource management, these programs were promoted also through the watershed management programs. Ongoing community managed programs that are focused on sustainable rural livelihoods through rejuvenation of natural resource base are one of the most suitable option to scale-up UTFI. The present paper discusses the UTFI methodology and implementation process, as well as attempts to identify links between ongoing watershed management practices for scaling-up opportunities though such programs.

2 Concept and Methodology

2.1 Broad-Scale Opportunity Assessment

Though the basic concept of UTFI was developed during a study on Chao Phraya River Basin, Thailand (Pavelic et al. 2012); the major ground test of practicality of the concept and impacts on natural and socio-eco-institutional components are being tested very recently (Pavelic et al. 2015).

The process of UTFI concept follows a framework which has primarily four steps for a successful implementation and receives the benefits at maximum at basin scale. The steps are:

- Planning, design, implementation and evaluation of demonstration trials.
- Socio-economic and institutional analysis Synthesis of the key research findings from the earlier activities for conceptual proof Up-scaling to basin level to attain maximum benefits.

The first step of UTFI implementation is to identify hydrogeologically suitable locations with characteristics suited for UTFI using scientifically-based methods and tools. Using geo-spatial technologies to identify ground water potential zone delineation of various permeable and impermeable layers and site suitability analysis for managed aquifer recharge (MAR) have been adopted by several researchers (Saraf and Choudhury 1998; Piscopo 2001; Murray and Mcdaniel 2003; Shankar and Mohan 2005; Jasrotia et al. 2007; Kumar et al. 2008; Chitsazan and Akhtari 2009; Yeh et al. 2009; Adham et al. 2010; Jha et al. 2010; Chenini et al. 2010; Kumar and Kumar 2011; Gontia and Patil 2012; Singh et al. 2013; Russo et al. 2014; Mallick et al. 2015; Samson and Elangovan 2015). Though most of the studies integrated several classified and weighted thematic layers, the classification methods and weighing factors for the layers differed. For the present study nine surface characteristics of the basin viz-drainage density, population density, geology, flood frequency, flood mortality and distribution, extreme rainfall events in a year, land use, slope and soil type and two subsurface characteristics groundwater level and transmissivity of the aquifer above and also economic losses due to floods were considered. Layers were analyzed in a GIS platform with a predetermined suitability index (SI) for optimal site selection.

Similar to previous studies that focused on broad-scale assessment of the potential of MAR (Smith and Pollock 2011; Alraggad and Jasem 2010; Chusanathas et al. 2010), by analyzing two primary aspects (1) nature of the subsurface formations, its capacity to accept, store and recover water; and (2) availability of sufficient water to meet the project objectives. UTFI also looks for socio-economic and institutional opportunities.

Originated from central Himalayas, the Ganga river system spreads across India (79%), Nepal (13%), Bangladesh (4%) and China (4%); and around 410 million people directly or indirectly depend on it (Bharati et al. 2011; Verghese 1993). It is one of the most populated (600 million) river basins in the world with a population density of 550 persons/km^2 (Rajmohan and Prathapar 2013). Regular inundation of low-lying areas during monsoon and occurrence of devastating floods in many parts of Ganga Basin is well-known. Apart from significant flow during monsoon, many Himalayan rivers are highly sediment-charged causing rising river bed and reduction in carrying (Tare et al. 2015). Rapid increase in population and economic growth have increased the food demand so the agricultural activities. The gross irrigated area in the Indian part of the basin is estimated at 23 mha and a major part (>65%) of the area is being supported by groundwater irrigation. The heavy reliance on groundwater has led to depletion across the northwestern Gangetic Plains, which includes much of Haryana, Delhi and wes tern Uttar Pradesh (Pavelic et al. 2015). Considering the dual prerequisites for UTFI approach i.e. flood hazard and groundwater depletion, Ganga Basin was identified as target region.

To further identify the suitable areas for UTFI literature based databases were developed that are focused on frequency and impact of floods, and groundwater development and potential zones for recharge Primarily the database covered two categories of the basin: surface and subsurface. The surface characteristics are (i) drainage density, (ii) population density, (iii) geology, (iv) flood frequency, (v) flood mortality and distribution, (vi) extreme rainfall events in a year, (vii) land use, (viii) slope and (ix) soil type. Other than two sub-surface characteristics, i.e. (i) groundwater level and (ii) transmissivity of the aquifer; economic losses due to floods were included in the selection process. Most of the data were acquired from public domains, which is presented in Table 1. All the databases were converted to

Table 1 Data list and sources

Data	Resolution	Source
Drainage density (SRTM)	90 m	Ganges River Basin Management Plan http://gisserver.civil.iitd.ac.in/grbmp/downloaddataset.aspx
Population density	2.5 by 2.5 min grid (≈5 km^2)	CHRR, CIESIN-Columbia University, and IBRD-WB http://sedac.ciesin.columbia.edu/data/collection/gpw-v3
Geology	1:5,000,000 and 1:10,000,000 scale sheets	Central Energy Resources Team, U.S. Geological Survey http://certmapper.cr.usgs.gov/geoportal/catalog/search/resource/details.page?uuid=%7B328C366A-6034-47D4-A4B7-47C3B7561994%7D
Flood frequency and spatial extent (MOD09 8-day composite time-series data; 2000–2011)	500 m	International Water Management Institute
Irrigated area maps of India	500 m	International Water Management Institute
Slope (SRTM)	90 m	Ganges River Basin Management Plan http://gisserver.civil.iitd.ac.in/grbmp/
Soil	1:5,000,000 scale	FAO-UNESCO Soil Map of the World http://www.fao.org/geonetwork/srv/en/metadata.show?id=14116
Extreme rainfall events (per year; 1998–2009)	Horizontal: 4 km; vertical: 250 m	Bookhagen (Unpublished data, 2010) http://www.geog.ucsb.edu/~bodo/TRMM/
Flood mortality and distribution (10 classes of increasing hazard; 1981–2000)	2.5 by 2.5 min grid (≈5 km^2)	CHRR, CIESIN-Columbia University, and IBRD-WB (2005) http://sedac.ciesin.columbia.edu/data/set/ndh-flood-mortality-risks-distribution
Groundwater level	Annual average groundwater level data district wise for the Indian states	District brochures, Central Ground Water Board, Government of India http://cgwb.gov.in/District_Profile/AP_districtProfiles.html

(continued)

Table 1 (continued)

Data	Resolution	Source
Transmissivity (m^2/day)	Average transmissivity data district wise for the Indian states	District brochures, Central Ground Water Board, Government of India http://cgwb.gov.in/District_Profile/AP_districtProfiles.html
Economic loss due to floods (10 classes of increasing hazard)	2.5 by 2.5 min grid ($\approx$5 km^2)	CHRR, CIESIN-Columbia University, and IBRD-WB (2005) http://sedac.ciesin.columbia.edu/data/set/ndh-flood-proportional-economic-loss-risk-deciles

Geographic Information System (GIS) layers with suitable geo-rectifications and analyzed in a GIS platform to determine a suitability index (SI). The data bases were classified in three major groups, (i) flood occurrence and impacts (FOI); (ii) capture and recharge (CR); and (iii) groundwater storage, use and demand (GSD) (Pavelic et al. 2015; Brindha and Pavelic 2016).

The selection process was broadly grouped into three phases: (i) pre-feasibility analysis, (ii) spatial data processing and (iii) determining Suitability Index (SI) to rank prospects for UTFI. For the first phase of selection, a Boolean logic based analysis was performed to meet the three conditions, i.e. (i) Risk of flooding during the wet season, (ii) Risk of drought/demand for water during the dry season and (iii) Risk to human lives and economic loss from floods and droughts (Pavelic et al. 2015; Brindha and Pavelic 2016). After excluding non-viable areas it was observed that nearly 13% of the inner Ganga Basin is not suitable for UTFI, whereas about 67% has high to very high potential. The Indian states of Uttar Pradesh and Bihar were found to have comparatively higher prospects for UTFI (Brindha and Pavelic 2016). Finally Ramganga sub-Basin was chosen because of relatively easy to access, contrasting SI values and institutional support (Fig. 1).

A major part of the Ramganga sub-Basin falls in Uttar Pradesh state, which is severely affected by both floods and drought. According to State Disaster Management Authority, out of the 24.09 mha geographical area of the State of UP about 7.3 mha is flood prone. The estimated loss to crops, houses and livestock is to the tune of USD 66.5 million annually apart from loss of human lives (SDMA 2016). The stare heavily depends on ground water 67% of total irrigated area is fed from aquifers and 80% of the total recharge occurs through monsoon rainfall.

2.2 *Pilot Site Selection*

From sub-basin level to focusing on village level site is primarily based on the physical suitability Based on the SI rank, presence of host structure (i.e. retention

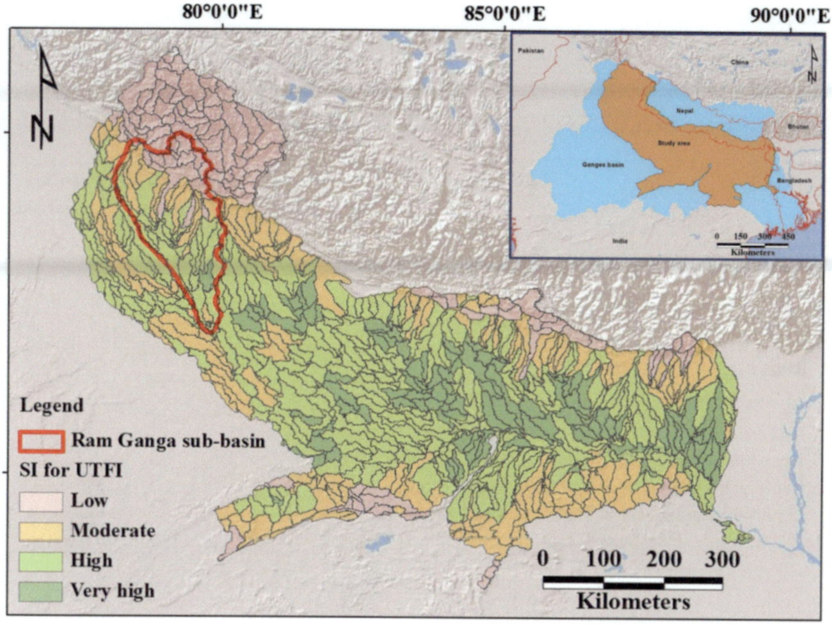

Fig. 1 Study area with different SI for UTFI (*Source* Brindha and Pavelic 2016)

pond), proximity of natural or manmade channels and accessibility; two adjacent meso-scale watersheds of the size of 248 and 295 km^2 were selected. Both the watersheds fall in the administrative boundary of Rampur district of Uttar Pradesh. Along with district-level datasets such as Central Ground Water Board (CGWB), Department of Agriculture and Cooperation (AGRICOOP) and watershed level information, a set of ground level indicators were selected to identify the pilot site. The primary indicators for the pilot site selection process were:

(i) Occurrence and frequency of flooding,
(ii) Depth and long-term trend of water level,
(iii) Proximity of source (canal/river) to host structure (ponds),
(iv) Size and number of the ponds,
(v) Ownership (community/government or private), and
(vi) Accessibility from transport network.

After a detailed analysis of supplementary data in the target watersheds, 10 villages were selected for further study. Based on the field observations and public domain satellite data (Google Earth) all the sites/villages were compared and ranked.

During the site selection process, extensive field campaign was carried out to cross check all the indicators such as physical practicality to connect source (canal/stream) in host structure, tentative cost of modifications, willingness of community through focused group discussions (FGDs), access to site from nearest

Table 2 Site summary

Pro	Con/work to be done
Site 1 (28° 47.026′ N, 79° 12.054′ E)	
Gram panchayat owned	Being used as a waste water disposal pond
Next to canal	New drain need to be constructed to divert waste water from 14 households
Easy access by highway	Irrigation channel need to be reconstructed next to highway
Not in specific use	
District administration supportive	
Villagers are enthusiastic	
Ideal site for outreach as close to road	
Groundwater level is deep and some shallow wells in the village had dried out. The cropping intensity is high. Due to low well yields as a result of depletion farmers have switched to low water intensive crops in some cases	
Site 2 (28° 47.004′ N, 79° 11.971′ E)	
Gram panchayat owned	Being used as a sewage pond
District administration supportive	Major drain network need to be constructed to divert sewage water from pond
Villagers are enthusiastic	Not easy to divert water from canal
Site 3 (28° 46.928′ N, 79° 11.990′ E)	
Gram panchayat owned	Being used as a sewage pond
District administration supportive	Major drain network need to be constructed to divert sewage water from pond
Villagers are enthusiastic	Difficult to divert water from canal

transport network to move heavy machinery (e.g. earth mover) and present use of host structure (Table 2). The village for pilot site was selected as Jiwai Jadid after a detailed site verifications among three candidate sites. A village meeting was conducted to discuss the project goal and to seek their active participation.

2.3 *Local Setup and Additional Site Details*

The village, Jiwai Jadid is located approximately 30 km from Rampur city and close to river Pilakhar. With a total population of 1814 including 291 children (Census 2011), Jiwai Jadid basically has an agrarian community. With a widely varied temperature from summer (43–30 °C) to winter (25–05 °C), and receives ∼900 mm rainfall/annum which occurs mostly in monsoon season spreading between July and September. Though flooding is not disastrous every year, floods in the upstream part causes minor to moderate damages when compared to other parts of Uttar Pradesh. There is an unmet irrigation demand during the summer months i.e. from April to May, sometimes extending up to June. A canal flows from west to east along the north of the village with a total annual average discharge of 378 cusecs (270 cusecs Rabi 2015–16 and 108 cusecs Kharif 2014–15). Much of the irrigation demand for water intensive crops such as paddy, sugarcane, and wheat are met by groundwater which is also evident by plethora of production wells. Additionally, recent additions in summer paddy cultivation also increased the use of groundwater. Soil in this area is silty-sand and silty-clay with thick layer of clay extending to depths of 3 or 6 m from the ground surface. Presently, the village does not have a sewerage system and according to 2011 census 13.7% of total 313 household have septic tanks. The wastewater from the households is disposed in three ponds located in the village. One of these ponds is renovated for the pilot trial and a concrete drain has made to bypass the drain water from ~12 households (which were using this pond) so that the domestic wastewater and sewage do not mix and contaminate the freshwater pond from which recharge will take place.

2.4 Pilot Site Design and Implementation

A wide spectrum of techniques is in practice which is being implemented to harvest excess surface runoff and recharge the aquifer basically based on the terrain and local hydrogeology. Surface water impoundments such as percolation tank is one of the most preferred method because it requires small technological interventions, and cheaper (CGWB 2013) and can be managed by community. However, where tap layer is impermeable or clayey—additional intervention is required, such as recharge shaft. It was observed in the pilot area that first 3–6 m of formation is sticky clay (Fig. 2). Though one of the goal of UTFI research approach is to use community for site preparation and maintenance, to achieve the maximum benefits from 2015 recharge season a recharge construction firm was hired for faster land shaping and construction of recharge structures.

In consultation with key stakeholders such as local farmers, Irrigation Department, research partners from Indian Council of Agricultural Research (ICAR), and implementing agency; the design of the recharge site was finalized, the pond base was excavated with an average depth of 2 m with a maximum depth of 2.41 m. The excavated soil was used to strengthen the embankment of the pond and fodder (Napier grass) was planted for slope stabilization. Next a total number of 10 gravity-fed recharge wells (Fig. 3) were constructed. PVC pipes of 150 mm inner diameter and slotted at the base were sunk into the bottom of the recharge well. Brick walls were constructed around the recharge wells and filled with 2–6 mm

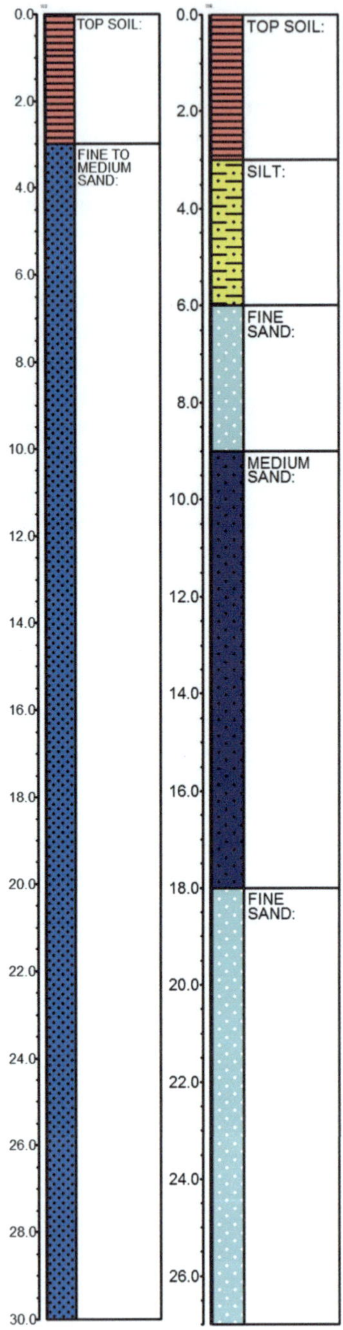

Fig. 2 Lithology of two recharge-wells in the extreme east and west of the site

Fig. 3 Status of the UTFI site **a** before and **b** after implementation

pea-gravels as filter material (Fig. 4). Out of 10 recharge wells, 5 wells are of 3 m and there remaining are of 1.5 m diameter. The first set of (3 m diameter) recharge wells have a depth of 30 m and rest (1.5 m) are 24 m deep. The purpose of constructing two different types of recharge wells is to understand their recharge rate and cost/benefit analysis in the later stage. One of the each type recharge wells were modified with a higher brick wall, where the water flow in the recharge wells can be controlled using a control to estimate the recharge rate. Before diverting the canal/flood water to the pond, a silty chamber was constructed on the north-west corner of the pond to lower the sediment load of canal/flood water. After every recharge season this stilling chamber needs to be cleaned.

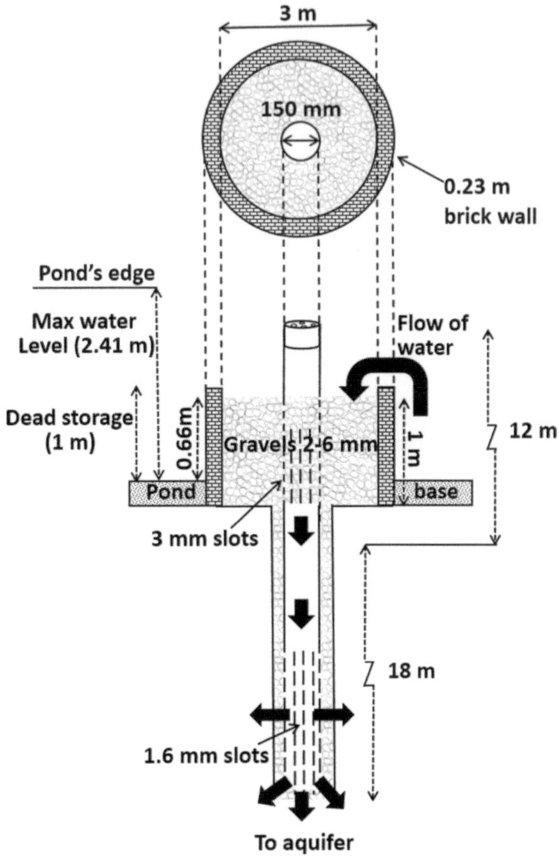

Fig. 4 Design of the recharge well (30 m)

3 Scaling-up Opportunities

Detailed study of the 2010 flood event showed that to reduce 50% flow across sub-basin, on average a flow of 1741 m^3 ha^{-1} needed to be intercepted, stored and recharged and it could be up to 2377 m^3 ha^{-1} in an extreme situation. For an alluvial terrains like Ramganga flood plain, surface infiltration method could be a viable option where the unsaturated zones are permeable. For this method, considering literature recharge value which is minimum infiltration rate of 10 m per wet season, around 2% of the land area would need to be dedicated to recharge interventions across the entire in the Ramganga sub-basin. In case of recharge wells, it is estimated that a well density of 0.17 is required for each hectare considering a well can recharge 100 m^3 day^{-1} which was calculated from previous studies conducted in Uttar Pradesh (MacDonald et al. 2015). Thus the sub-basin requires a large numbers of recharge sites and for a practical model for putting UTFI into practice at a scale greater than the pilot scale involves strong stakeholder acceptance and its integration into ongoing government programs that are focused on livelihood improvement and natural resources management. The ongoing government programs/schemes that are close to UTFI in terms of objectives and implementing guidelines, are discussed below.

(i) *The Integrated Watershed Management Program (IWMP)* is one of the best approach to harvest (excess) rainwater to improve livelihood by providing community managed simple low-tech solutions and mitigate adverse flood conditions. However, it's a bottom-up approach, whereby the user's group themselves suggest their work program where minimal or no technological intervention is required. First launched during 2009–10, the IWMP program is being implemented as per Common Guidelines for Watershed Development Projects 2008. Presently it has been implemented under Prime Minister Krishi Sinchayee Yojna (Watershed Development Component) or WDC-PMKSY. Having a cost norms from $92.3 to $184.6 per ha in plains and $230.8 per ha in difficult/hilly areas, the projects under this program can be implemented in flexible time period of 4–7 years. From 2009 to 2014, a total fund of $1.67 billion (utilized $1.65 billion) has been released Within the Ram Ganga command area, IWMP is being implemented in several districts of The primary objective of these IWMP projects are to address the problems of (i) moisture stress-drought conditions, (ii) groundwater depletion/deep groundwater table, poor quality of groundwater, (iii) excess runoff, (iv) lack of irrigation facilities, (v) low productivity of crops, (vi) low income of the households, (vii) low wages etc.

(ii) *Introduced in September 7, 2005, the Mahatma Gandhi* National Rural Employment Guarantee Act (MNREGA) made paradigm shift from the previous wage employment programs. Being a rights-based approach, MNREGA makes the government legally accountable for providing employment to those who demand it. The primary objective of MNREGA is to provide 'wage employment opportunities' to create 'sustainable rural livelihoods'. Apart from improving livelihood approach, MNREGA emphasizes also on 'rejuvenation of natural resource base i.e. augmenting productivity and supporting creation of durable assets' and 'strengthening rural governance through decentralization and processes of transparency and accountability'. Over the years MNREGA is showing an encouraging trend which includes rise in household income, reduction in distress migration, improvement in groundwater agricultural productivity and cropping intensity. However, having a ceiling of $10.8 thousands per project (with a possibility of higher budget reviewed by an expert committee), MNREGA alone might not be the best option to implement UTFI. Nevertheless, as per the directives of Ministry of Rural Development, convergence between IWMP and MNREGA is crucial and mutually beneficial as 70% of the work taken by MNREGA are related to soil and water conservation and both the programs are introduced to improve livelihood conditions.

(iii) The Cabinet Committee on Economic Affairs has recently introduced Pradhan Mantri Krishi Sinchayee Yojana (PMKSY) which has been formulated by merging ongoing schemes viz. Accelerated Irrigation Benefit Programme (AIBP) of the Ministry of Water Resources, River Development & Ganga Rejuvenation (MoWR, RD&GR), Integrated Watershed Management Programme (IWMP) of Department of Land Resources (DoLR) and the On Farm Water Management (OFWM) of Department of Agriculture and Cooperation (DAC). In PMKSY the 'watershed' component is being managed by DoLR and MoRD. Based on the DoLR report of Uttar Pradesh, in 2014–15 under IWMP or WDC-PMKSY a total amount of \$286,000 was allocated for new farm ponds and \$98,000 was allocated for percolation tanks and for groundwater recharge structure. However, the achievement under these heads was sub-optimal as only \$232,300 and \$1700, respectively were utilized and can be substantially improved through UTFI interventions.

From the perspective and operational mode, UTFI in IWMP or WDC-PMKSY are well suited. One of the primary objectives of IWMP or WDC-PMKSY is to develop a cluster of micro-watersheds in a holistic manner, which is compatible with UTFI's synergistic approach of addressing the issues of groundwater depletion and flood mitigation at basin level. As IWMP is being implemented by the "District Rural Development Agency" through Village Watershed Committees (VWC), it provides an excellent opportunity for the community to be involved from the beginning and contribute in planning and execution as well as receive benefits from the project over the short and long term.

The process of up scaling may not be very smooth and few challenges that might affect the actual process are listed below:

(i) Community level: Lack of information regarding the threat of ground water depletion and without direct financial benefits it is difficult to motivate and engage the community.
(ii) Technical level: Shortage of qualified community member to operate the site and take decision in critical situations.
(iii) Physical environment level: Uncertainties of surface flow might reduce the recharge rate, as the sites will be often connected from canal or natural channels.
(iv) Policy level: Programs such as MGNREA has a cap of maximum of 40% for materials (of total project cost) which might not be sufficient for a recharge well site.
(v) Institutional level: Shift from present practices and lack of willingness to implement at lower administrative level.

However, all of these challenges are manageable when the stakeholders are fully engaged in the program and district management is involved in problem appreciation, site selection, design of the intervention and its satisfactory implementation. Capacity building of the stakeholders through classroom and on-site interactive sessions, and development of comprehensive guidelines and operating procedures shall be an important component for successful implementation of the program. UTFI holds the key to successfully manage the present resource variability and also take care of the future climate induced changes in water cycle for agriculture, livelihoods and food security.

4 Conclusions

Although India has a long history of storing and recharging runoff waters through community participation, the concept and application of Underground Taming of Floods for Irrigation (UTFI) is significantly different from the previous approaches. By drawing from high water flows during the monsoon and storing this water underground through modified village ponds, this approach not only secures water for irrigation during the dry season, but also reduces flood hazards locally and in downstream areas. Apart from improved livelihoods from enhanced irrigation, UTFI could be an effective means of disaster risk reduction when implemented at a larger scale. Though it is too early to conclude from limited observations from the 2015 recharge season, the preliminary results from the pilot site shows an improvement in the local groundwater resources. Additionally, opportunities and risks of UTFI are being studied to provide the necessary degree of confidence for scaling up. As observed from the results of hydrological modeling outputs, for the Ramganga basin a large number of UTFI sites are required to have a significant impact in sub-basin scale. Linking UTFI with ongoing livelihood enhancement and natural resource management programs under the Govt. of India holds the key to implementation at a larger scale to reap the benefits with strong involvement of communities.

Acknowledgements The authors would like to gratefully acknowledge the supports and contributions from Central Soil Salinity Research Center (Lucknow), Krishi Vigyan Kendra (Rampur) and Rampur District Administration.

References

Adham MI, Jahan CS, Mazumder QH, Hossain MMA, Haque AM (2010) Study on groundwater recharge potentiality of Barind tract, Rajshahi District, Bangladesh using GIS and remote sensing technique. J Geol Soc India 1:432–438

Alraggad M, Jasem H (2010) Managed Aquifer Recharge (MAR) through surface infiltration in the Azraq Basin/Jordan. J Wat Res Prot 2(12):1057–1070

Bharati L, Lacombe G, Gurung P, Jayakody P, Hoanh CT, Smakhtin V (2011) The impacts of water infrastructure and climate change on the hydrology of the upper Ganges river basin. International Water Management Institute, Colombo, p 36 (IWMI Research Report 142). doi:10.5337/2011.210

Brindha K, Pavelic P (2016) Identifying priority watersheds to mitigate flood and drought impacts by novel conjunctive water use management strategies. Env Earth Sci 75(5):1–17

Census India (2011) The Registrar General & Census Commissioner, Government of India. http://www.censusindia.gov.in/2011census/population_enumeration.html. Accessed 20 Aug 2016

Central Ground Water Board (2007) Manual on Artificial Recharge of Groundwater. Ministry of Water Resources, Government of India

Central Ground Water Board (2013) Master plan for artificial recharge to ground water in India. Ministry of Water Resources, Government of India

Chenini I, Mammou AB, May ME (2010) Groundwater recharge zone mapping using GIS-based multi-criteria analysis: a case study in central Tunisia (Maknassy basin). Wat Res Manag 24:921–939

Chitsazan M, Akhtari Y (2009) A GIS-based DRASTIC model for assessing aquifer vulnerability in Kherran Plain, Khuzestan, Iran. Wat Res Manage 23:1137–1155. doi:10.1007/s11269-008-9319-8

Chusanathas S, Uppasit S, Munyou S, Intarasut T, Pholkern K, Srisuk K (2010) Application of GIS techniques for determining suitable areas for managed aquifer recharge in the Lower Ping-Yom River Basin, Thailand. In: Zuber A, Kania J, Kmiecik E (eds) Proceedings of the 38th IAH Congress on Groundwater Quality Sustainability, Krakow, pp 1961–1972

Gontia NK, Patil PY (2012) Assessment of groundwater recharge through rainfall and water harvesting structures in Jamka microwatershed using remote sensing and GIS. J Ind Soc Rem Sens 40(4):639–648

Jasrotia AS, Kumar R, Saraf AK (2007) Delineation of groundwater recharge sites using integrated remote sensing and GIS in Jammu district, India. Int J Rem Sens 28(22)

Jha MK, Chowdary VM, Chowdhury A (2010) Groundwater assessment in Salboni Block, West Bengal (India) using remote sensing, geographical information system and multi-criteria decision analysis techniques. Hydrogeol J 18(7):1713–1728

Kaledhonkar MJ, Singh OP, Ambast SK, Tyagi NK, Tyagi KC (2003) Artificial groundwater recharge through recharge tube wells: a case study. IE (I) J–AG 84:28–32

Kumar B, Kumar U (2011) Ground water recharge zonation mapping and modeling using Geomatics techniques. Int J Environ Sci 1(7):1671

Kumar GM, Agarwal AK, Bali R (2008) Delineation of potential sites for water harvesting structures using remote sensing and GIS. J Ind Soc Rem Sens 36:323–334

Kumari R, Sharma B, Singh R, Singh RM, Tewari RK (2014) Estimation of groundwater recharge using well recharging unit in Parasai-Sindh watershed of SAT region of India. Ind J Ecol 41 (2):252–256

MacDonald AM, Bonsor HC, Taylor R et al (2015) Groundwater resources in the Indo-Gangetic basin: resilience to climate change and abstraction. British Geological Survey, Nottingham, Open Report OR/15/047, p 63

Mallick J, Singh CK, Al-Wadi H, Ahmed M, Rahman A, Shashtri S, Mukherjee S (2015) Geospatial and geostatistical approach for groundwater potential zone delineation. Hydrol Process 29:395–418

Murray J, Mcdaniel PA (2003) Development of a GIS database for ground-water recharge assessment of the Palouse Basin. Soil Sci 168(11):759–768

National Disaster Management Authority (2016) http://www.ndma.gov.in/. Accessed 20 Aug 2016

Pavelic P, Brindha K, Amarnath G, Eriyagama N, Muthuwatta L, Smakhtin V, Gangopadhyay PK, Malik RPS, Mishra A, Sharma BR, Hanjra MA, Reddy RV, Mishra VK, Verma CL, Kant L (2015) Controlling floods and droughts through underground storage: from concept to pilot implementation in the Ganges river basin. International Water Management Institute (IWMI), Colombo, p 33 (IWMI Research Report 165). doi:10.5337/2016.200

Pavelic P, Srisuk K, Saraphirom P, Nadee S, Pholkern K, Chusanathas S, Munyou S, Tangsutthinon T, Intarasut T, Smakhtin V (2012) Balancing-out floods and droughts: opportunities to utilize floodwater harvesting and groundwater storage for agricultural development in Thailand. J Hydrol 470–471:55–64

Piscopo G (2001) Groundwater vulnerability map explanatory notes: castlereagh catchment. NSW Department of Land and Water Conservation, Parramatta

Rajmohan N, Prathapar SA (2013) Hydrogeology of the Eastern Ganges Basin: an overview. International Water Management Institute (IWMI), Colombo, p 42 (IWMI Working Paper 157). doi:10.5337/2013.216

Russo TA, Fisher AT, Lockwood BS (2014) Assessment of managed aquifer recharge site suitability using a GIS and modeling. Ground Water. National Ground Water Association. doi:10.1111/gwat.12213. (http://es.ucsc.edu/~afisher/CVpubs/pubs/Russo2014_GW_PV-MAR.pdf)

Samson S, Elangovan K (2015) Delineation of groundwater recharge potential zones in Namakkal District, Tamilnadu, India using remote sensing and GIS. J Ind Soc Rem Sens 43(4):769–778

Saraf AK, Choudhury PR (1998) Integrated remote sensing and GIS for groundwater exploration and identification of artificial recharge sites. Int J Rem Sens 19(10):1825–1841

Shankar MNR, Mohan G (2005) A GIS based hydrogeomorphic approach for identification of site-specific artificial-recharge techniques in the Deccan Volcanic Province. J Earth Syst Sci 114(5):505–514

Singh P, Thakur J, Singh UC (2013) Morphometric analysis of Morar River Basin, Madhya Pradesh, India, using remote sensing and GIS techniques. Environ Earth Sci 68:1967–1977

Smith AJ, Pollock DW (2011) Assessment of managed aquifer recharge potential using ensembles of local models. Ground Water. doi:10.1111/j.1745-6584.2011.00808.x

State Disaster Management Authority (2016) Disaster Management Department, Govt. of Uttar Pradesh. http://rahat.up.nic.in/. Accessed 20 Aug 2016

Tare V, Bose P, Roy G (2015) Ganga River Basin Management Plan, Consortium of 7 Indian Institute of Technology (IITs). https://nmcg.nic.in/writereaddata/fileupload/13_GRBMP%20-%20MPD.pdf. Accessed 20 Aug 2016

The Associated Chambers of Commerce & Industry of India (2016) http://www.assocham.org/newsdetail.php?id=5678. Accessed 20 Aug 2016

United Nations International Strategy for Disaster Reduction (2015) http://www.preventionweb.net/english/hyogo/gar/2015/en/home/data.php?iso=IND. Accessed 20 Aug 2016

Verghese BG (1993) Harnessing Himalayan rivers: regional cooperation in South Asia. In: Iyer RR (ed), Konark Publishers, New Delhi

Yeh HF, Lee CH, Hsu KC, Chang PH (2009) GIS for the assessment of the groundwater recharge potential zone. Environ Geol 58:185–195

Impact Assessment of Demonstrative Project on Artificial Recharge to Groundwater in Gangavalli Block, Salem District, Tamil Nadu

A. Subburaj, E. Sampath Kumar, S. Suresh, V. Elanchelian and K. Rajarajan

Abstract A total of 41 nos of artificial recharge structures (ARS) were constructed in Gangavalli block, Salem District, Tamil Nadu. The area is located in the central parts of the state and the block is marked with groundwater over exploitation. The project was executed to demonstrate the efficacy of the ARS in the aquifer type. The contributing aquifers are made up of weathered zone and underlying fractures within granite gneiss and charnockites. The area is depending on groundwater resources for all utilisation and the abstraction structures are dug well and borewell. The efficacy of the ARS constructed in the area has been assessed by studying its impact on hydro-geological scenario and agricultural practices.

Keywords Artifical recharge · Groundwater · Vedose zone · Hydrogeology Infiltration · Aquifers

1 Introduction

The component of water that infiltrates to below the surface and gets stored in the pore spaces years together under various conditions called groundwater. Groundwater is extracted for various purposes, mainly for irrigation, followed by drinking and replenished naturally mainly from precipitation, called recharge to aquifer system. The degree of recharge to groundwater depends upon the source of water, contact time between surface water and soil and characteristics of the vadose zone, through which the water percolates downwards before joining the water table. Because of various reasons, the contact time is often very less and at many a time, the source water is inadequate over space and time. In order to increase the contact time, the source water is stopped at many places by constructing ARS which abets

A. Subburaj (✉) · V. Elanchelian · K. Rajarajan
Central Ground Water Board, SECR, Chennai, India
e-mail: subbucgwb@yahoo.com

E. Sampath Kumar · S. Suresh
Central Ground Water Board, NH-IV, Faridabad, India

D. Saha et al. (eds.), *Clean and Sustainable Groundwater in India*,
Springer Hydrogeology, https://doi.org/10.1007/978-981-10-4552-3_13

increased infiltration and recharge to groundwater. An improvement on groundwater resource due to the construction of artificial recharge structures, particularly check dam has been studied by many authors (Mudrakartha 2003; Muralidharan 2007; Alderwish 2010; Rajarajan et al. 2012; Subburaj et al. 2015 and Venkateswaran et al. 2015). In the present study, the recharge to groundwater due to ARS is assessed by various ways, like change of the water level, increase of pumping hours and change in cropping pattern.

2 Study Area

The study area falls in the Gangavalli block of Salem district lying between North Latitudes 11° 00′ 30″ and 11° 58′ 30″ and East Longitudes 77° 39′ 0″ and 78° 50′ 30″ forming part of survey of India topographical maps Nos. 58E and I. The total geographical area of the block is 410 km^2 and has 27 revenue villages and 14 panchayats. The location map is given in Fig. 1. Major part of the block is covered by red in situ soil and forest soil in areas under reserved forest. In addition, black cotton soil covers a small patch on the north-eastern part and sandy-loamy mixed soil as a small patch on the south-western part of the block. The prominent geomorphic units are (1) Structural hills, (2) Valley fill, (3) Pediments and (4) Buried Pediments. An ephemeral stream, Suvedha Nadi drains the central part

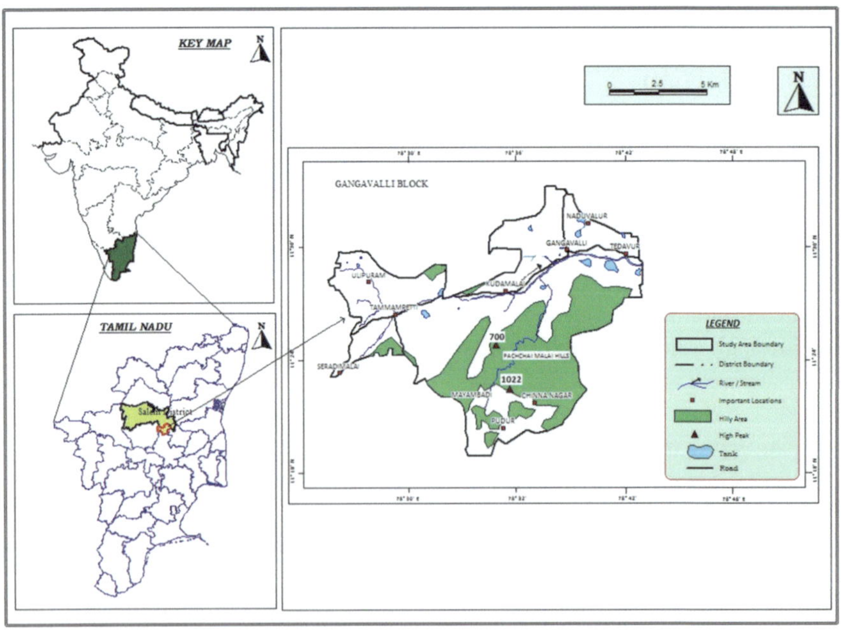

Fig. 1 Location map of study area

of the block is the major water course. The area is characterised by dentritic drainage. The block receives the rain under the influence of both south-west and north-east monsoons. The north-east monsoon chiefly contributes to the rainfall, while the south-west monsoon rainfall is highly erratic. Rainfall data from two stations over the period from 1901 to 2003 have been analysed revealed normal annual rainfall in the block varies from about 800 to 1600 mm. It is observed that the chances of receiving normal rainfall are minimum (40–50%) around Attur, located close the northern boundary and maximum (50–60%) around Pachamalai in the southern boundary of the block.

3 Hydrogeology of the Area

Gangavalli block is underlain entirely by Archaean Crystalline formations (Palanisamy et al. 1989) with Recent alluvium (Fig. 2). The major part of the block is underlain by charnockite and pyroxene granulite occurring as bands trending NE–SW. Hornblende biotite gneiss occurs as patches on the north-western part. Recent alluvium is seen along the river courses and in the intermountain valleys. In general, aquifer system can be divided into two, viz. weathered residuum and fractures below within the hard rocks. The thickness of weathered zone varies from 2.5 to 28 m. Two to five sets of fractures are encountered within 200 m bgl, with yield ranging from 10 to 840 lpm. The depth of the dug wells in the area is ranging from 4.00 to 38.00 m bgl with yield ranging from 35 to 140 lpm and maximum up to 250 lpm in alluvium. The borewells recorded transmissivity values ranging from 1 to 250 m^2/day. The storativity values range from 4.3×10^{-3} to 9.6×10^{-5} indicating semi-confined nature of the deeper fractures. Chemical quality of groundwater is fresh. The electrical conductivity remains within 750 µS/cm in about 40%, moderately fresh (751–2250 µS/cm) in about 55% of the wells, and in 5% of the wells, it ranges between 2251 and 3000 µS/cm revealing that groundwater is of locally brackish at places.

4 Surplus Run-Off

The volume of non-committed surface water resources available in the watershed, where the study area is located, is the difference between the annual run-off in the entire watershed and the committed volume required for filling up the surface water bodies. Based on the computations, the non-committed surface water resource available has been worked out as 20.47 MCM. This volume is considered harvestable by artificial recharge structures at suitable locations.

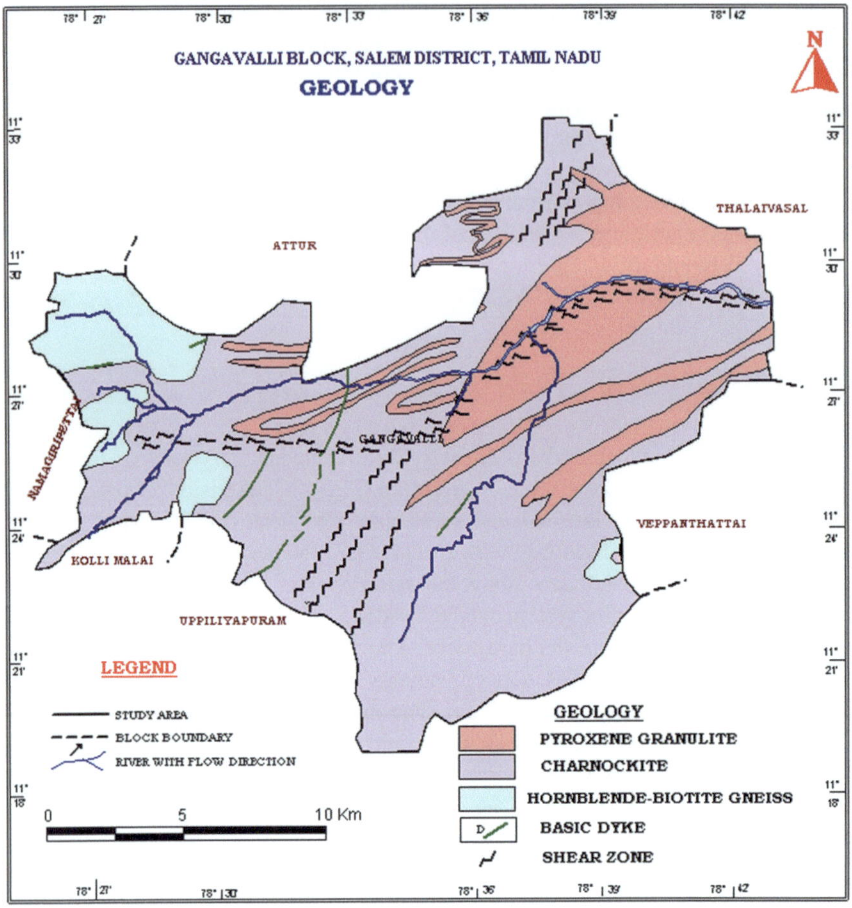

Fig. 2 Geology map of study area

5 Artificial Recharge Structures

A total of 41 artificial recharge structures have been constructed, including 22 check dams, 17 percolation tanks and 2 desilting of existing ponds. All the constructions have been executed during 2006–08. Recharge borewells also constructed in the check dams and Percolation tanks to transfer the water in deeper fractures. The structures were constructed by Tamil Nadu Water Supply and Drainage Board (TWAD Board), Agricultural Engineering Department (AED) and Public Works Department (PWD). The locations of artificial recharge structures constructed in Gangavalli block are shown in Fig. 3.

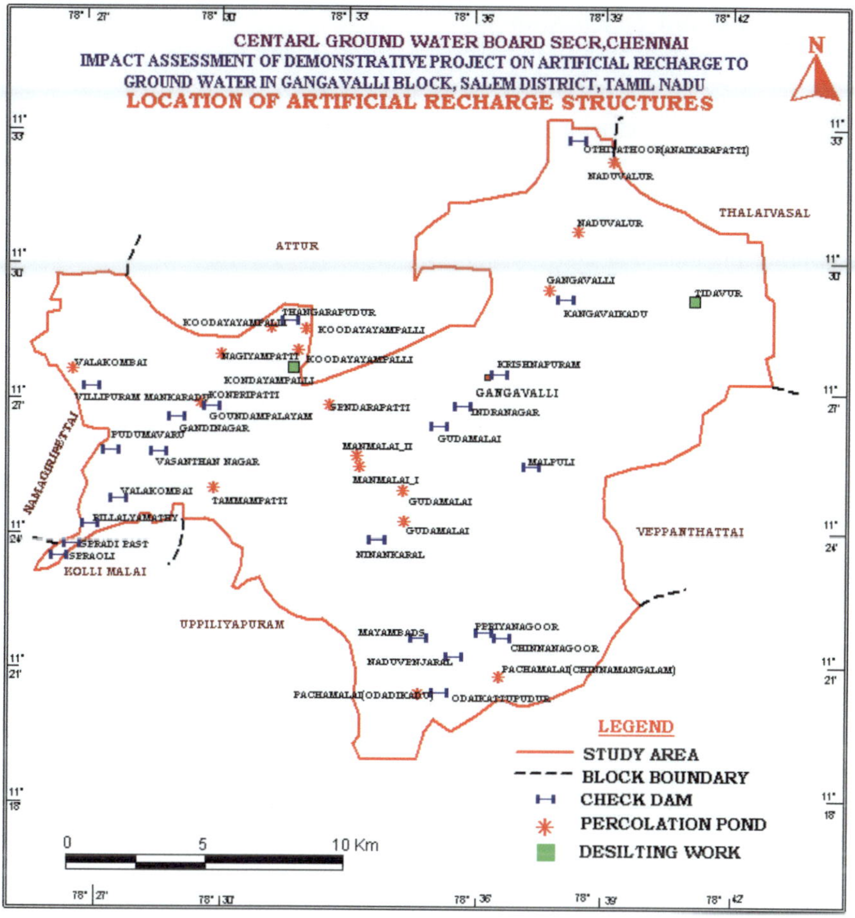

Fig. 3 Location of ARS constructed in the study area

6 Harvested and Recharged Water

On the basis of rainfall pattern and the run-off generated from the catchment, it is taken that there will be three to five fillings of the ARS in each year. A life of 20 years has been considered for each structure and accordingly, quantum of water that can be harvested in the life span of each structure has been computed. In accordance with the GEC methodology, only 50% of the water harvested has been considered for the recharge to groundwater. Based on this, 893,938 m^3 can be harvested in a year and 17,878,751 m^3 can be harvested during the life of the ARS. The quantum of water that could be recharged in a year is 489,618 m^3 and during the life of these structures would be 9,792,367 m^3.

7 Results and Discussion

The impact of the ARS on groundwater regime has been studied based on the water-level changes, cropping pattern and changing in pumping hours. The construction of structures was commenced in 2006 and completed in 2007. Monitoring 41 nos of observation wells, established in the vicinity of the recharge structures, was carried out pre-construction and post-construction of the ARS.

During pre-monsoon 2007, shallowest water level recorded as 2.25 m bgl at Mayambadi and the same well has shown 1.65 m bgl in the same season in 2008. The deepest water level recorded in May 2007 is 27.72 m bgl at Naduvalur West, which recorded 24.42 m bgl in May 2008 after construction of ARS. The rise in water level during pre-monsoon 2007 and 2008 is in the range of 0.6 m to 4.05 m bgl. The water-level fluctuation map (Fig. 4) indicates that the 2–4-m water level rise in northern and western parts, while in other parts less rise is observed. The maximum rise occurred at two places, namely Odaikattu pudur and Gandhi Nagar, which is a mark of better efficiency of structure (Plot 1). The minimum raise of 0.60 m observed at Mayambadi which reflects the poor efficiency of the structures (Plot 2). At places, well located near to ARS is become an auto-flowing (Photo 1).

Analyses of post-monsoon (January) data reveal that the shallowest levels in 2007 is 1.87 m bgl at Mayambadi, whereas in 2008, the level has been encountered at 0.85 m bgl at Ninankarail. The deepest levels of 20.79 and 14.60 m bgl are observed in Naduvalur West in both the years (Plot 3), revealing maximum rise in

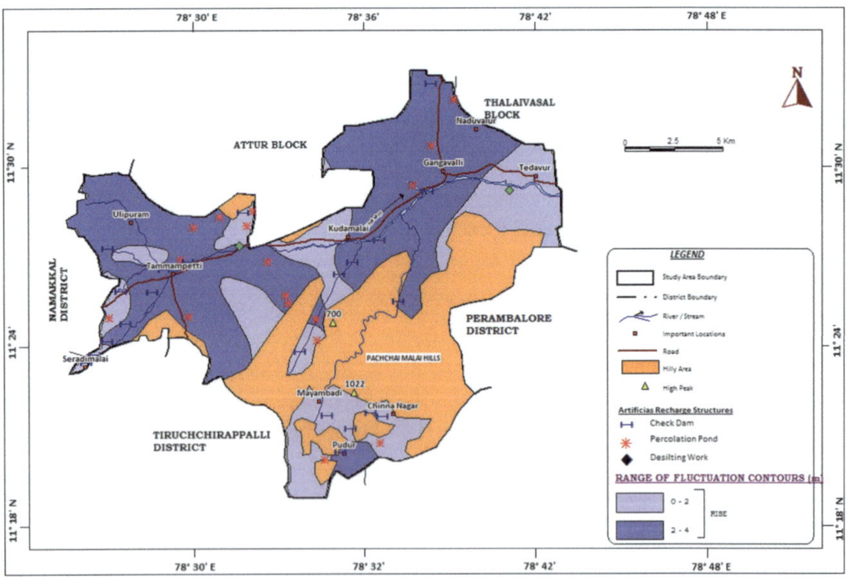

Fig. 4 Water-level fluctuation (May–2007 and 08)

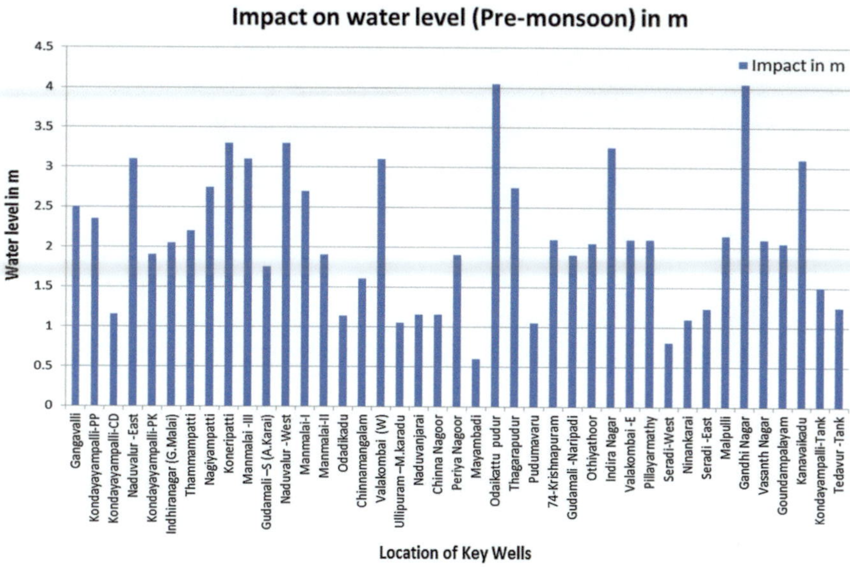

Plot 1 Water level of May–2007 and 08 (pre-monsoon)

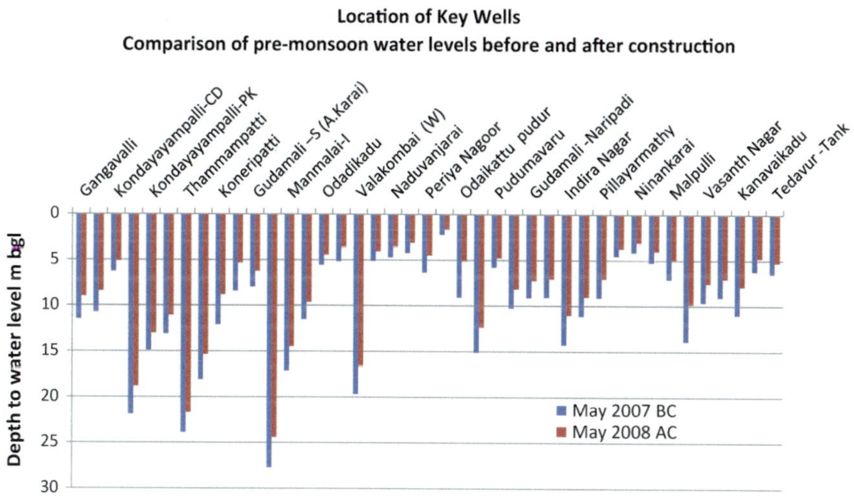

Plot 2 Rise of water level of between May 07 and 08 (pre-monsoon)

water level in the area. The minimum rise in water level has been observed as 0.72 m at Mayambadi. The average rise in water level in post-monsoon 2007 and 2008 is 2.51 m. Water-level partially maximum area is showing 2–4-m water level rise (Fig. 5), while in the northern parts >4 mts rise is occurring. The maximum rise in water level is reported as 6 mts (Plot 4).

(a) Pre construction of ARS (b) Free flow after construction of ARS

Photo 1 Borewell located near ARS

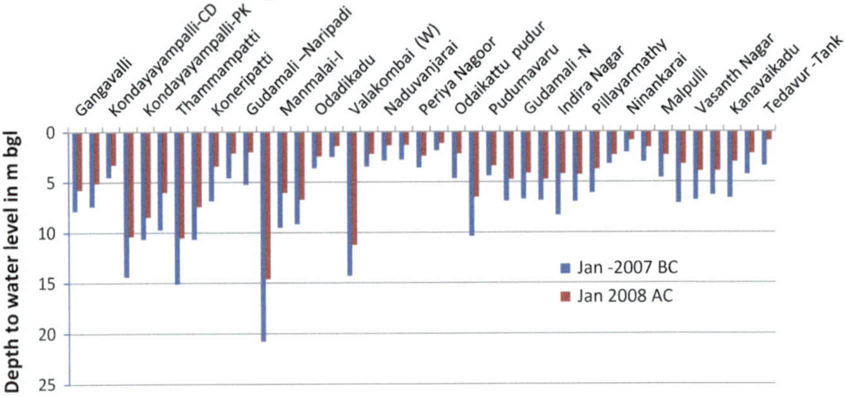

Plot 3 Water level of January 2007 and 2008 (post-monsoon)

The total area of wet crops grown during 2006–2007, before construction of the ARS, was 203 acres which increased to 457 acres, during 2007–2008. The range of increase in cropped area against individual structures varies from 1 acre at Chinnamangalam to 30 acres at Tedavur. It is more significant in the areas where desilting of pond was carried out. Large quantum of desilting resulted in increased storage of surface water resource, which can be directly applied for irrigation. Due to this, the area under cultivation of wet crop has increased from 20 to 30 acres. The percolation tank and check dam with recharge borewell has impacted in has helped in change in cropping pattern due to increased groundwater resource. When the recharge wells are constructed with rainwater harvesting structures, there is significant increase in groundwater resource, which is reflected in increase of cropped area. The area increased ranges from 5 to 10 acres when recharge wells are also

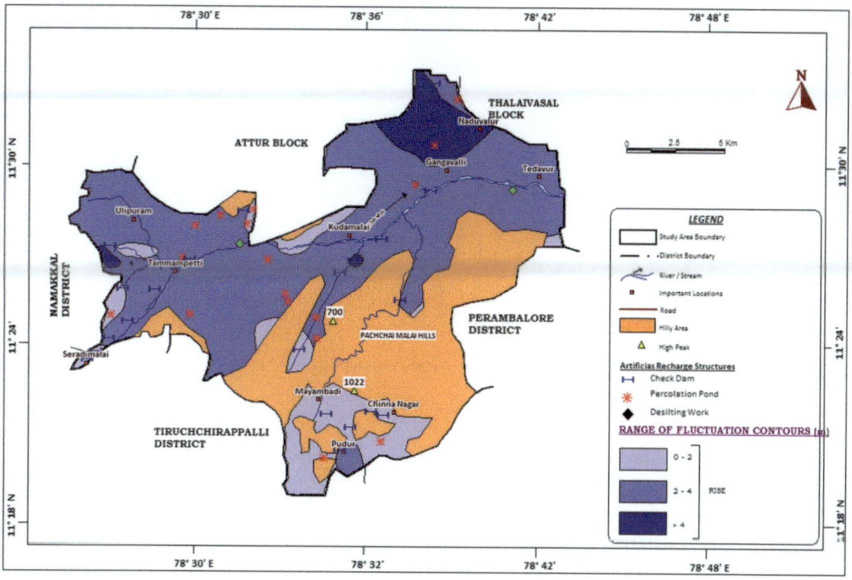

Fig. 5 Water-level fluctuation of January 2007 and 2008 (post-monsoon)

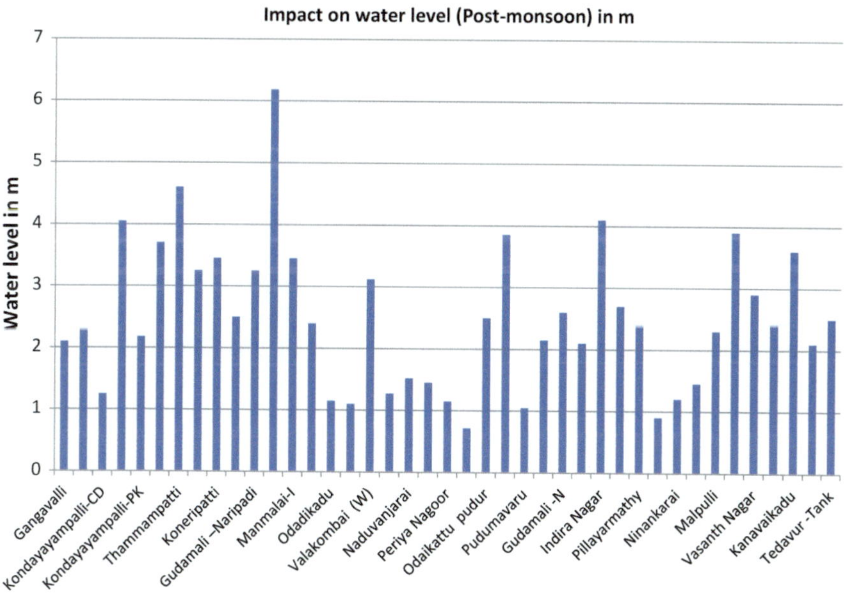

Plot 4 Rise of water level of between January 2007 and 2008 (post-monsoon)

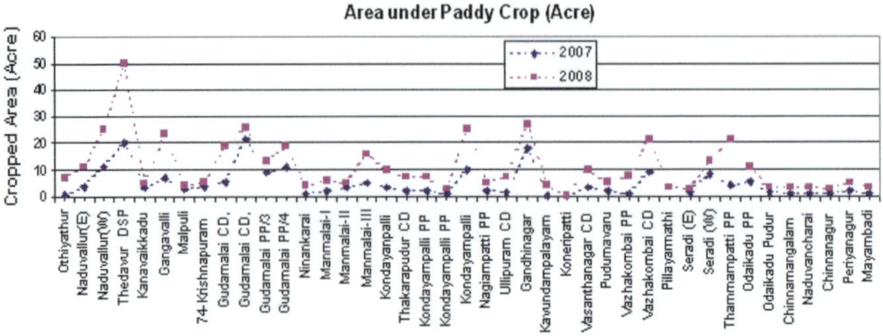

Plot 5 Increase in cropping area before and after construction of recharge structure

Photo 2 Photographs showing increase in paddy cultivated area in vicinity of artificial recharge structure

constructed. In case of percolation tank and check dam constructed without recharge well, the cropped area increased from 3 to 6 acres (Plot 5 and Photo 2).

The pumping hours of the 302 nos borewells working in the area have been collected during the research. It ranges from 100 to 1000 h/year at pre-construction of ARS, while in post-construction, the pumping hours range from 210 to 1400 h/year. The highest increase in pumping of borewell is observed in the north-east and central parts underlain by pediplains. The southern part formed by the moderately and highly dissected hill is showing the marginal increase in pumping hours. The borewells are maximum, influenced by the percolation tanks with recharge well having an average pumping hour of 1400 h/year. The check dams have also influenced the pumping hours significantly. The highest pumping hour in such case is observed as 1200 h/year. In borewell located in highly dissected and moderately dissected landforms influenced by the check dam, the highest increase in pumping hour is observed 700 h/years (Plot 6).

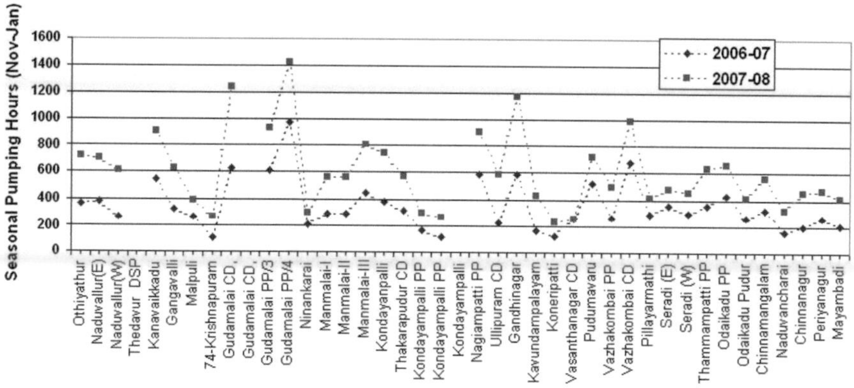

Plot 6 Increasing of pumping hours

8 Conclusion

The impact assessment of ARS constructed in the Gangavalli block was assessed and found that two places, namely Odaikattu Pudur and Gandhi Nagar shows maximum impact on water level rises. In both pre- and post-monsoon, the maximum area shows more than 2 mts water level rise. In the northern part, the level rise is more than 4 m. The land use practices show increase of wet crops area by 254 acres. The range of change in cropped area is recorded and found to be ranging from 1 to 30 acres in and around the ARS. The pumping hours of the well increased from 210 to 1400 h/year. The impact analysis clearly reveals positive response of artificial recharge on groundwater resources at places.

Acknowledgements The authors wish to place on record their sincere gratitude to Sri. K.B. Biswas, Chairman, and Dr. Dipankar Saha Member, Central Ground Water Board, New Delhi, for publication. The technical support rendered by officers from CGWB, Chennai, is gratefully acknowledged. The authors also express thanks to the Engineers from PWD, TWAD Board, and AED, Govt. of Tamil Nadu, for providing technical support during the construction of artificial recharge structures.

References

Alderwish AM (2010) Induced recharge at new dam sites—Sana'a Basin, Yemen". Arab Journal of geosciences, 3: 283–29

Mudrakartha S (2003) Augmenting groundwater resources by Artificial Recharge Research Site At Aravalli Hills, Gujarat, India. AGRAR Inception Report: Aravalli Case study, Gujarat, VIKSAT, Ahmedabad

Muralidharan D (2007) Evaluation of check-dam recharge through water-table response in ponding area. Curr Sci 92:10

Palanisamy V, GSI (1989) Report on the geology in parts of Attur valley and adjoining parts of Attur taluk of Salem district and Thuraiyur taluk, Tiruchirapalli district, Tamil Nadu

Rajarajan K et al (2012) Impact of artificial recharge structures and influencing factors in Hosadurga taluk, Chitdradurga district—a case study. In: Proceedings of national seminar on rain water harvesting and artificial groundwater recharge, BIS, New Delhi

Subburaj A et al (2015) Aquifer management plan for sustainable development in over exploited Gangavalli block of Salem district Tamil Nadu—a case study. Indian J Appl Res 6(2):56–61

Venkateswaran S et al (2015) Impact assessment of water table fluctuations in and around artificial recharge structures in Vaniyar sub basin of the Ponnaiyar river. South India Int J Recent Sci Res 6(7):5480–5486

Soil Infiltration Test in Hard Rock Areas—A Case Study

S.S. Vittala, G.R.C. Reddy, K.R. Sooryanarayana and G. Sudarshan

Abstract The study carried out in Ankasandra watershed covering on area of 375 km^2, falling in parts of Tiptur and C.N. Halli taluks of Tumkur district, Karnataka. Three major rock types identified in the area are gneisses, schists and granites belong to Archaean age. Groundwater is being extracted through borewells tapping fractures in depth ranges of 10–249 m below ground level. Over a period of time, due to increase in number of borewells and increasing groundwater extraction, the water level has gone down to more than 100 m bgl at certain location. The soil type in the area is fine red soils followed by clayey skeletal and loamy skeletal soils, and the rainfall is the main source of recharge. Twenty infiltration tests were conducted at various soil types using double-ring infiltrometer to understand the infiltration characteristics. The results of studies revealed that highest infiltration rates were observed at Madapurahattithanda, Kaval and Siddaramanagara sites characterised by sandy roils and lowest rates were noticed at Halkurike, Bommanahalli, Bhairanayakanahalli, Kuppur and Sasalu sites located on tank beds, pointing towards, need for desiltation of the tanks so that the tanks contribute to recharge also.

Keywords Over exploitation · Water level · Infiltration · Hard rock Karnataka

1 Introduction

Recharge of groundwater is a significant component of nature's hydrologic cycle in which precipitation that falls on the ground surface percolates through the sub-soil to ultimately reach the groundwater passing through the unsaturated zone (Sridevi et al. 2013; Saha et al. 2011). Soil plays a vital role in filtration of groundwater and

S.S. Vittala (✉) · G.R.C. Reddy · K.R. Sooryanarayana · G. Sudarshan
Central Ground Water Board, South Western Region, Ministry of Water Resources, River Development and Ganga Rejuvenation, Government of India,
27th Main, 7th Cross, H.S.R. Layout, Sector-1, Bengaluru 560102, India
e-mail: srisrivittala@gmail.com; vittala_99@rediffmail.com

D. Saha et al. (eds.), *Clean and Sustainable Groundwater in India*,
Springer Hydrogeology, https://doi.org/10.1007/978-981-10-4552-3_14

thus recharging the aquifer. After touching the ground, the rainwater entering into the soil by infiltration and thereafter percolates down and finally joins the groundwater. Infiltration rate is the physical property of soil, which determines the velocity of downward movement of surface water into the ground which is dependent on soil matrix, i.e., grain size, texture, loose of compactness, interconnectivity of pore spaces, moisture, slope or topography, temperature. In dry soil, the water infiltrates rapidly called as the initial infiltration rate. As more water replaces the air in the pores, the water from the soil surface infiltrates more slowly and eventually reaches a steady rate known as the final infiltration rate. The infiltration rate helps in estimating groundwater recharge with respect to rainfall (Subrahmanyam et al. 2003; Sridevi et al. 2013; Saha et al. 2013). In semi-arid regions, the spatial patterns of infiltration are governed by geomorphology and vegetation pattern (Horton 1933). Field studies in arid regions have also shown that the spatial variability of infiltration is related to the geomorphic setting as distinguished by inter-drainage area, topographic depression and drainage area (Scanlon et al. 1999; Loague and Gander 1990; Tricker 1981). Spatial patterns of the infiltration capacity are crucial for the determination of run-off generating areas especially in semi-arid regions where infiltration excess run-off development (Hortonian flow) is a dominant process (Horton 1933). Several researchers have attended infiltration test using double-ring infiltrometer method in hard rock and other areas (Johnson 1963; Turner 1963; Youngs 1991; Berndtsson and Larson 1987). In the state of Karnataka, as per the classification based on agricultural capability, the soils are grouped as red soils, laterite soil, black soils, alluvio–colluvial soils, brown forest soils and coastal laterite and alluvial soils (ICAR 1980; NBSS&LUP 1998). The details of the district wise distribution of soils and its characteristics including infiltration parameters with respect to the Karnataka State are briefly given in Table 1. The objective of this study is to determine the infiltration rates and characteristics of different types of roils, underlain by various rock types.

Table 1 Distribution of soils and its characteristics in Karnataka State

S. No.	Soil name	District covered	Soil properties	Infiltration characteristics
1	Black cotton soil	Belgaum, Bijapur, Gulbarga and Bidar and parts of Raichur, Chitradurga and Bellary	In texture, these soils vary from loamy to clays. Generally, they are neutral to alkaline in reaction, calcareous and well supplied with bases such as Ca, Mg and K. These soils are known to get self-ploughed due to their swelling and shrinking properties with change in their moisture content	Poor infiltration (0.75–2.5 cm/h)

(continued)

Table 1 (continued)

S. No.	Soil name	District covered	Soil properties	Infiltration characteristics
2	Lateritic soil	Malnad and coastal area of Uttara Kannada, Dakshina Kannada and part of Dharwad, Chikmagalur and Hassan	Lateritic soils are in the advanced stage of weathering, highly leached, poor in bases and very acidic in reaction. The moisture retentivity of the soil is very poor, and they become hard during summer month. These soils contain adequate quantities of organic matter.	Lateritic soils: moderate to good infiltration (3.6–19.62 cm/h)
3	Red and red loamy soils	Shimoga, Chikmagalur, Hassan, Mysore and Kodagu	These soils are light textured from sandy or gravely to loams with poor aggregating ability. They are poor in bases and acidic to neutral in reaction.	Red soils: moderate to good infiltration (2.6–3.8 cm/h) Red loamy soils: moderate to good infiltration (2–12 cm/h)
4	Coastal alluvials	Uttara Kannada and Dakshina Kannada	The surface soil is generally grey, yellow or light brown, the intensity of the colour increases with depth. These soils are acidic in nature, low in cation exchange capacity and bases.	Alluvial soils: good to very good infiltration
5	Dark brown clayey soil	Uttara Kannada, Dakshina Kannada Kodagu and Mysore	They are clayey, low bases and rich in organic matter as the surface soil receives the decomposition product of the virgin forest	Clayey soils: poor to very poor infiltration rate (0.5–1 cm/h)
6	Mixed red and black soils	Belgaum, Bijapur, Dharwad, Raichur, Bellary and Chitradurga	Black soil seen in the lowland and valleys have properties resembling those of medium black soil. These are productive under good management practices.	Mixed soils: moderate to good

Source www.raitamitra.kar.nic.in, www.agropedia.iitk.ac.in and www.cgwb.gov.in

2 Study Area

The Ankasandra watershed is covering an area of 375 km^2 and experiencing the over-exploitation of groundwater deep groundwater levels. The area receives about 680 mm annual rainfall (Vittala et al. 2013). It is located in parts of Tiptur and

C.N. Halli taluks of Tumkur district of Southern India. The watershed falls in the Survey of India toposheets no. 57 C/7 and 57 C/11, lying between north latitudes 13° 16′ 15″ to 13° 25′ 45″ and east longitudes 76° 23′ 45″ to 76° 38′ 40″. The watershed is drained by 1st to 5th order streams which forms part of Vedavathi sub-basin of Krishna basin. The drainage pattern is dendritic to sub-dendritic in nature with flow direction from south to north, ultimately joining the Torehalla stream (Fig. 1).

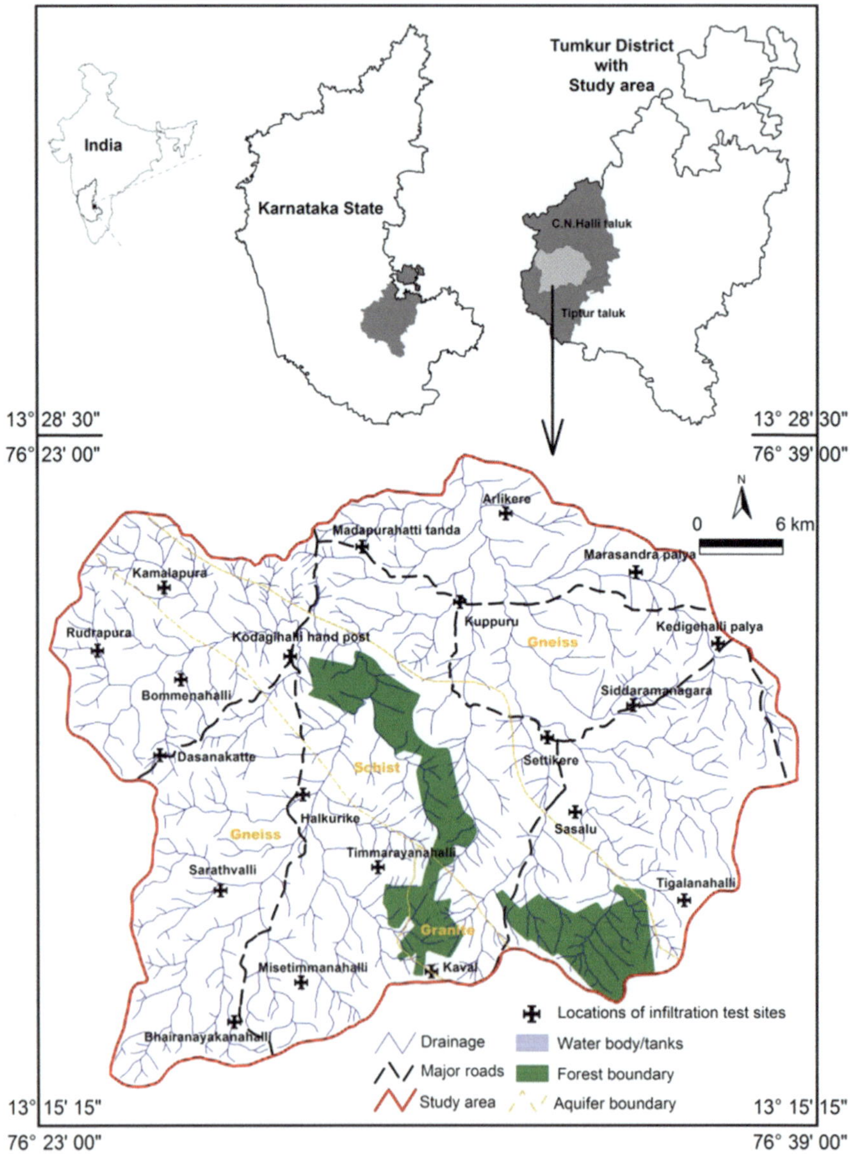

Fig. 1 Location map with infiltration test sites

Three principal rock types are identified, as gneisses, schists and granites, which belong to Archaean age. Majority of the area is occupied by gneisses covering about 270 km^2, 72% of the total area, followed by schist which covers 94 km^2, 25% of above the area. The schist bends trend in NW–SE direction cutting across the entire area. A small part of 11 km^2 is covered by granite intrusive in the southern part. The swarm of dolerite dykes is also noticed all over the area. The groundwater is being extracted mostly through borewells in the area. Due to increase in number of borewells over a period of time mostly for irrigation purpose, the depth to groundwater levels has gone deeper and deeper and found to be more than 100 m bgl at certain locations (Huchanahatti village, Sarathavalli, Ankasandra villages, etc.) (Vittala et al. 2013). The weathered zone which is 5–40 m thick is found to be desaturated. Groundwater occurs in fractured system in semi-confined to confined condition. Groundwater in small pockets around Halkurike village and other area also occurs in phreatic condition where the depth to water level is about 5–10 m bgl.

3 Methodology

The infiltration tests were conducted during non-rainy seasons using double-ring infiltrometer (ASTM 2003) in 20 different locations distributed all over the study area (Fig. 1), of which 18 tests were carried out in gneissic aquifers and two were in schistose aquifer. A double-ring infiltrometer with a diameter of 0.6 m for the outer ring and 0.3 m for the inner ring placed one inside the other, and both the rings were driven in the soil up to the depth of 5 cm as shown in Fig. 2. The rings were driven without tilt or too much disturbance of the soil and with constant annular

Fig. 2 Field photograph showing conducting infiltration test using double-ring infiltrometer

space between the rings in all directions (Raju 1986; Karanth 1978, 1987; Subrahmanyam et al. 2003). It was made sure that the bottoms of both the rings (underground) are at the same depth, and the rings are easily accessible and that water can easily be added over a period of hours. The water is poured inside both the rings simultaneously till to the brim and maintained that water levels are same in both inner and outer rings at a same time (Fig. 2). The lateral spread of water from outer ring was reduced the error by allowing vertical downward movement of inner ring water. The water column height in inner and outer ring was maintained constantly throughout the duration of the experiment. Using the established interval times and from a fixed reference, measurements were taken and recorded the water level drop in the inner ring at each interval with the help of stop timer. When the water level reached the bottom level, the rings were and the water level continued to drop again. This process was continued till stabilized infiltration rate was attained at each site, and all precautions were taken to minimize the error. Infiltration rates in the inner ring were recorded using a stop watch at 1-min interval for the first 10 min, every 2 min from 10 to 20 min, every 5 min from 20 to 50 min, every 10 min from 50 to 100 min. Then every 20-min intervals were kept till completion of test. The stabilized infiltration rate was calculated using the following formula suggested by Horton (1933).

$$F_t + F_c + (F_o - F_c)\, e^{-kt}$$

where,

F_t Infiltration rate cm/h at time 't' in mins
F_o Initial rate of infiltration (cm/h)
F_c Final (constant) rate of filtration (cm/h)
k Constant depending on soil and vegetation

$$K = \frac{Ln(F_1 - F_c) - Ln(F_2 - F_c)}{t_2 - t_1}$$

$$F_o = F_c + \frac{F_1 - F_c}{e^{-kt1}}$$

$$F_o = F_c + \frac{F_2 - F_c}{e^{-kt2}}$$

where,

F_1 Infiltration rate (cm/h) at time 't_1'
F_2 Infiltration rate (cm/h) at time 't_2'

The graph infiltration rate curves showing three representative soil characteristics are shown in Fig. 3. The spatial distribution of final infiltration rates is shown in Fig. 4. The details of sites, initial infiltration rates, final infiltration rates, cumulative depth of infiltration and duration of test obtained from the studies are given in Table 2.

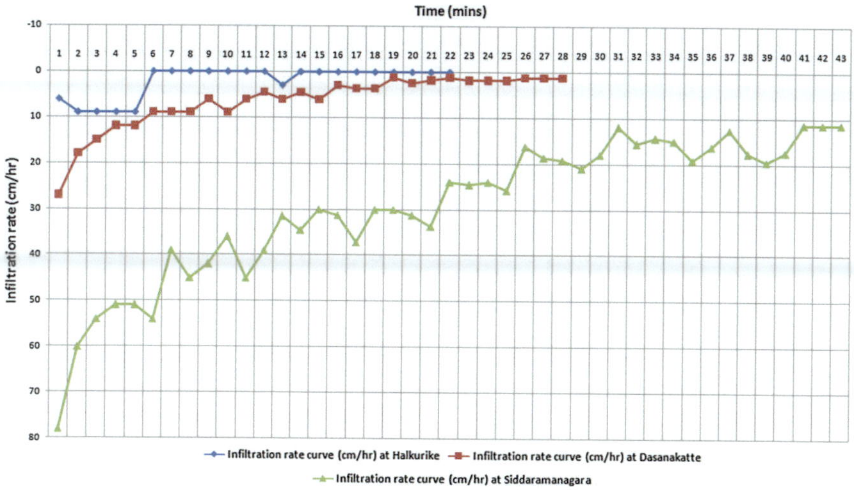

Fig. 3 The infiltration rate curves showing three representative soil characteristics

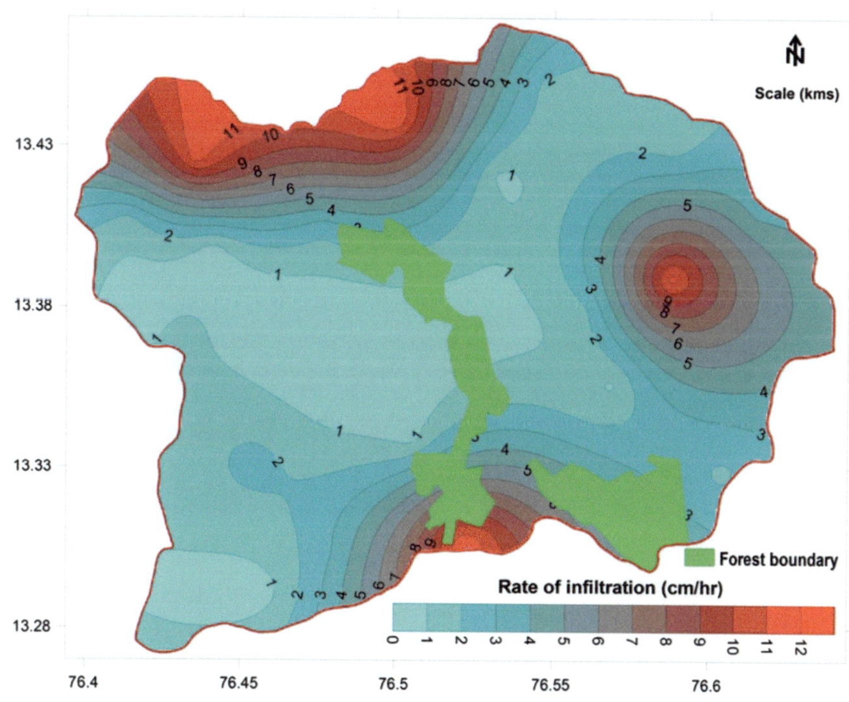

Fig. 4 The spatial distribution of final infiltration rates

Table 2 Results of infiltration test obtained from the studies

Sl. No.	Site name	Taluk	Initial infiltration rate (cm/h)	Final infiltration rate (cm/h)	Cumulative depth of infiltration (cm)	Duration (mins)	Geomorphology	Aquifer
1	Bhairanayakanahalli	Tiptur	6.0	0.60	1.50	110	Tank bed	Gneiss
2	Madapurahattithanda	C.N. Halli	93.0	12.00	110.20	380	Cultivated land	Gneiss
3	Kaval	Tiptur	60.0	12.00	86.20	300	Cultivated land	Gneiss
4	Sarathavalli	Tiptur	15.0	2.40	9.20	160	Tank bed	Gneiss
5	Halkurike	Tiptur	6.0	0.10	0.80	60	Tank bed	Gneiss
6	Misetimmanahalli	Tiptur	18.0	3.00	13.10	210	Cultivated land	Gneiss
7	Kodagihalli hand post	Tiptur	30.0	2.40	11.50	170	Cultivated land	Schist
8	Kamalapura	C.N. Halli	36.0	11.40	29.90	150	Uncultivated land	Schist
9	Timmarayanahalli	Tiptur	27.0	0.75	4.75	140	Uncultivated land	Gneiss
10	Rudrapura	Tiptur	9.0	1.50	5.30	200	Uncultivated land	Gneiss
11	Dasanakatte	Tiptur	27.0	1.20	6.40	140	Cultivated land	Gneiss
12	Bommanahalli	Tiptur	24.0	0.45	4.70	280	Uncultivated land	Gneiss
13	Arlikere	C.N. Halli	36.0	1.80	9.20	200	Uncultivated land	Gneiss
14	Marasandrapalya	C.N. Halli	18.0	1.50	5.70	160	Cultivated land	Gneiss
15	Kuppuru	C.N. Halli	18.0	0.90	5.70	200	Cultivated land	Gneiss
16	Kedigehallipalya	C.N. Halli	66.0	3.90	20.30	240	Uncultivated land	Gneiss
17	Siddaramanagara	C.N. Halli	78.0	11.40	94.80	270	Cultivated land	Gneiss
18	Settikere	C.N. Halli	54.0	1.80	9.90	160	Tank bed	Gneiss
19	Sasalu	C.N. Halli	39.0	0.90	8.70	260	Uncultivated land	Gneiss
20	Tigalanahalli	C.N. Halli	36.0	1.80	13.20	160	Uncultivated land	Gneiss

4 Results and Discussion

The objective of the study is to evaluate subsurface hydraulic variability of a complex hard rock watershed to predict the recharge in the watershed using infiltration rates. The infiltration tests were carried out in various soil types to study the infiltration characteristics. The test results reveal that the rate varies from place to place. The higher initial infiltration rates are noticed at Madapurahattithanda (93 cm/h), Siddaramanagara (78 cm/h) and Kaval villages (60 cm/h), whereas the lower initial infiltration rates are noticed at Halkurike and Bhairanayakanahalli (6 cm/h) and Rudrapura villages (9 cm/h). The higher final infiltration rates are also observed at Madapurahattithanda and Kaval villages (each 12 cm/h), and Siddaramanagara (11.4 cm/h). The lower final infiltration rates are noticed at Halkurike (0.1 cm/h), Bommanahalli (0.5 cm/h), Bhairanayakanahalli (0.6 cm/h), Kuppur and Sasalu sites (0.9 cm/h) (Fig. 4). Besides, the higher cumulative depth of infiltration was found to be 110.2 cm in 380 min at Madapurahattithanda, followed by 94.8 cm at Siddaramanagara in 270 min and 86.2 cm at Kaval in 300 min. The lower values cumulative depth of infiltration is noticed at Halkurike with 0.8 cm after 60 min, followed by 1.5 cm at Bhairanayakanahalli sites in 110 min. There is no much variation noticed in the results among the aquifers.

Interestingly, the lowest infiltration rates are noticed at tank beds at Halkurike and Bhairanayakanahalli due to the presence of clay and silt as observed in field, whereas in Madapurahattithanda, Kaval and Siddaramanagara sites, the infiltration rates are high because of sandy soils. The wide variation in infiltration rates is due to characteristics of soil at surface, soil hydraulic properties and local hydrogeological conditions in the study area. The infiltration rate is higher in cultivated lands, moderate to low in non-cultivated lands and low to least in tank bed areas. In Halkurike site (tank bed), the low infiltration rate is due to fine textural soil and may also be the result of shallow water table condition. It is also recommended that the suitable strategy can be planned for artificial recharge studies wherever necessary for the sustainable development of these over-exploited aquifers in the area.

5 Conclusions

The Ankasandra watershed is located in parts of Tiptur and C.N. Halli taluks of Tumkur district experiencing the over-exploitation of groundwater, deep groundwater levels, low rainfall, resulting in a drought conditions. The area represents typical hard rock terrains of Southern India. Three major rock types identified are Gneiss, Schists and Granites. Infiltration tests were conducted at 20 locations in various soil types distributed all over the study area during non-rainy seasons using double-ring infiltrometer. Based on the results obtained from initial infiltration rates, final infiltration rates, cumulative depth of infiltration and duration of test, it is revealed that the lowest infiltration rates are noticed at tank beds whereas at place like Madapurahattithanda, Kaval and Siddaramanagara sites, the infiltration rates are high because of sandy soils. Low infiltration rate at Halkurike site may also be the result of shallow water table condition and presence of sheet rocks. It is noted that there is no much variation in the results observed in the test sites located at different rock types.

Acknowledgements The authors are grateful to K.B. Biswas, Chairman, K.C. Naik, Member (RGI), Dipankar Saha, Member (SAM) and E. Sampath Kumar, Member (SML) & Member Secretary (CGWA) of Central Ground Water Board, CHQ, Faridabad, for their permission to carry out the study. The authors expressed their sincere thanks to Sh. K.M. Vishwanath, Regional Director and colleagues of CGWB, South Western Region, Bengaluru, for their valuable suggestions rendered during the course of the study. The critical comments and suggestions by the anonymous reviewer for improving the manuscript are duly acknowledged.

References

ASTM (2003) Standard test method for infiltration rate of soils in field using double-ring infiltrometer. D-3385-03, ASTM International, 100 Barr Harbor Drive, PO Box C700, West Conshohocken, PA 19428-2959, United States. Accessed from www.astm.org

Berndtsson R, Larson M (1987) Spatial variability of infiltration in a semi-arid environment. J Hydrol 90:117–133. Accessed from http://dx.doi.org/10.1016/0022-1694(87)90175-2

Horton RE (1933) The role of infiltration in the hydrologic cycle. Trans Amer Geophys Union 14:446–460

ICAR (1980) Hand book of agriculture. In: Jaiswal PL, Wadhwani AM (eds) 5th edn. Indian Council of Agriculture Research, New Delhi

Johnson AI (1963) A field method for measurement of infiltration. U.S. Geological Survey Water-Supply Paper 1544-F, pp 4–9

Karanth KR (1978) Studies of hydrological parameters of ground water recharge in water balance computations in Andhra Pradesh. Hydrol Rev Contrib IHP CSIR New Delhi 4(1–4):65

Karanth KR (1987) Ground water assessment, development and management. Tata McGraw-Hill Publishing Company Ltd, Delhi, pp 21–26

Loague, K Gander GA (1990) Spatial variability of infiltration on a small rangeland catchment. Water Resour Res 26:957–971. Accessed from http://dx.doi.org/10.1029/WR026i005p00957

NBSS&LUP (1998) Soils of Karnataka for optimizing land use. National Bureau of Soil Survey and Land Use Planning, Publication, p 47

Raju TS (1986) Infiltration process and techniques of measurement. Bhu-Jal News. Quart J Central Ground Water Board 3:8–11

Saha D, Sinha UK, Dwivedi SN (2011) Characterization of recharge processes in shallow and deeper aquifers using isotopic signatures and geochemical behavior of groundwater in an arsenic-enriched part of the Ganga Plain. Appl Geochem. doi:10.1016/j.apgeochem.2011.01.003

Saha D, Dwivedi SN, Roy, G.K, Reddy DV (2013) Isotope-based investigation on the groundwater flow and recharge mechanism in a hard-rock aquifer system: the case of Ranchi urban area, India. Hydrogeol J. doi:10.1007/s10040-013-0974-3

Scanlon BR, Langford RP, Goldsmith RS (1999) Relationship between geomorphic settings and unsaturated flow in an arid setting. Water Resour Res 35:983–999. Accessed from http://dx.doi.org/10.1029/98WR02769

Sridevi PD, Sarah S, Gandolfi JM, Ahmed S (2013) Relevance of hydrological parameters in artificial recharge sites selection and assessment in a granitic aquifer. Int J Hydrol Sci Technol 3(3):270–288

Subrahmanyam K, Marechal JC, Brue I.D, Ledoux., E, Ahmed, S and DeMarsily, G (2003) Geological Investigation and Hydraulic Tests for Aquifer Characterization. Scientific Report, vol 2, Indo-French Collaborative Project (2013-1), Technical Report No. NGRI-2003-GW-410

Tricker AS (1981) Spatial and temporal patterns of infiltration. J Hydrol 49:261–277. Accessed from http://dx.doi.org/10.1016/0022-1694(81)90217-1

Turner AK (1963) Infiltration, runoff and soil classifications. J Hydrol 1(2):129–143

Vittala SS, Reddy GRC, Najeeb KM (2013) Data gap analysis for ground water surveys: a case study. J Geol Soc India 82(10):440–441

Youngs EG (1991) Infiltration measurements—a review. Hydrol Process 5(3):309–319

Peoples' Participation for Sustainable Groundwater Management

Bishwadeep Ghose, Harshvardhan Dhawan, Himanshu Kulkarni, Uma Aslekar, Siddharth Patil, M.V. Ramachandrudu, Bakkareddy Cheela, Yogesh Jadeja, Brijen Thankar, Ravi Chopra, Anil Gautam and Eklavya Prasad

Abstract India is the largest user of groundwater in the world. Over 85% of rural domestic water, around 48% of urban domestic water, and 60–70% of agriculture water are groundwater dependent. This has resulted in the overexploitation and acute depletion of the resource in many parts of the country. Despite the manifold short- and long-term consequences of such dependence on a fast depleting and critical resource, India has made little headway in its regulation or conservation. While groundwater exhibits the qualities of a classic common pool resource—those of subtractability and excludability, in reality, it has largely been treated as private property. Much of the problem lies in the juxtaposition between the public and common pool nature of groundwater and its rampant private, atomized, and unregulated extraction. As the volume diminishes and quality deteriorates, lack of regulation and appropriate management can lead to both inter- and intra-use conflicts with considerable political and socio-economic impacts. Therefore, in order to conserve this resource it is imperative to shift away from a paradigm of private groundwater development to a more sound system of groundwater management. This paper argues that despite the relative invisibility of groundwater and the complexities that surround its governance, decentralized management options offer better solutions for long-term sustainability of the resource and ensure social equity.

B. Ghose (✉) · H. Dhawan
ARGHYAM, 599, 12th Main, HAL IInd Stage, Indiranagar, Bangalore 560008, India
e-mail: bishwadeep@arghyam.org

H. Kulkarni · U. Aslekar · S. Patil
Advanced Centre for Water Resources Development and Management (ACWADAM), Pune, India

M.V. Ramachandrudu · B. Cheela
Watershed Support Services and Activity Network (WASSAN), Secunderabad, India

Y. Jadeja · B. Thankar
Arid Communities and Technologies (ACT), Bhuj, India

R. Chopra · A. Gautam
People Science Institute (PSI), Dehradun, India

E. Prasad
Megh Pyne Abhiyan (MPA), Delhi, India

D. Saha et al. (eds.), *Clean and Sustainable Groundwater in India*,
Springer Hydrogeology, https://doi.org/10.1007/978-981-10-4552-3_15

Firstly, it argues for a hydrogeological foundation for groundwater management. Secondly, given the decentralized nature of aquifers, community participation is essential for the sustainability of this resource. Finally, it highlights the urgent need for policy initiatives that recognize the common pool nature of groundwater and facilitate bottom-up innovations that reflect the local geological and socio-economic specificities of the resource Sengupta 2015.

Keywords Groundwater · Participatory management · India · Hydrogeology Aquifer

1 Dependence on Groundwater in India

From historical perspective, water management in India can be broadly categorized under three distinct phases (Shah 2008). First, the pre-1800s or the Mughal period, which saw predominantly irrigation through wells and tanks. Although supported by state patronage, the governance of water was based on traditional knowledge systems and followed a decentralized system of governance. During the period between 1800 and 1970, the colonial and post-colonial periods, the model for groundwater management was through state-supported schemes, focused on enhancing the irrigation potential. It provided a major impetus to cash crop production and made the country food secure. Post-1970, with the proliferation of borewells and small pumps, irrigation became atomized and hugely groundwater dependent. The institutional mechanisms were retained under the state and policies followed a 'command and duty' regime.

Presently, the dependence on groundwater remains strong with irrigation for >85% of all groundwater extraction (MDWS 2016). It caters to 85% of the rural and nearly 48% of the urban drinking water needs. The overwhelming dependence on groundwater has put aquifers in several regions under a lot of stress (CGWB 2013).

2 Current Paradigm and Challenges

The atomized use of groundwater posits several challenges. The invisibility of groundwater does not lend itself to easy quantification and consequent management options. Its ownership is tied to land rights, which is in conflict with its nature as a common pool resource stored in aquifers and shared between a multitude of users. It also attracts competition and raises the potential for future conflicts. The problem analysis per se is neither new nor unknown, but needs to be understood from the perspective of bottlenecks to efficient groundwater management (Figs. 1, 2, and 3).

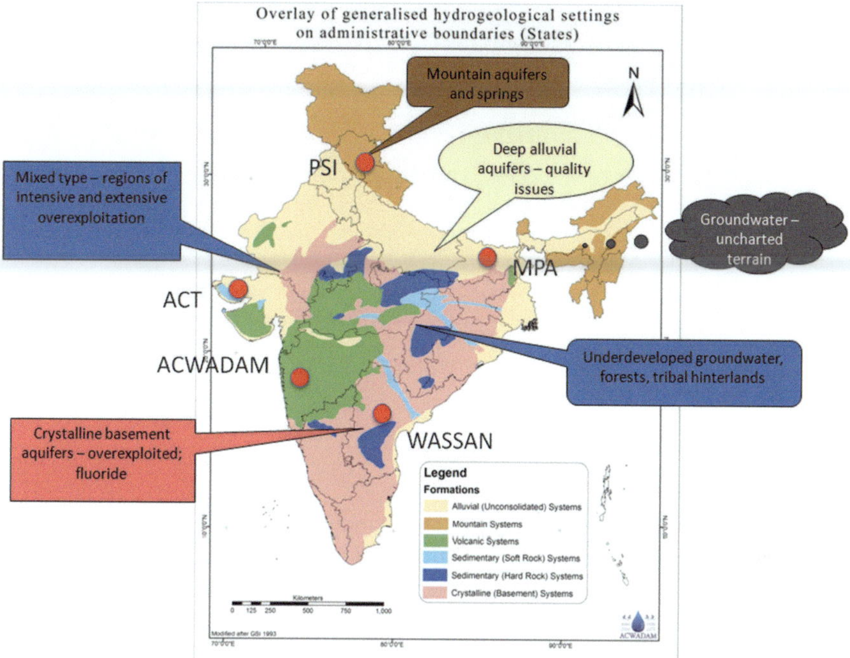

Fig. 1 Typology of hydrogeological settings in India (modified after COMMAN 2005; Kulkarni et al. 2009)

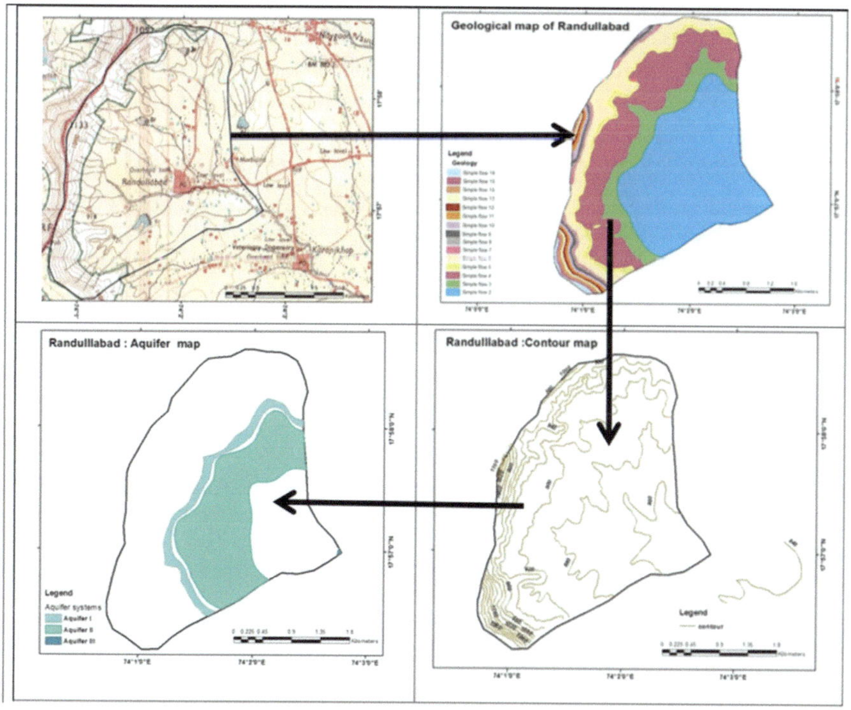

Fig. 2 Process of aquifer delineation in Randullabad (PGWM ACWADAM 2012)

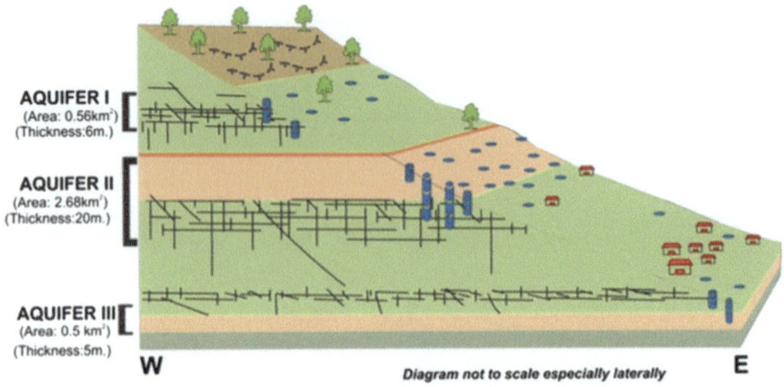

Fig. 3 Aquifer systems in Randullabad (PGWM ACWADAM 2012)

2.1 Development Versus Management

Till very recently the slant of looking at groundwater was more on development rather than on management. There is a continued emphasis on engineering solutions focused on source development rather than overall and sustainable management of the resource. The annual slip back of villages brought under safe drinking water supply is of around 40% (MDWS 2006) due to sources drying up or other issues like quality deterioration, points towards the inadequacy of this approach. Managing demand, especially in water-stressed areas, continues to be a political hot potato.

Beside large areas under overexploitation and unbridled, abstraction has sprouted the issue of quality deterioration, like flour decontamination in several states and arsenic contamination of large stretch of Indo-Gangetic plains and its sub-basins (CGWB 2013). Salinity ingress in coastal areas, nitrate, iron, and other heavy metal contaminations are also emerging as serious issues along with the potential bacteriological contamination of groundwater through increased toilet coverage. As per the data available, approximately 59% of the total districts in the country show groundwater vulnerability and have issues related to either water quality, quantity, or both (Figs. 4, 5, and 6).

2.2 Issues of Data and Institutions

There are multiple government departments that are engaged in different aspects of managing groundwater with very little coordination between them. The compartmentalization between the supply-side ministries and departments (Department of Land & Records, Department of Forest, CGWB, CWC, Department of Rural development, Department of Drinking Water Supply) and demand-side (Agriculture, Energy, Industry, Urban, etc.) renders has a common vision for water management very difficult. The Pollution Control Board, which plays an

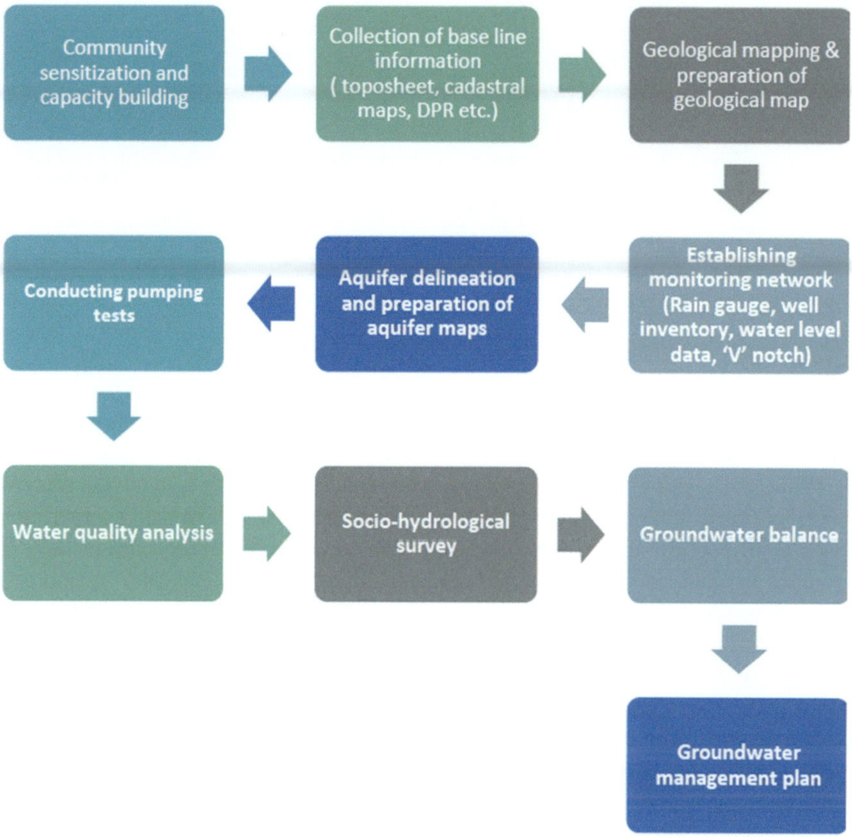

Fig. 4 Process of participatory groundwater management (PGWM ACWADAM 2014)

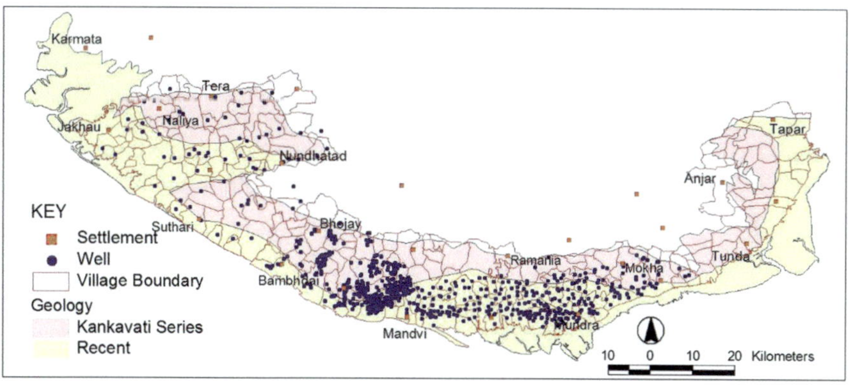

Fig. 5 Villages based on Kankavati Sandstone (PGWM ACT 2014)

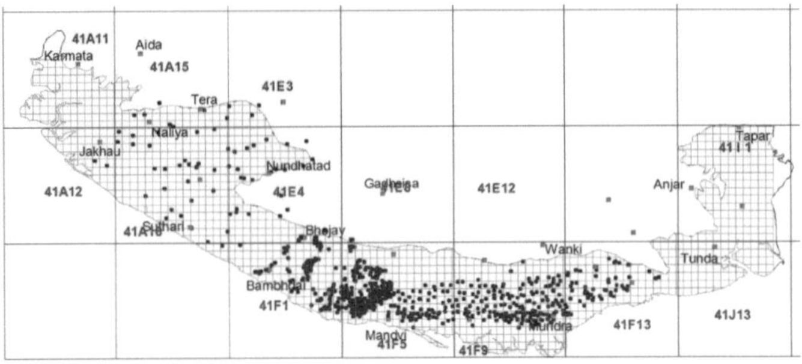

Fig. 6 Existing well on monitoring network grid (PGWM ACT 2014)

important role in monitoring and regulation of effluent discharge into ground and surface water, is at the periphery of groundwater management.

While each agency has its own mechanisms and repositories of data, the use and users of such data are not predefined. For example, CGWB's groundwater assessment reports are possibly the only data sets available on various uses of groundwater and are very useful. However, it suffers from both a lack of granularity and a time lag in publishing the reports that make decision-making based on the data relatively inaccurate. From a user point of view, say a farmer or even the collector of a district, it may not be very easy to find data related to groundwater, surface water, irrigation, and weather at a village or block level, making it difficult to develop decision support systems for allocation of water for various users. The water quality data on bacteriological and chemical contaminations are collected by different departments and are not integrated in a common platform. The fragmented institutions and processes raise some questions related to the use of data for the management of groundwater as a resource, like the mapping of aquifers through National Aquifer Management Programme (NAQUIM) taken up by CGWB that eventually should lead to participatory groundwater management.

Two questions arise out of this. One, can the water data be crowd sourced from citizens, which will enhance the density of the data to be collected? Second, can the citizens or managers be enabled to engage with the data that help them in understanding the resource better and therefore partake in the management of the resource? The use of technology in collecting these data can make compilation and analysis in real time and useful for management compared to what is currently the practice (Figs. 7, 8, and 9).

2.3 Groundwater Ecosystem Connection

Our knowledge is improving on the interplay between forests, wetlands, springs, and wells as a part of the same ecosystem. The changes in one element of the ecosystem have profound impact on the others. Upstream–downstream issues in the

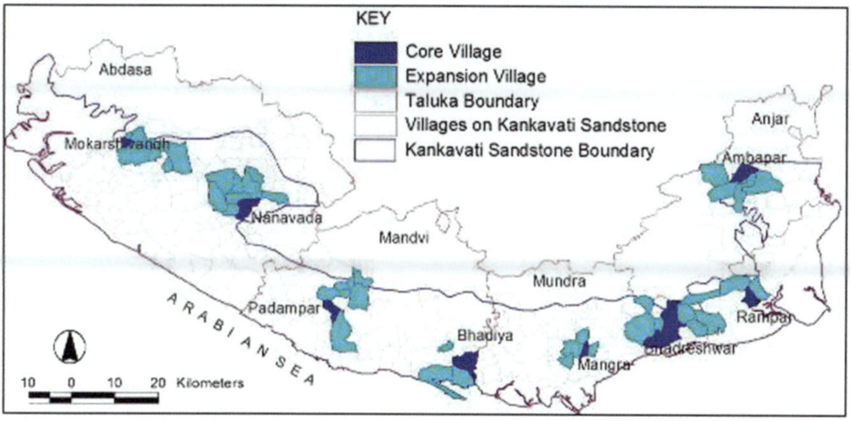

Fig. 7 PGWM demonstration clusters (PGWM 2014)

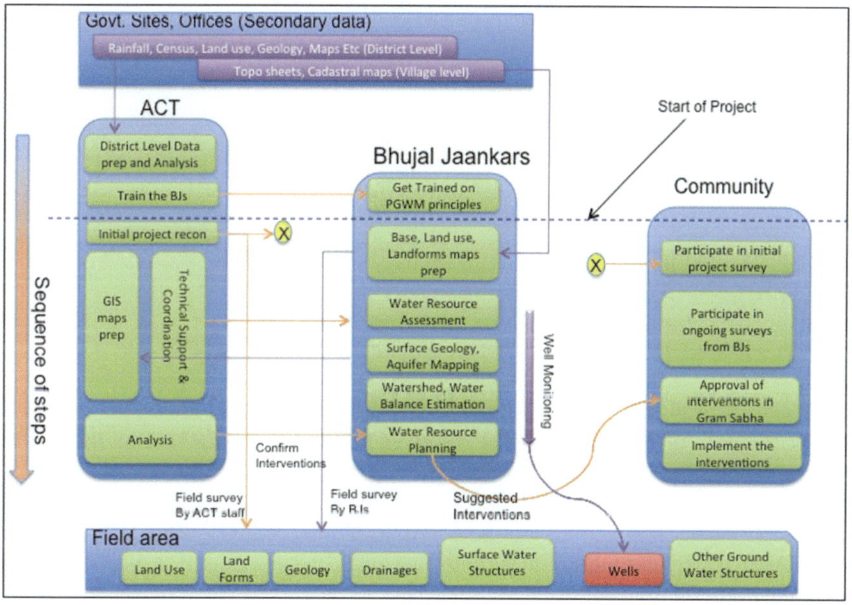

Fig. 8 PGWM process adopted in Kachchh area: activities, participants, and interactions (PGWM ACT 2016)

case of springs and rivers, effect of overabstraction by borewells on base flows, ground and surface water interplay in the flood plains, the possible connection between forest fires and groundwater recharge, ownership of the sources of groundwater, etc., are some of the issues that need to inform the local management and governance. It is challenging for any centrally driven approach to respond to the diversity of our ecosystems. On the other hand, traditionally every subculture

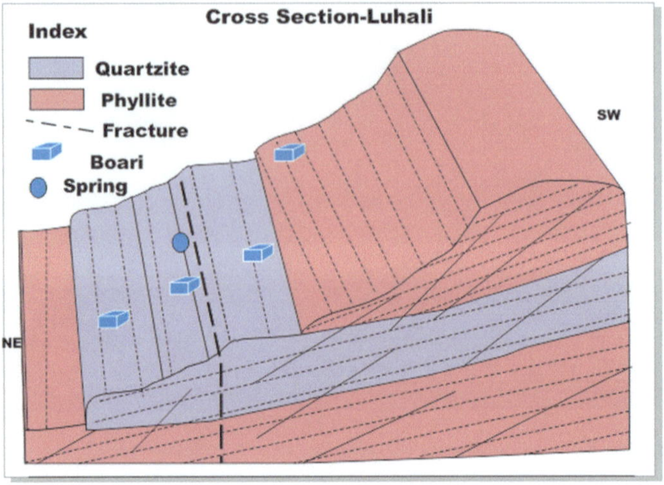

Fig. 9 Cross section of Luhali village (PGWM PSI 2012)

across regions had a good understanding of its ecology. As we would see in the cases below, traditional knowledge when meshed with modern science and technology helps in developing contextually appropriate systems and practices of groundwater management.

2.4 Human Resource: Issues and Challenges

For a state-driven approach, there seems to be an acute mismatch of manpower at all levels, both, in terms of quality and quantity. Perspective building for resource management at the top to skills required for mapping and managing aquifers seems to be woefully inadequate. These in turn have implications on efficient management of the resource. Inadequacy of physical infrastructure like good-quality water testing laboratory hampers timely detection and mitigation of water quality problems. With many institutions presiding over the management of groundwater and the poor coordination between them has led to an overall lack of accountability within the system (Figs. 10 and 11).

While most of the policies and programmes have been espousing peoples' participation for managing groundwater, the statist approach has created a patronage system that undermines the agency of the citizens in urban as well as rural areas in playing a critical role in management of the resource. The work on ground informs about the dire need of shifting from a services-centric to a more resource-centric approach. The emergence of citizens' groups both in urban and rural areas has demonstrated a qualitative improvement in resource management outcomes, including the critical aspects of equity and sustainability. However,

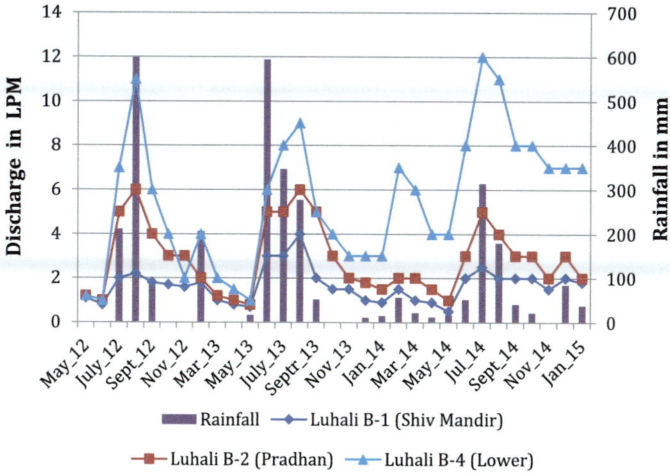

Fig. 10 Luhali spring hydrograph (PGWM PSI 2015)

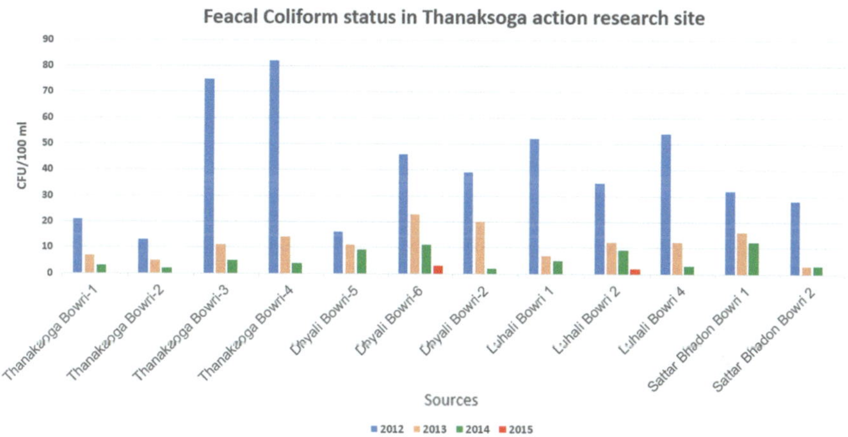

Fig. 11 Springs water quality improvement hydrograph (PGWM PSI 2015)

citizens' actions are more often triggered through natural causes such as scarcity or some health hazards due to poor water quality. Besides, it is not very easy to incentivize peoples' participation through public policy instruments.

3 The Contours of Groundwater Management

Arghyam has been supporting work around water and sanitation over the last decade across India. The learning has informed its work on groundwater management. Some empirical principles have evolved from the experience of our partners' work on the ground. With the only input for any water being rainfall, the question becomes that of capturing this water for deferred use. Rainwater stored in reservoirs, dams, or ponds are more common to be seen but cost more as compared to storing the same in natural underground aquifers, where there is less evaporation and reach through natural recharge. However, understanding the boundaries and characteristics of aquifers is not very easy and neither is the quantification of groundwater, both of which requires quite an understanding of hydrogeology.

Much of our collective learning stems from the understanding of nature of groundwater as a resource and our communities' historical understanding accrued through the interaction with this resource. The aquifer boundaries within the rock layers make groundwater a renewable but finite resource. Given that the resource supports several uses and users, it exhibits the twin characteristics of subtractability and excludability for any Common Pool Resource (CPR). And as for any CPR, exceeding its carrying capacity leads to problems of scarcity or even quality problem.

Having a unit of management, in this case the aquifer assists in defining the problem statement for that unit and helps in monitoring if the objectives of management are being achieved. The problem statements could be related to scarcity, quality, access or a combination of all three. This approach also helps in looking beyond the source to understanding the resource. This helps in bringing supply augmentation and demand management as integral parts of the management plan.

The work has demonstrated that communities come up with their own management protocols once their connection with the resource improves. Resource understanding requires the science of hydrogeology to be broken down for the communities to assimilate better, which in turn is meshed with their own traditional knowledge. The practice has shown that the communities can use their own agency to develop water security plans that are a combination of supply-side earth construction and demand-side interventions through a set of decisions defining the usage of water for multiple uses and users. The principle of subsidiary applies for decision-making. Decisions or protocols once agreed upon are formalized in the Gram Sabha. Over time, the panchayats or water user groups have been able to mobilize resources from existing government schemes for implementing the interventions.

The participatory groundwater management (PGWM) programme has helped in building a cadre of para-professionals who are now adept in collecting data, developing maps, and facilitating water balance exercises with the communities that helps in water budgeting. The data collected are either used for developing decision support systems or for evidence building. In most cases, it requires the support of an intermediary NGO to analyse the data, which is shared among the community members for decision-making.

It is important to engage with the community leadership, particularly women. The interventions are planned keeping in mind the issues of equity and sustainability. The protocols developed work more on the basis of social sanctions and do not necessarily have a legal anchorage. The endeavour has been to institutionalize these management protocols in institutions of local governance.

Given the complexities of institutional structures at the state level, the panchayat becomes the focal point of convergence for various programmes, schemes, and interdepartmental coordination as well as implementation. The state needs to create a conducive environment where negotiations on allocations, incentives, and regulations can take place. It can help formalize the management protocols developed by communities themselves. More importantly, it should be able to resource the water security plans developed by the communities.

Ensuring that the decision-making is decentralized to the level of the panchayat and enabling the communities to manage their resource has turned out to be efficient and effective than the purely statist approach. The lack of manpower, especially at the last mile itself acts as a deterrent. It has been observed through our work that thorny issues such as prohibiting new borewells to be sunk, moving away from cash crops, and sharing of borewells become relatively easy through community engagement and for the resource management. Conflicts related to water sharing, if any, can be settled at the lowest levels of governance. This process derives strength from the provisions in the 73rd and 74th amendments and helps in deepening the democratic processes in this respect.

However, externalities weigh heavy on this work and it is still not clear about the long-term sustainability of such efforts. For example, any change in the catchment areas of springs due to deforestation or forest fires can reduce the discharge of springs; price or energy incentives for cash crops in a water-stressed region may lead to the disregard of social protocols; additional complexities arise for peoples' participation, when the same aquifer is shared between multiple villages, pushing the management boundaries beyond the confines of administrative units. In Himalayan region, where the recharge and discharge areas can follow a valley-to-valley approach, rather than a ridge-to-valley approach can also make peoples' participation for management of the resource a bit more complex.

Despite its limitations, the impact of the PGWM programme has been significant. Some are illustrated in the following partner case studies, representing critical hydrogeological typologies of India.

4 Case Studies

4.1 Hard Rock Areas—Western India

Droughts, drinking water scarcity, depletion of groundwater levels, and water quality are the main water-related issues of Maharashtra. In the state, the rainfall ranges from 400 to 4000 mm; however, over the last three years there has been a deficit rainfall resulting in consecutive cycle of droughts. This has led to severe shortage of ground and surface water in the region. Groundwater is considered as the lender's last resort, and hence, there is high dependence on this common pool resource, which is under threat due to increased exploitation of the aquifers.

However, a few villages in the state have managed to tide over this vicious cycle of crisis. In these villages, local communities used scientific principles along with their traditional knowledge to manage groundwater in a democratic, equitable, and sustainable manner.

In the drought-prone villages of Muthalane in Pune District and Randullabad in Satara District of Maharashtra, a three-year-long watershed development project undertaken with PGWM principles brought the village back from the brink of drinking water scarcity to becoming a water-sufficient village. The project involved geological mapping, recharge of regional aquifers, testing of water quality, and establishing usage protocols for drinking and irrigation. Drilling of borewells was banned, and 90% of wells in the village were used on a sharing basis as farmers took turns to irrigate their lands. Groundwater recharge and discharge areas were demarcated.

As a result of these interventions, groundwater levels have improved and local water structures have been revived. The impact of the programme is seen in improved kharif productivities, improvement in irrigation and water use efficiency, improved equitability particularly for farmers, and improved drinking water security.

The chief reason this programme succeeded was because of the involvement of the community. Over a period of three years from 2011, there was a significant increase in groundwater recharge despite rainfall being below normal. This was because of the watershed development undertaken by the village water and sanitation committee that facilitated construction of check dams in natural recharge areas. A sense of ownership among the farmers helped them take decisions together, such as diversifying crops based on the availability of water, using low-powered pumps, and adopting micro-irrigation techniques for effective use of water and reduce abstraction considerably.

4.2 Mixed Sedimentary Systems—Kachchh

Groundwater is the only reliable and critical source for drinking water in arid and semi-arid regions both for rural and urban areas. Surface water, particularly in such regions, becomes a supplementary source, given its scarcity and seasonality. Groundwater, apart from drinking water security, also supports other natural resource-based livelihoods such as agriculture, animal husbandry, pastoralism. It has enabled industrial development in regions that would otherwise have seemed impossible due to the scant availability of surface water seasonal sources. Groundwater is increasingly getting stressed that adversely affects the quality, quantity, and sustainability of the resource.

The sites are located in Kachchh area which is underlain by the Tertiary and Jurassic sedimentary formations—mainly sandstones—in the Kankavati river basin and in Kamaguna-Vatacchad watershed. It is a mixed-type region with intensive and extensive overexploitation, contamination (geogenic and bacteriological), and salinity ingress into the freshwater aquifer systems in the coastal parts, which has directly impacted the traditional drinking water sources.

The participatory groundwater management at 18 locations in this hydrogeological typology setting has been studied and monitored for last 5 hydrogeological cycles. On the basis of which groundwater management plans were made and implemented in these villages, which helped in achieving water security particularly drinking water.

The programme has now expanded to four areas, spread across Kankavati Sandstone in Bhuj and Nakhatrana Taluka (sedimentary: multi aquifer system); Wagad Sandstone in Mandvi, Mundra, and Anjar Taluka (sedimentary: single aquifer system); and north Gujarat Alluvium in Kheralu Taluka (unconsolidated sediments: multi aquifer system), that includes aquifer mapping and their characterization, and a participatory groundwater management plan. The programme has also scaled up the efforts towards achieving water security for drinking water needs and developed participatory decision support tools that will guide the interventions and protect groundwater resource from exploitation in the sedimentary and alluvial systems, which also include a shared aquifer system. A groundwater monitoring network through community participation has been established in this region to facilitate understanding of the resource, plan interventions, and develop groundwater management protocols based on decision support tools.

The programme has demonstrated how the communities or group of stakeholders relying on an aquifer can reduce water stress and conflicts by adopting PGWM principles. Through various interventions, these stakeholders have understood and demonstrated an equitable and sustainable sharing mechanism based on the area's hydrogeology. It has incorporated aquifer management into mainstream watershed and drinking water projects.

ACT's work across Kachchh District and several locations across India showcase PGWM-based water security can be achieved. It also demonstrates cadre building at various levels (para-hydrogeologists or Bhujal Jankars), key stakeholders resource understanding, community-based groundwater management and its governance supported by scientific evidences.

4.3 Mountain Typology—Himalayan Aquifers and Springs

Springs, which occur where groundwater table intersects the surface and provide safe, perennial drinking water, feed rivers, and anchor entire ecosystems. Springs are a source of common pool resource, i.e. groundwater. But this vital resource is under threat due to environmental degradation, increasing water demand, and climate change. Spring discharge and quality is declining due to rampant drilling in the catchment, unregulated abstraction changing land use patterns, ecological degradation, poor sanitation, population pressure, and climate variations.

Studies carried out in the Himalayan region in the Sirmour District, Himachal Pradesh. The area lies in the Proterozoic rock formations of the Lesser Himalayan region. In the Himalayan mountain systems, local aquifers are found over a large region that feed springs and streams. The aquifers are often fed by recharge from distant locations. Dependency for drinking water is high on springs and spring-fed streams than on wells in this region.

Springs require a different approach of management. The conventional ridge-to-valley approach does not necessarily work in these regions. The springshed approach looks at valley-to-valley interventions and has demonstrated positive results. The major interventions would need to focus on springs rejuvenation and understanding the mountainous aquifer systems. Based on the implementation of interventions, springshed recharges area protection, social fencing, developing community for springs/groundwater management through the various institutions like van panchayats, and mitigating quality issues based on the hydrogeological interventions.

A pilot action research initiated in five villages, viz. Thanakasoga, Luhali, Dhyali, Dandor, and Sattarbhadon of Thanakasoga Panchayat in Sirmour District, has developed an understanding of mountain aquifers, its augmentation, and management using physical, vegetative, and social measures through community participation. These interventions have protected and regenerated springs, which have led to an improvement in the availability of water for the communities especially during the lean months of the year. Social fencing of the recharge area and its protection from open defecation, cattle grazing, deforestation, and land use change is one of the major outcomes of this programme which has resulted in significant water quality improvement for all the springs in the pilot locations.

PSI has demonstrated that recharge treatment area for springs is on an average 2 ha, as compared to more than 10 ha used by traditional watershed management approaches. Thus, the springshed approach has potentially reduced overall costs by 80%. Interventions indicate that springshed management requires an investment around Rs. 30,000 per ha. This investment includes the cost of essential infrastructure, human resource development, and knowledge sharing costs.

4.4 Crystalline Basement—Peninsular India

In the absence of knowledge of local aquifers that support local governance/management practices at community level, both government and individuals seem to be working at cross-purposes and result in a water crisis especially that of drinking and agriculture water.

The area under groundwater-based irrigation systems is increasing even those parts where traditionally surface water-based irrigation systems used to be popular. This is observed in Andhra Pradesh, Telangana, and other parts of Peninsular India too. Similarly, as groundwater abstraction has increased, the quality of drinking water in several sources has deteriorated. There is also an increase and acceleration of water markets and quick-fix solutions (through reverse osmosis plants) in several parts of peninsular India.

Studies have been taken up in Mahbubnagar and Ranga Reddy District in Telangana and Anantapur District of Andhra Pradesh (Fig. 12). These hard rock aquifers are heterogeneous and overexploited with a water quality problem of excess fluoride. There is high dependency on groundwater for both drinking water and agricultural purposes on the homogeneous aquifer systems. The various degrees of groundwater extraction have lead to intense competition around depth of wells and borewells.

The work carried out in this location under PGWM aimed to demystify the processes/protocols to enhance the application of knowledge in aquifer-based groundwater governance systems within mainstream public investment projects for enhancing water security at local level (with a clear focus on drinking water security and irrigation purposes).

In the villages identified in these districts for study, many farmers—both with and without access to groundwater, were facing an increasingly precarious livelihood situation. The farmers were brought together through a system of voluntary compliance. It meant rather than individuals digging new borewells, existing borewells were linked through a network of distribution pipes to provide access to water for all parties involved. The scope was to create a sustainable model for resource sharing and groundwater management.

The villages were to pool the groundwater from farmers who had borewells and share it with other farmers, who do not have access to water, thus providing critical

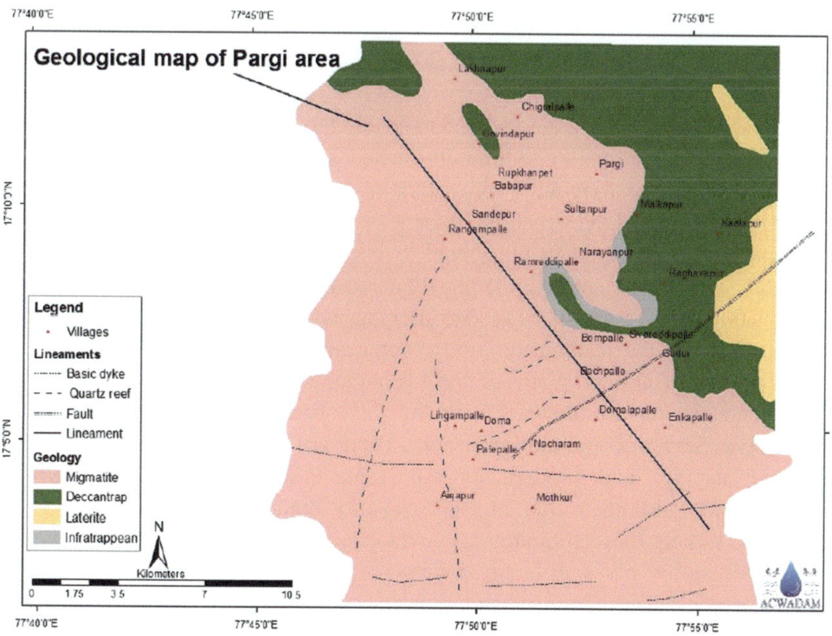

Fig. 12 Geological map of area around Pargi, Ranga Reddy District, Andhra Pradesh (after Geological Survey of India 1995)

irrigation to the rainfed crops and achieving water security with specific focus on drinking water. It has successfully demonstrated this groundwater sharing through borewells pooling in more than 100 villages and introduced the PGWM concept with significant change in agricultural cropping pattern based on participatory water budgeting exercises (Fig. 13).

Trainings and capacity building exercises for NGOs and government officials also organized to spread the knowledge and skills related to aquifer management and participatory groundwater management. It has demonstrated that large-scale replication of good practices is feasible through cadre building/training inputs to partners (village level to facilitators) on PGWM-related issues.

This study demonstrates social regulations evolving from within community with technological innovation like borewell pooling led to sustainable groundwater management and drinking water security.

4.5 Alluvial Typology—Flood Plains

Floods in north Bihar are a recurring disaster and one of the most serious problems faced during floods is the inability to access safe drinking water. Groundwater is the only source of drinking water, for scattered habitations that dot the flat landscapes of flood-prone areas of north Bihar. While considering the region as water abundant, groundwater quality remains a serious issue to be addressed. Iron, arsenic, and biological contamination is widespread. Clearly, in such areas, the quantity of groundwater is of secondary importance as compared to accessing good-quality water.

Study taken up in these areas focused to enable the communities to shift from the contaminated hand pumps to community-owned dug wells as sources of drinking water. To add to the longevity and safety of the dug well, these communities engaged with design of the structure, hygiene, sanitation, and solid and liquid waste management at the individual household levels. The acceptance of phaydemandshauchalay (ecological sanitation), phaydemandsokta (banana circle), and phaydemand compost (compost) was encouraging and also went a long way in understanding and barricading the groundwater and sanitation nexus.

PGWM in alluvial flood plains of north Bihar aims to understand and address the high dependency on groundwater with the same principles and protocols for its management. Several communities in north Bihar are using geogenic—chemically (iron and arsenic) and biologically contaminated—water from hand pumps. Under PGWM guidelines, training, and capacity building, action research work commenced to help these communities develops an aquifer-based understanding of groundwater and evolves different options of long-term engagements with this common pool resource. Training and capacity building on key groundwater modules specifically designed for alluvial flood plains helped the communities, government officials, students, NGOs, and INGOs to understand various aspects of groundwater quality and quantity in these regions.

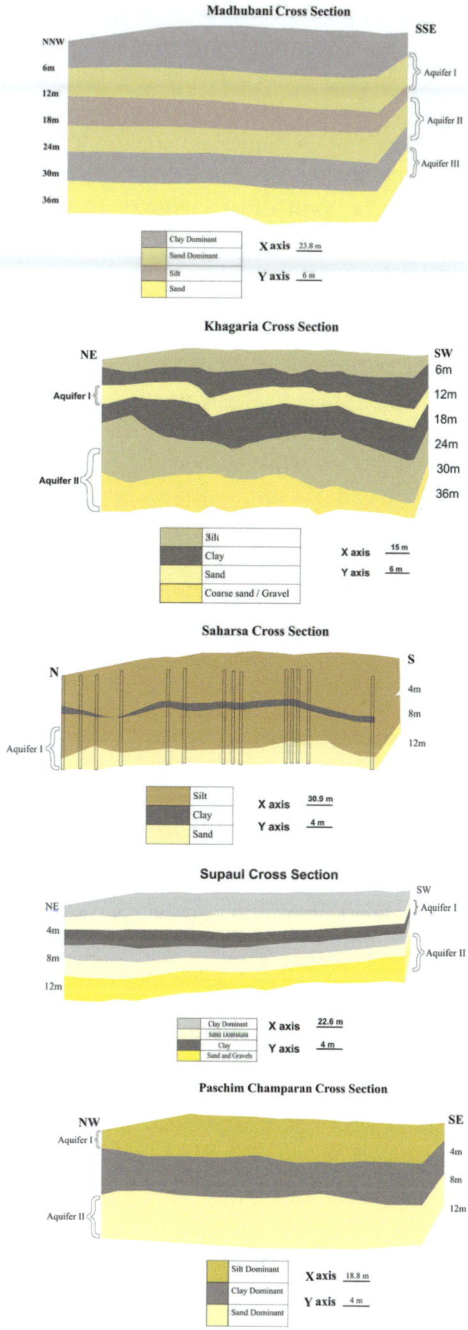

Fig. 13 Alluvial flood plain typologies in North Bihar (PGWM MPA and its partners 2016)

Alluvial flood plains of north Bihar area have unique typology of groundwater setting. It is a part of the great Indo-Gangetic plains, made up of sequences of loose unconsolidated sediments up to hundreds of metres in thickness, spread across regional expanses of land. Alluvial settings provide a very complex set of characteristics for groundwater studies. The storage capacities of such settings are exponentially higher than that of hard rock regions. High storage and high precipitation in north Bihar (approximately 1300 mm) and in the catchments of the Himalayan rivers in Nepal ensure that this region is persistently flood prone, with perennial surface water flow and shallow groundwater table. In addition, iron and arsenic contamination is widespread, but the distribution is poorly understood, and biological contamination is omnipresent.

The application of PGWM in the flood plains of Bihar has helped in understanding of groundwater hydrogeology of the aquifers including its sub-typologies. It also has evolved a nuanced understanding of floods in north Bihar and also its characterization while associating with water quality issues and possible solutions. It has facilitated in understanding the diversity within groundwater systems (availability, access, and contamination) in the alluvial flood plains and provided flood analysis and classified them under the following five sub-themes (sub-typologies): inside embankments, outside embankments, flash floods, general floods, and trans-boundary aquifers.

5 Conclusions

The PGWM principles, premised on groundwater as common pool resource, have now been demonstrated to be working across several regions and hydrogeological typologies. The approach invariably empowers the local communities, as managers of the local groundwater resource to make informed choices for achieving overall water security. The development and implementation of the water security plans is also a demonstration of bottom-up planning that fosters the agency of the communities with better outcomes for equity and sustainability. Peoples' participation encourages more collaboration with the government, yet less dependence on them for management. From passive recipients of water and sanitation services, communities have become active managers of the resource, in all the above cases. The community–government interaction also becomes much more informed and meaningful through this approach.

The outcomes of this approach go beyond mere drinking water security to resilience building of communities against climate variances such as rainfall, drought, and floods. The limitations of state's interventions for regulating groundwater use through punitive measures can be circumvented to a large extent by enabling communities to make informed choices—as has been demonstrated above through the instances of banning and regulating the depth of new borewells, shifts in cropping pattern, sharing of borewells, etc.

Peoples' participation is not new, and through this approach, the effort is to rediscover its contemporary relevance, meshed with a scientific temper. However, there are larger questions related to how peoples' participation would work with heterogeneous communities, in shared aquifer systems and its ability to deal with externalities such as urbanization, energy incentives, and other land use changes. The current phase of PGWM programme looks at collaborations with the state at various levels for scaling up the practice by using public investments and human resources of the state. Various technological options for data collection, collation, and analysis for decision support systems are being explored that could assist in accelerating the scaling up of the programme.

The draft model groundwater bill 2016 reflects several of the foundational principles of PGWM. The already expanding practice of PGWM across many states could pave the way for those state governments to change laws and frame policies informed by the draft model groundwater bill 2016, the draft national water framework bill 2016 as well as the proposed institutional reforms. Apart from the conjunctive use of surface as well as groundwater, the policies may provide the much needed legal backing for the community-based protocols and help in adjudication for conflict resolution and allocations for water sharing between various users of the common pool resource.

The programme's platform has also brought on board researchers and think tanks that are examining various legislations such as the Easement Act, 1882, relevant judgments of the supreme court forest policies and environmental acts, some of which support and some of which contradict the principles and practice of the PGWM programme. The need for responsive decentralized resource sensitive water governance is clearly being felt. A growing number of a community, NGOs, funders, academia, and governments at different levels are becoming a part of this collaborative effort that seeks to understand and manage the groundwater resources more sustainably and equitably.

Acknowledgements PGWM programme has been jointly developed by Arghyam and its programme partners, Advanced Centre for Water Resources Development and Management (ACWADAM), Arid Communities and Technologies (ACT), People's Sciences Institute (PSI), and Watershed Support and Services and Action Network (WASSAN) in 2010, who have been pioneers in promoting this programme. The authors acknowledge the efforts of the study group and team members working in these organizations for their valuable inputs, support during this PGWM programme work since January 2010. We are also grateful to the rural communities whose adoption of these principles and practice of PGWM have provided rich insights and learning from across the PGWM locations in various hydrogeological typologies. We sincerely acknowledge the support received from various levels of government, national, state, district, block, and panchayats for providing resources, knowledge, and guidance to the implementation of the programme. We would also like to thank all our colleagues at Arghyam, specially Sundar M Senthilnathan for editing the document and for their valuable inputs and comments on various drafts of this paper.

References

ACWADAM, Pune (2014) Groundwater situation and dependency. Action research in participatory groundwater management: a report on six action research sites. Compiled by ACWADAM, Pune and Supported by Arghyam, Bangalore. pp 8

Department of Drinking Water Supply, Government of India (2006) Summary of nation-wide statistics from Rajiv Gandhi Drinking Water Mission. DDWS, Ministry of Water Resources, GoI, New Delhi

Kulkarni H, Vijay Shankar PS, Krishnan S (2009) Synopsis of groundwater resources in India: status, challenges and a new framework for responses, Report submitted to the Planning Commission. ACWA Report ACWA/PC/Rep—1, Government of India, 97 p

Planning Commission (2012) Report of the steering committee on water resources and sanitation for Twelfth Five Year Plan (2012–2017). Government of India, New Delhi

Sengupta MB et al (2015) Groundwater efficient use and peoples' participation for its sustainable management. CGWS publication

Shah T (2008) Taming the anarchy: groundwater governance in South Asia. Resources for the Future Press, Washington

Shah M (2013) Water: towards a paradigm shift in the Twelfth Plan. Economic and Political Weekly XLVIII, 40–52 (Special Article)

Geospatial Analysis of Recharge of Groundwater and Irrigation Draft from Different Aquifer Systems of Rajasthan, India

Bidyut Kumar Bhadra, Rakesh Paliwal and S. Srinivasa Rao

Abstract *Groundwater* over-exploitation for agricultural crops caused adverse impact on the sustainability of the resources of Rajasthan. Western arid to central semi-arid regions of the state shows a large spital variation in annual rainfall. The study of groundwater elevation zones during pre-monsoon, post-monsoon, and post winter irrigation (January to March) seasons shows a steady declining of groundwater level of places such as Osian (1.93 m/year) and Piprali (1.50 m/year) which is attributed to excessive groundwater draft due to irrigation in Rabi season. It has been predicted that most of the tube wells in Osian and Piprali may get dry completely by 2048 and 2068, respectively. However, during the same period Khamnor area shows rise in groundwater level at the rate of 1.19 m/year. This variation in groundwater level is attributed to the nature of different aquifer systems viz. alluvium in Piprali, sandstone in Osian, and gneisses in Khamnor area. Implementation of water harvesting structures shows significant improvement in recharge of groundwater in Osian and Khamnor area underlain by hard rock aquifer system. Geospatial analysis of recharge and draft from groundwater during 2008–2009, 2010–2011, and 2011–2012 shows significant changes due to rise and fall in groundwater levels. Over-exploitation of groundwater has developed a non equilibrium stage between recharge and irrigation draft. In contrary, a positive impact on groundwater recharge in Khamnor area is evident due to higher rainfall and hard rock aquifer with secondary porosity (joints/fractures). Unless preventive measures are taken, the gap between groundwater recharge and draft will increase many folds in future.

Keywords Groundwater recharge · Irrigation draft · Aquifer system
Geographic information system · Rajasthan

B.K. Bhadra (✉) · R. Paliwal · S. Srinivasa Rao
Department of Space, Regional Remote Sensing Centre (West), NRSC/ISRO, Jodhpur 342003, Rajasthan, India
e-mail: bkbhadra63@gmail.com

D. Saha et al. (eds.), *Clean and Sustainable Groundwater in India*,
Springer Hydrogeology, https://doi.org/10.1007/978-981-10-4552-3_16

1 Introduction

Rajasthan is the largest state in India covering an area of 34.22 million hectares and shares only 1.15% of the country's water resources. The state is predominantly agrarian as the livelihood of 70% of its people depends on agriculture-based activities. In the past 50 years, a threefold increase in human population has put tremendous pressure on the fragile water and fertile land resources of Rajasthan. The estimated annual per capita water availability in the state was 840 m^3 in 2001 which is expected to fall to 439 m^3 by 2050. Over the several years, a continuous demand of groundwater for irrigation has been observed in the country as also in Rajasthan (Karanth 1987; Allison et al. 1994; Aswathanarayana 2001). Though low rainfall is observed, indiscriminate tapping of aquifers is observed in various areas of Rajasthan, resulting in decline of groundwater level significantly in the last couple of decades that brought down into several over-exploited zones of the State.

There are physical methods to estimate groundwater recharge. The water table fluctuation method using storage parameter (specific yield in case of unconfined aquifer) is rated as dependable and widely used, and because of various factors, like its accuracy, it can be easily used and is of low cost for adopting in semi-arid areas (Sophocleous 1991; Sukhija et al. 1996; Scanlon et al. 2006; Aksever et al. 2015; Lakshmamma et al. 2015; Shahid et al. 2015). The fluctuation of hydraulic heads is used to assess groundwater recharge condition, but also helps in assessing withdrawal or draft, efficiency of water use, crop productivity, groundwater flow regime and management, water balance studies etc. (Asano and Cotruvo 2004; Crosbie et al. 2005; Gundogdu et al. 2000; Healy and Cook 2002; Radhakrishna 2003; Scott and Shah 2004; Marechal et al. 2006; MoWR 2009; Saha et al. 2009; Sharif and Ashok 2011). Extraction of groundwater is assessed by multiplying the mean yield and running hours of a well in a year (Chatterjee and Purohit 2009; MoWR 2009). Ravikumar et al. (2005) have adopted the groundwater irrigated area and trend of groundwater level in a long-term basis for groundwater draft assessment in Palar basin, Tamil Nadu. Jahan et al. (2010) have shown a progressive decline in hydraulic head owing to rapid rise in demand from irrigation in Bangladesh. The variation in groundwater level in fractured aquifer system has been studied by Reddy (2012) in parts of Anantapur district and by Saha and Agrawal (2006) in Deccan trap terrain in Pune district. Their hydrograph study shows steady decline in hydraulic head in a range of 0.50–2.91 m/year. Various models have been developed and adopted to assess draft of groundwater resources in the country. They are like adopting regional groundwater flow model in south India, covering upper part of Bhima River Basin (Surinaidu et al. 2012), estimating potentiality of recharge in consolidated formations in Aravalli (Bhuiyan et al. 2009) and also, hydro-economic optimization model in order to reduce over-extraction in California (Tulare Basin) (Harou and Lund 2008). Reckless withdrawal can also result in adverse impact on ecology and environment, besides economic and social consequences, and also deterioration of groundwater quality (Karanth 1987; Das 2008; Bhadra et al. 2013), potable water shortage (SRSAC 1999) and frequent occurrence of droughts particularly in semi-arid areas (Shiferaw et al. 2008). Recently, recharge and draft for irrigation in parts of Rajasthan is discussed by Bhadra et al. (2016a, b).

Modern tools such as Geographic Information System (GIS) and satellite data are used to assess land and water resources aided by a wide array of satellite data with various temporal and spatial resolution (Gupta 1991; Khan et al. 2008; Skidmore 2010). In the present study, spatial maps have been prepared for recharging groundwater and its extraction adopting interpolation techniques in a GIS environment. The temporal relationship between water level rise and fall owing to recharge and extraction for irrigation has been produced by the declining groundwater level trend. An attempt has been made to assess the positive impact of structures constructed for harvesting water on enhancing the groundwater resource by recharge using geospatial analysis.

2 Hydrogeology of the Area

In the present study, three key areas have been selected based on unique aquifer systems viz. Piprali (620.28 km^2) in Sikar district, Osian (589.24 km^2) in Jodhpur district, and Khamnor (413.47 km^2) in Rajsamand district in Rajasthan (Fig. 1). However, detailed hydrogeological analysis was done using village-wise groundwater data of Piprali (70 villages), Osian (33 villages), and Khamnor (88 villages) area, respectively. Due to gently undulating topography with availability of aquifers, more than 75% area of Piprali and Osian is represented by agricultural land in comparison with Khamnor area (32% only) which represent a hilly topography where agricultural lands are confined to the valley portions only. In Rajasthan, Kharif crops (cotton, vegetables, maize, bajra, groundnut, and castor) are cultivated under rain-fed irrigation while the Rabi crops (mustard, wheat, garlic, onion, and cumin) are grown using groundwater. Generally, flood irrigation is used; however, for better efficiency of irrigation, few cultivators used drip irrigation to grow fruits.

The three areas taken up for the presents study represent distinct aquifer characteristics like alluvium in Piprali, sandstone in Osian, and gneiss in Khamnor area. The general stratigraphy with major rock types of Piprali, Osian and Khamnor area is given in Table 1. The Piprali area is characterized by two major formations viz. Older alluvium and Aeolian sand of Recent age, underlain by quartzite/schist beds of Alwar Group (Delhi Supergroup).of Precambrian age (Pareek 1984; GSI 1980, 1985, 1993, 1997). Unconsolidated to semi-consolidated sand, silt, clay, kankar with sand, and gravel lenses form Older Alluvium. They form potential aquifers in the area. The major rock types of Osian area are from Jodhpur Group of Marwar Supergroup (L to U Proterozoic) which overlies basement rock of Malani Rhyolite. Jodhpur sandstone forms the main aquifer underlying a thin (5–10 m) layer of Quaternary sediments. The major rock types of Khamnor area are gneisses of Bhilwara Supergroup of Precambrian age (GSI 1997), which are comprised of porphyritic and non-porphyritic gneissic complex associated with aplite, amphibolite, schists, and augen gneisses.

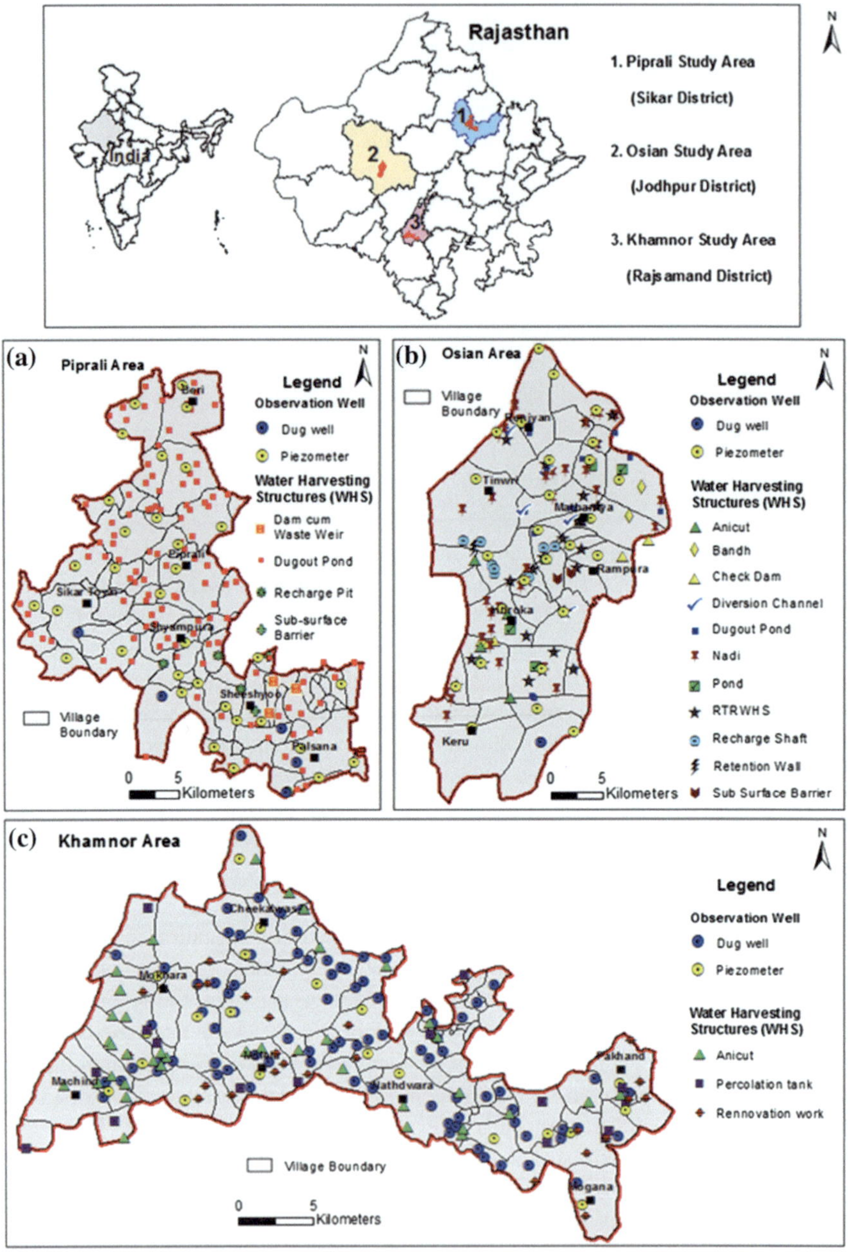

Fig. 1 Location map of the study area with different aquifer systems in Piprali (alluvium), Osian (sandstone), and Khamnor (gnesisses) of Rajasthan. Water harvesting structures (WHS) in respective areas are shown in different symbols

Table 1 Stratigraphic succession of Piprali, Osian, and Khamnor area, Rajasthan (After Pareek 1984; GSI 1980, 1985, 1993 and 1997)

(a) Stratigraphy around Piprali area, Sikar district			
Quaternary to recent			**Older alluvium** and aeolian sand
Delhi Supergroup (lower to upper proterozoic)	Alwar Group		Massive quartzite/schist
(b) Stratigraphy around Osian area, Jodhpur district			
Quaternary to recent	Windblown sand, alluvium, pebbles, gravel and kankar		
Marwar Supergroup (lower to upper proterozoic)	Nagaur Group (75–500 m)	Tunklian Formation	Grits, pebbly sandstone, siltstone and clay
	Bilara Group (100–300 m)	Pondlo Formation	Chert, cherty dolomite and limestone with interbedded siltstone and claystone
		Gotan Formation	Limestone, dolomitic limestone and dolomite
		Dhanapa Formation	Limestone, dolomite, chert, cherty dolomite
	Jodhpur Group (125–240 m)	Girbhakar Formation	Gritty and pebbly sandstone with few interbeds of sandy shale
		Sonia Formation	**Sandstone**, shale and minor chert, cherty dolomite and limestone
	Pokhran Bolder Bed		
Basement rock—Malani Rhyolite			
(c) Stratigraphy around Khamnor area, Rajsamand district			
Lower to middle proterozoic	Delhi Supergroup	Gogunda Group	Quartzite, schist, Calc-silicate rock, Hornblende schist
	Intrusives	Dharwal granitoid	Granite and gneisses
Lower proterozoic	Aravalli Supergroup	Nathdwara Group	Quartzite, Bi schist, Hbl schist, Dolomitic marble, gneisses and migmatites
		Udaipur Group	Chl. Phyllite, Gt Bi Schist, Dolomitic marble, Quartzite, Hbl. Schist, Meta tuffs
		Debari Group	Quartzite, Basic volcanics, Dolomite, Feldspathic schist, Conglomerate
Archaean	Bhilwara Supergroup	Mangalwar Complex	Quartzite, Pegmatite and **Composite gneisses**

3 Materials and Methods

Studies were conducted in the study region using groundwater level data of observation wells (OW) in Piprali (39), Osian (26), and Khamnor (106). The measurements were taken in pre-monsoon, post-monsoon, and post-irrigation. The respective months are from April to June, October to December, and January to March, in the years 2004–2005 to 2011–2012 (*Source* Ground Water Department, Government of Rajasthan, Jodhpur). These datasets were used to estimate groundwater recharge and irrigation draft which have been correlated with the corresponding annual rainfall data collects from Govt. of Rajasthan. To augment groundwater level, Ground Water Department (GWD), Govt. of Rajasthan has constructed a large number of Water Harvesting Structures (WHS) during 2008–2009 viz. nos in Piprali (121 dugout ponds, 3 weir, 3 recharge pits, and 2 subsurface dams), 118 no. in Osian (36 nadi, 19 dugout ponds, 29 roof top rain Water harvesting structure 5 anicuts, 2 bandhs, 5 check dams, 4 ponds, 10 recharge shafts, 5 diversion channel, 1 retention wall, and 2 subsurface dams) and 81 no in Khamnor area (38 anicuts, 16 percolation tanks, 13 roof top rain Water, and 14 renovation layers). For long-term trend analysis, monthly rainfall data of the past 17 years (1995–2011) have been collected. Normally June, July, and August are the rainy months in western Rajasthan with a temporal variation in rainfall within a month.

Initially, depth to groundwater levels was converted to elevation in respect of the mean sea level (msl). To represent the seasonal variation in groundwater level, the groundwater elevation data was processed by the Kriging interpolation method. The kriging is an advanced geo-statistical technique that generates an estimated surface from a scattered set of points with z-values (Oliver 1990; Chai et al. 2011). The general equation for interpolation technique is presented by Eq. (1).

$$Z(S_0) = \sum_{i=1}^{N} \lambda_i Z(S_i) \quad (1)$$

where

- $Z(S_i)$ measured value at the *i*th location
- λ_i an unknown weight for the measured value at the *i*th location
- S_0 prediction location
- N number of measured values

The weight of λ depends on the distance between the measured points and the prediction location as well as on the overall spatial arrangement of the measured points. This method is adopted in many countries to analyze the spatial variation of levels of groundwater (Kumar and Remadevi 2006; Gundogdu and Guney 2007; Kamińska and Grzywna 2014). ArcGIS 9.3.1 software, a geo-statistical tool is used for interpolation of point data. Overlay analysis was carried out after converting interpolated maps i.e., raster to vector form, asses the variation in groundwater fluctuation owing to draft for irrigation and recharge.

Groundwater level variation owing to recharge from rainfall has been assessed by subtracting post-monsoon level from pre-monsoon level. In a same way, decline in level owing to draft for irrigation has been assessed by deducting post-irrigation level (in Rabi season) from post-monsoon level. Interpolated maps for groundwater fluctuation are represented by recharge and draft for irrigation. The variation in level fluctuation owing to extraction and recharge before the construction of WHS (2008–2009) and after the construction of WHS (2011–2012) has also been assessed using interpolated maps.

4 Results and Discussion

4.1 Analysis of Rainfall Data

The annual rainfall during 1901–2011 was 462.9 mm at Piprali, 257.4 mm at Osian, and 514.9 mm at Khamnor Fig. 2. Variation of mean annual rainfall indicates transition of isohyets from arid region (<400 mm rainfall) in Osian to semi-arid (>400 mm rainfall) in Piprali and Khamnor area. However, annual rainfall pattern in last 17 years (1995–2011) reveals erratic rainfall. Higher rainfall (above normal) has been observed in 1996, 2001, 2008, and 2010 in all the three areas. Exceptionally, Khamnor area has experienced highest rainfall (1004 mm) in 2005. Similarly, lowest rainfall years have been observed in 2000, 2002, and 2008 which were likely to be the drought year.

4.2 Analysis of Lithological Logs

In the present study, litholog data have been analyzed for Piprali (25 logs), Osian (19 logs), and Khamnor (21 logs). Out of these, only 6 representative logs have been selected for detailed analysis in each of the study area (Fig. 3). The depth of the wells for lithological logs varies from 166 to 193 m in Osian, 75–130 m of Khamnor and 36–90 m in Piprali. Most logs from Piprali area are represented by very fine to coarse sand except Palsana where quartzite/schist has been encountered at a depth of 30 m. Alternate layers of sandstone and shale with varying thickness (113–190 m) represents aquifer system in Osian. Barring a thin soil cover (0–5 m), weathered and fractured gneissic aquifer (up to 113 m bgl) has been deciphered in Khamnor area.

4.3 Analysis of Seasonal Groundwater Level Data

The trend line of groundwater elevation data for all the three seasons (pre-monsoon, post-monsoon, and post-irrigation) in the past eight years (2004–05 to 2011–12) is shown in best fit lines (Fig. 4). In Piprali and Osian area, constant declining phase is revealed in spite of inclining trend in rainfall during the same period. In contrary, the best fit line in Khamnor area shows rising trend which is commensurate with the

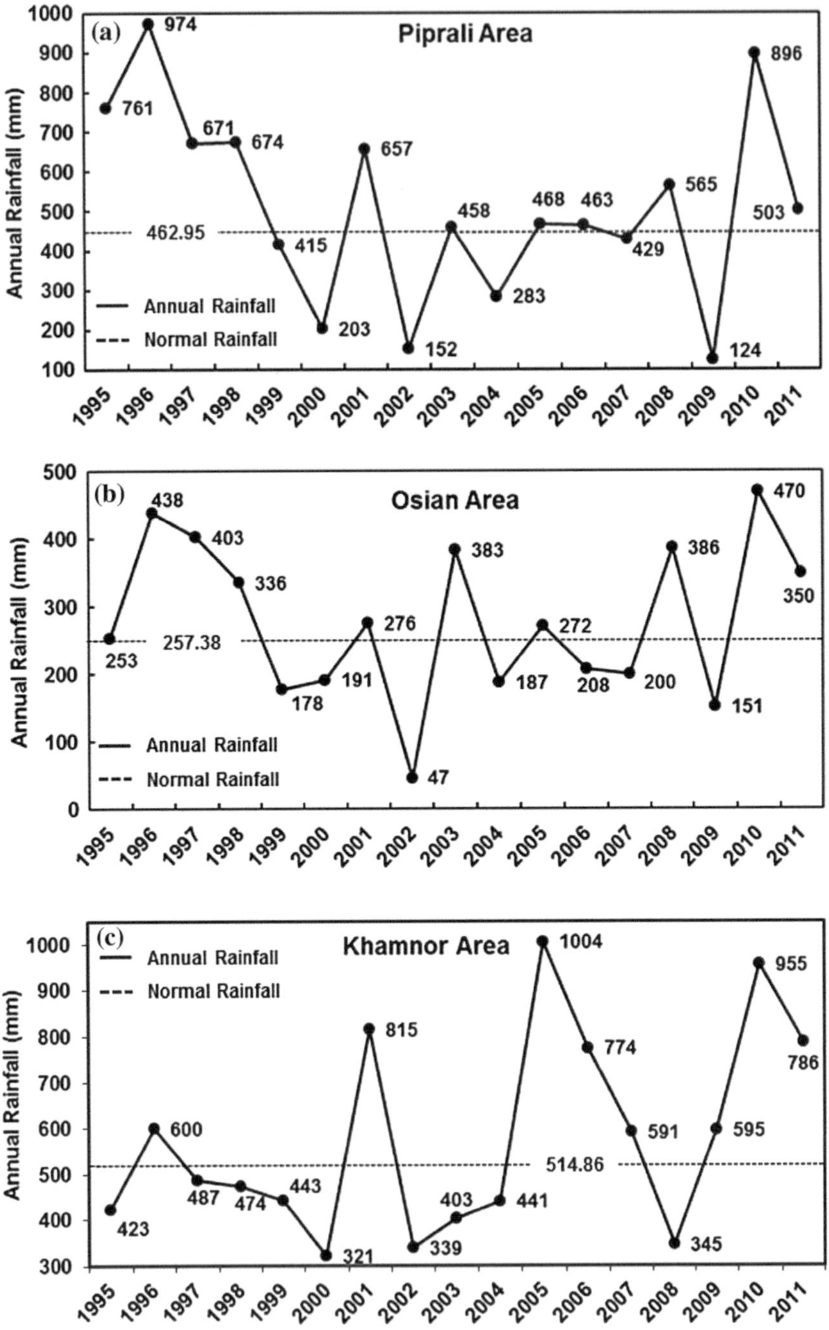

Fig. 2 Annual rainfall pattern during 1995–2011 in **a** Piprali, **b** Osian, and **c** Khamnor study area

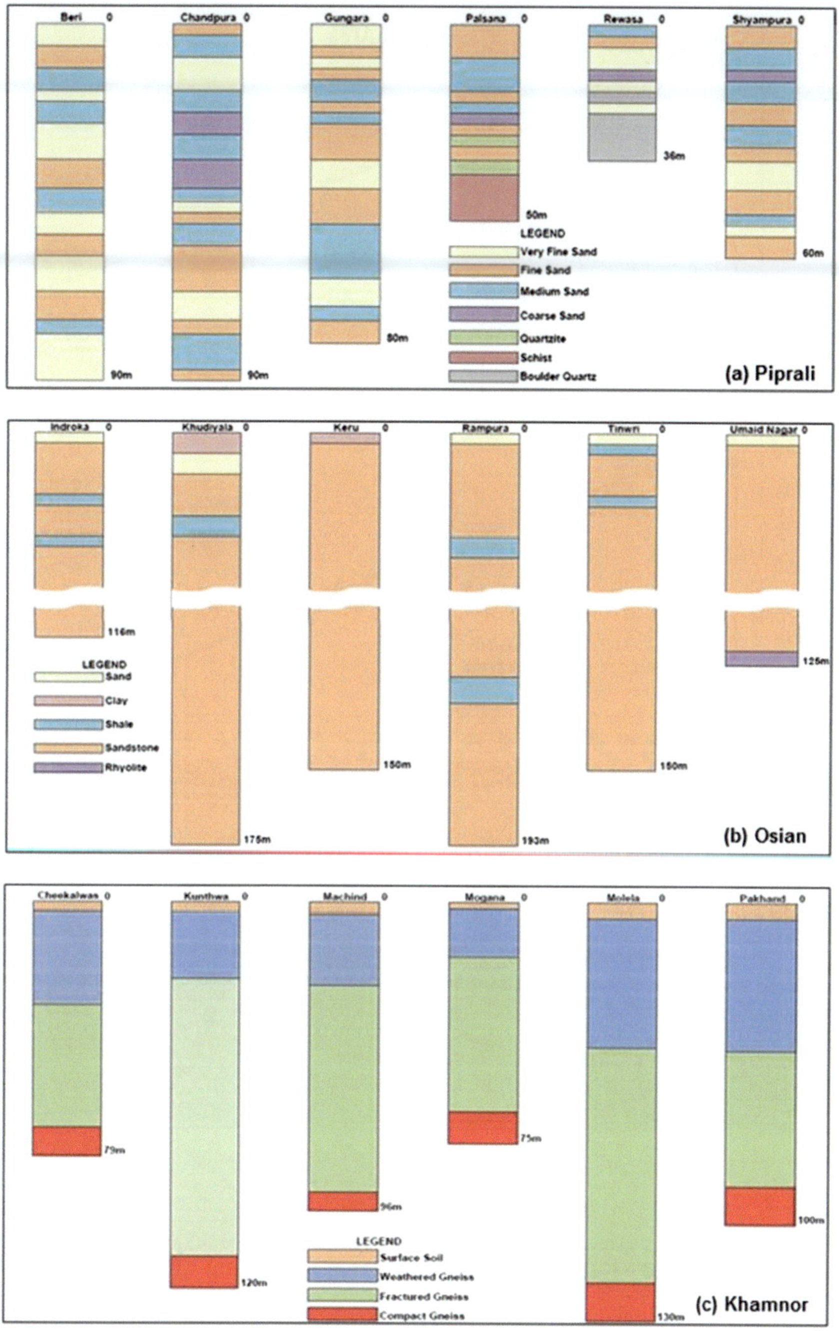

Fig. 3 Representative lithologs showing variation in aquifer materials at different depths in **a** Piprali, **b** Osian, and **c** Khamnor area

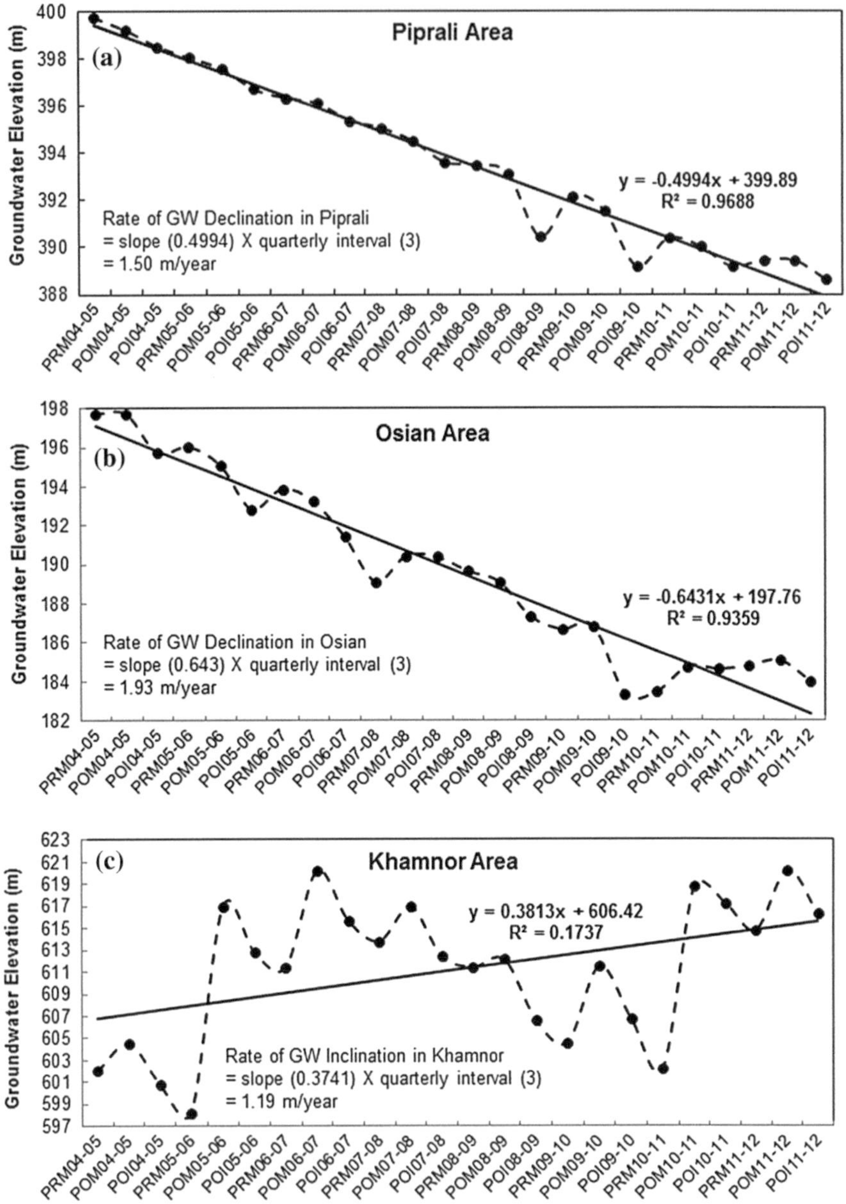

Fig. 4 Groundwater elevation trends in different seasons from 2004–2005 to 2011–2012 of **a** Piprali, **b** Osian, and **c** Khamnor study area

increasing rainfall pattern. Equations of the best fit lines for Piprali, Osian, and Khamnor area are represented by the following linear equations viz.

$$\text{For Piprali,}\quad y = -0.4994x + 399.89 \tag{1}$$

$$\text{For Osian,}\quad y = -0.6431x + 197.76 \tag{2}$$

$$\text{For Khamnor,}\quad y = 0.3741x + 606.49 \tag{3}$$

where y = Groundwater elevation (m) and x = Groundwater observation period.

The above analysis shows a steady declining of groundwater level in Piprali (1.50 m/year) and Osian (1.93 m/year) which is attributed to excessive groundwater draft for irrigation. It has been estimated that the deepest wells in Osian (193 m) and Piprali (130 m) area may get dry completely by 2048 and 2068, respectively, unless appropriate measures are taken. However, Khamnor area shows rising trend in level at the rate of 1.19 m. The variation in the bed of groundwater level is attributed to aquifer types viz. alluvium in Piprali, sandstone in Osian, and gneisses in Khamnor area.

4.4 *Temporal Geospatial Analysis of Groundwater Recharge and Irrigation Draft*

(a) **Groundwater Recharge**—Groundwater recharge has been assessed from the water level fluctuation between pre-monsoon and post-monsoon seasons. It has been observed that recharge commensurate with high rainfall and limited to shallow aquifers only due to good infiltration capacity of alluvium (Piprali), sandstone (Osian), and fractured/weathered gneisses (Khamnor). Deeper aquifers do not show much variation in groundwater fluctuation. Gneissic terrain shows faster and higher groundwater recharge than sandstone and alluvium terrain. Spatial and temporal variation (2009–2010, 2010–2011, and 2011–2012) in recharge pattern in Piprali, Osian, and Khamnor area are shown in Fig. 5. It has been observed that groundwater level in most of the villages in Piprali area fall up to 1 m during 2009–2010 which is attributed to low rainfall (124 mm). With the increase in rainfall in 2010 and 2011, the level has risen to >1 m in the southeastern and western extremity. But at the same time, groundwater recharge remains same in the north and central part of the study area. Thus, southeastern part of Piprali area indicates very good recharge zones in comparison with northern or central part the reason of which may be shallow basement rocks (Similarly, majority of the villages in the north of Osian area shows drastic fall in groundwater recharge (<1 m) during 2009–2010 due to poor rainfall (151 mm). As the rainfall intensity increases in 2010 (470 mm)

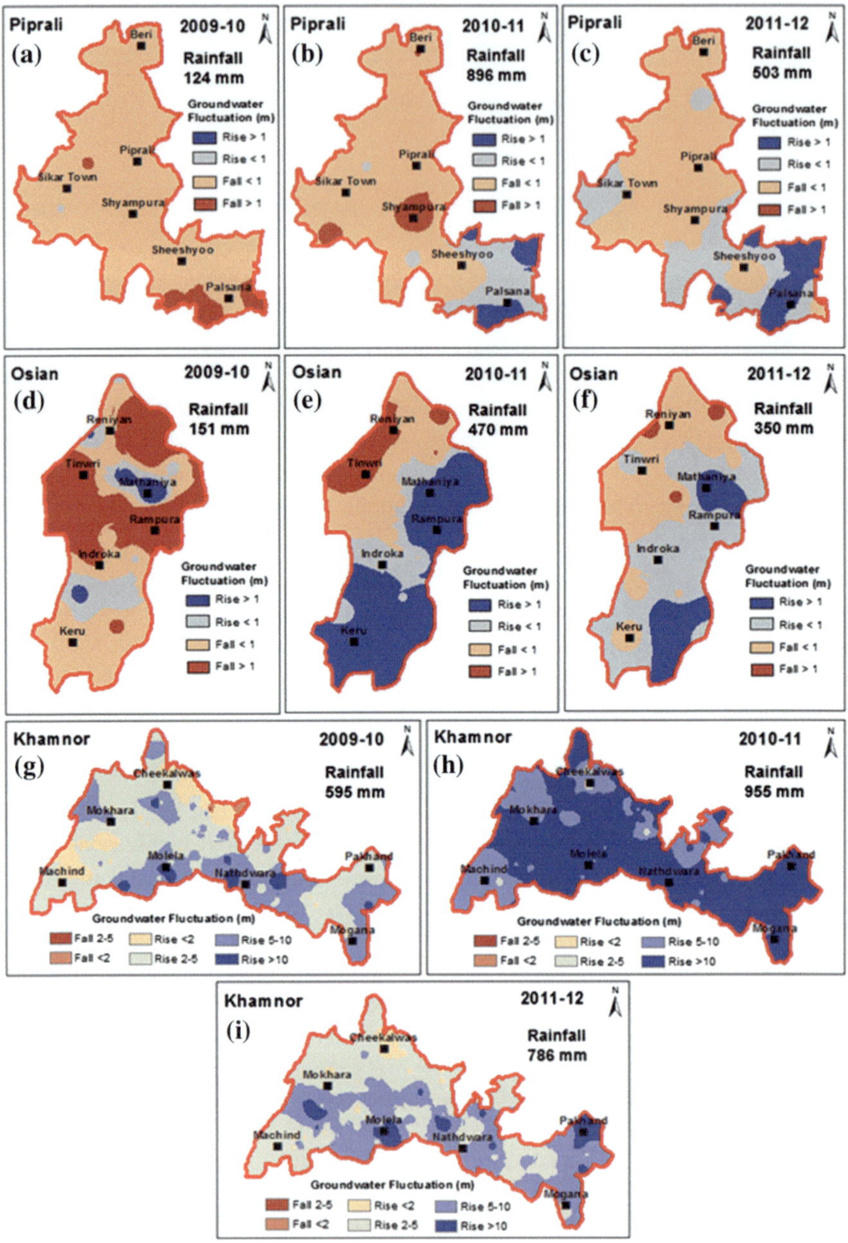

Fig. 5 Spatial maps of groundwater fluctuation due to recharge during 2009–2010, 2010–2011 and 2011–2012 in **a** Piprali, **b** Osian and **c** Khamnor area

and 2011 (350 mm), the recharge also increases (>1 m) in the eastern and southern parts of the study area. However, the recharge remains depleted (fall <1 m) in the northern villages. Thus, southern and eastern parts of the study area indicate very good recharge zones in comparison with northern part of Osian area. Faster recharge in southern part may be attributed to the presence of fractured sandstone aquifer. While comparing the recharge characteristics, large part of southeastern Khamnor area shows much higher groundwater recharge (>10 m) due to high rainfall intensity in 2010 (955 mm) and 2011 (786 mm) in comparison with 2009 (595 mm). Khamnor area acts as a good recharge potential zone which is attributed fractured and weathered gneissic aquifer.

(b) **Groundwater draft**—Groundwater draft due to irrigation in ideal season has been analyzed by groundwater fluctuation data between post-monsoon and post-irrigation groundwater level between 2004–05 and 2011–2012. In spite of wide variation in rainfall pattern, the study area experience significant irrigation draft (1–4 m in Osian and Piprali and 2–10 m in Khamnor area). Spatial map of post-monsoon and post-irrigation fall in during 2009–2010 shows >2 m fall in majority of the villages in Piprali area (Fig. 6), which possibly due to low rainfall in 2009 (124 mm). However, the situation has been improved considerably in 2010–2011, when the fall is <1 m due to good amount of rainfall in 2010 (896 mm) and 2011 (503 mm). Field verification with farmers' feedback reveals that that groundwater abstraction is quite extensive because of continuous sprinkler irrigation for Rabi crops. In the last 8 years (2004–05 to 2011–12), the rise in groundwater draft in Piprali area is quite alarming especially during 2008–2009 and 2009–2010. The less rainfall and groundwater draft during Rabi season has led the practice of abandoning crop fields by the local farmers after cultivating continuos for two years. Similarly, excessive withdrawal of groundwater (fall in water level 2–4 m) has been observed in the northern part of Osian during 2009–2010, possibly due to low rainfall (151 mm). The situation has been improved during 2010–2011 due to higher rainfall (470 mm) in western part of Osian. While at Rampura and Mathania area critical condition (fall >4 m) during 2009–2010 to 2011–2012 which is attributed to excessive groundwater abstraction in Rabi season. Geospatial analysis of Khamnor area also reveals excessive withdrawal of groundwater (>5 m) during 2009–2010 in the southeastern villages, with marginal improvement during 2010–2011 due to excessive rainfall (955 mm). Groundwater draft situation again deteriorated in 2011–2012 due to low rainfall (786 mm).

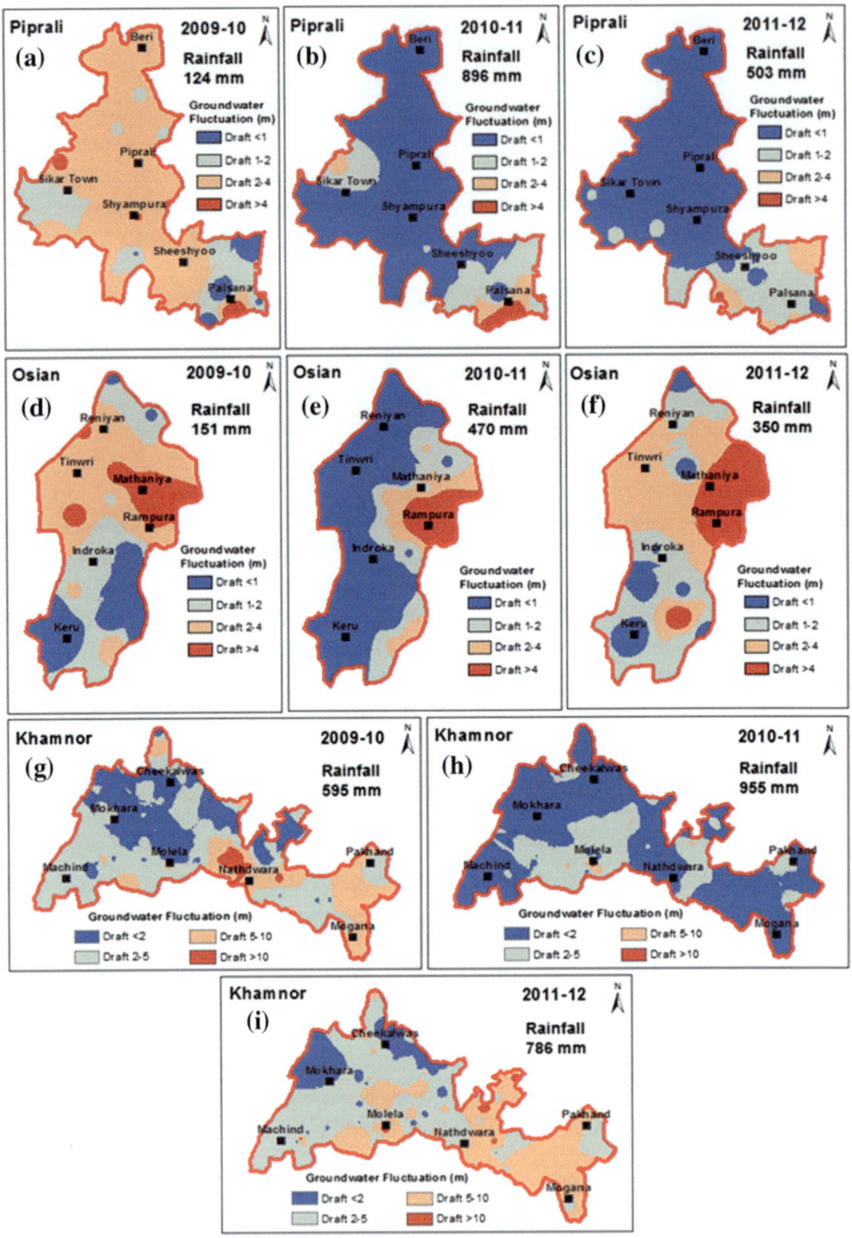

Fig. 6 Spatial maps of groundwater fluctuation due to irrigation draft during 2009–2010, 2010–2011 and 2011–2012 in **a** Piprali, **b** Osian, and **c** Khamnor area

4.5 *Impact of Water Harvesting Structures on Groundwater Regime*

A large number of WHS has been constructed in Piprali, Osian, and Khamnor area during 2008–2009. The impact of these WHS is assessed by comparing seasonal groundwater level fluctuations between 2008–2009 (pre-WHS) and 2011–2012 (post-WHS). It has been observed that groundwater level in most of the villages of the Piprali area has declined to the tune of 0.67–10.35 m in pre-monsoon, 0.3–9.36 m in post-monsoon, and 0.1–9.1 m in post-irrigation periods, respectively. Exceptionally, positive change in groundwater level (0.2–1.47 m) has been observed in the southeastern part of Piprali area. Similarly, in Osian area also exhibits a declining trend in groundwater level in different seasons in the range of 0.05–17.05 m during pre-monsoon, 0.2–16.5 m during post-monsoon, and 0.8–18.85 m during post-irrigation periods. A few villages in southern part of Osian show positive change in groundwater level (0.6–8.8 m). Thus, widespread negative groundwater fluctuation in Piprali and Osian area indicates that there is marginal on improvement in groundwater level due to construction of WHS; on the contrary, Khamnor area shows positive changes in most of the villages during 2008–2009 and 2011–2012 with average rise in level by 0.2–35.75 m in post-monsoon and 0.5–59.22 m in post-irrigation periods. The positive change in groundwater level may be attributed to recharge through fractures (secondary porosity) in hard rock terrain (sandstone in Osian and gneisses in Khamnor area).

4.6 *Spatial Change in Recharge of Groundwater and Draft for Irrigation Due to WHS*

For monitoring of WHS on groundwater fluctuation, spatial change analysis was done for the areas of groundwater recharge and irrigation extraction happened before and post-construction of WHS. The ‘before’ year is taken as 2008–2009 and the ‘after’ year is considered as 2011–2012.

(a) **Change in Groundwater Recharge**—The change matrix of recharge area from 2008–2009 to 2011–2012 in Piprali, Osian, and Khamnor areas is shown in Table 2a–c. In Piprali area, 513 km^2 (83%) of the recharge area that has observed during 2008–2009 is reduced to 379 km^2 (61%) in 2011–2012 under falling category (<1 m) of groundwater level which is a positive impact indicating groundwater remove (Table 2a). Similarly, under ‘rise’ category in 2008–09 (<1 m) the area increased from 10 to 27% in 2011–2012. The spatial map of Piprali area (Fig. 7a) shows significant increase in recharge area shows the change in recharge area between 2008–2009 and 2011–2012. It was observed that 162 km^2 (27%) of the recharge area in Osian, coming in falling groundwater level (>1 m) in 2008–2009, has been distributed to 7 km^2 (no

Table 2 Change matrix for groundwater recharge area before (2008–09) and after (2011–12) the construction of water harvesting structures in 2009–10

(a) Piprali area

Recharge category		Recharge area in (km^2)					
		Fall >1 m	Fall <1 m	Rise <1 m	Rise >1 m	Total	Area (%)
Groundwater recharge area (km^2)	Fall >1 m	–		2.69	Nil	28.83	–
	Fall <1 m	Nil			40.91	–	82.79
	Rise <1 m	1.19	Nil			–	10.25
	Rise >1 m	Nil	Nil	Nil	**14.32**	14.32	2.31
	Total	1.19	379.47	168.41	71.21	–	100.00
	Area (%)	0.19	61.18	27.15	11.48	100.00	

(b) Osian area

Recharge category		Groundwater recharge area in 2011–12 (km^2)					
		Fall >1 m	Fall <1 m	Rise <1 m	Rise >1 m	Total	Area (%)
Groundwater recharge area in 2008–09 (km^2)	Fall >1 m	**7.05**	106.73	45.37	2.73	161.88	27.47
	Fall <1 m	4.35	**88.98**	159.96	60.49	313.78	53.25
	Rise <1 m	Nil	21.17	**49.16**	28.34	98.67	16.75
	Rise >1 m	Nil	4.34	7.68	**2.89**	14.91	2.53
	Total	11.40	221.22	262.17	94.45	589.24	100.00
	Area (%)	1.93	37.54	44.49	16.03	100.00	

(c) Khamnor area

Recharge category		Groundwater recharge area in 2011–12 (km^2)							
		Fall >5 m	Fall 2–5 m	Fall <2 m	Rise <2 m	Rise 2–5 m	Rise >5 m	Total	Area (%)
Groundwater recharge area in 2008–09 (km^2)	Fall >5 m	**Nil**	Nil	Nil	Nil	Nil	3.02	3.02	0.73
	Fall 2–5 m	Nil	**Nil**	Nil	0.70	8.79	23.17	32.66	7.90
	Fall <2 m	Nil	Nil	**Nil**	8.70	104.68	64.50	177.88	43.02
	Rise <2 m	Nil	Nil	Nil	**3.05**	72.26	70.14	145.44	35.18

(continued)

Table 2 (continued)

(c) Khamnor area									
Recharge category		Groundwater recharge area in 2011–12 (km^2)							
		Fall >5 m	Fall 2–5 m	Fall <2 m	Rise <2 m	Rise 2–5 m	Rise >5 m	Total	Area (%)
	Rise	Nil	Nil	Nil	0.27	**22.77**	26.51	49.56	11.99
	Rise >5 m	Nil	Nil	Nil	Nil	0.20	**4.71**	4.91	1.19
	Total	Nil	Nil	Nil	12.71	208.71	192.04	413.47	100.00
	Area (%)	Nil	Nil	Nil	3.07	50.48	46.45	100.00	

Note The bold values indicate no change in groundwater recharge area between 2008–2009 and 2011–2012

change), 107 km^2 (fall <1 m), 45 km^2 (rise <1 m), and 3 km^2 (rise >1 m) in 2011–2012 which is a positive change (Table 2b). The spatial map of Osian area (Fig. 7b) shows significant increase in groundwater recharge area in 404 km^2 (68%) with negative change in recharge area only in 38 km^2 (6%), and no change in recharge area in 148 km^2 (24%). Due to higher groundwater fluctuation (rise >5 m), Khamnor area is divided into six categories (Table 2c). The entire recharge area of 178 km^2 (43%), under fall category of <2 m, in 2008–2009 has been improved by rising groundwater level viz. 9 km^2 (rise <2 m), 105 km^2 (rise 2–5 m), and 64 km^2 (rise >5 m), respectively. Spatial map of Khamnor area shows significant positive changes (92%) in groundwater recharge (Fig. 7c). Thus, Piprali, Osian, and Khamnor area shows an overall improvement in groundwater recharge between 2008–2009 and 2011–2012 due to construction of WHS in 2009–2010.

(b) **Change in Groundwater Draft**—The change matrix of groundwater draft in Piprali and Osian area are generated for different categories of falling groundwater level between 2008–2009 and 2011–2012 i.e., <1, 1–2, 2–4 and >4 m (Table 3a, b). In Piprali area, significant change has been observed due to drastic reduction in draft area (positive changes) from 93 to 6% under category of 2–4 m before and after construction of WHS. The matrix values show that 577 km^2 draft area in 2008–2009, under 2–4 m, has been transformed to 435 km^2 in <1 m category and 119 km^2 in 1–2 m category, respectively. This positive change may be attributed to higher rainfall in 2010–2011 as well as withdrawal of less groundwater for irrigation. Spatial interpolated maps of Piprali area show total positive changes in 581 km^2 area (94%) due to irrigation draft between 2008–2009 and 2011–2012 (Fig. 8a). Similarly, change matrix in Osian area also shows positive changes due to variation in 2–4 m category (246 km^2) in 2008–09 to 1–2 m category (38 km^2) and <1 m category (20 km^2) in 2011–12 (Table 3b). However, significant negative impact is shown is shown under falling categories of <1 m (146 km^2) in 2008–2009 to

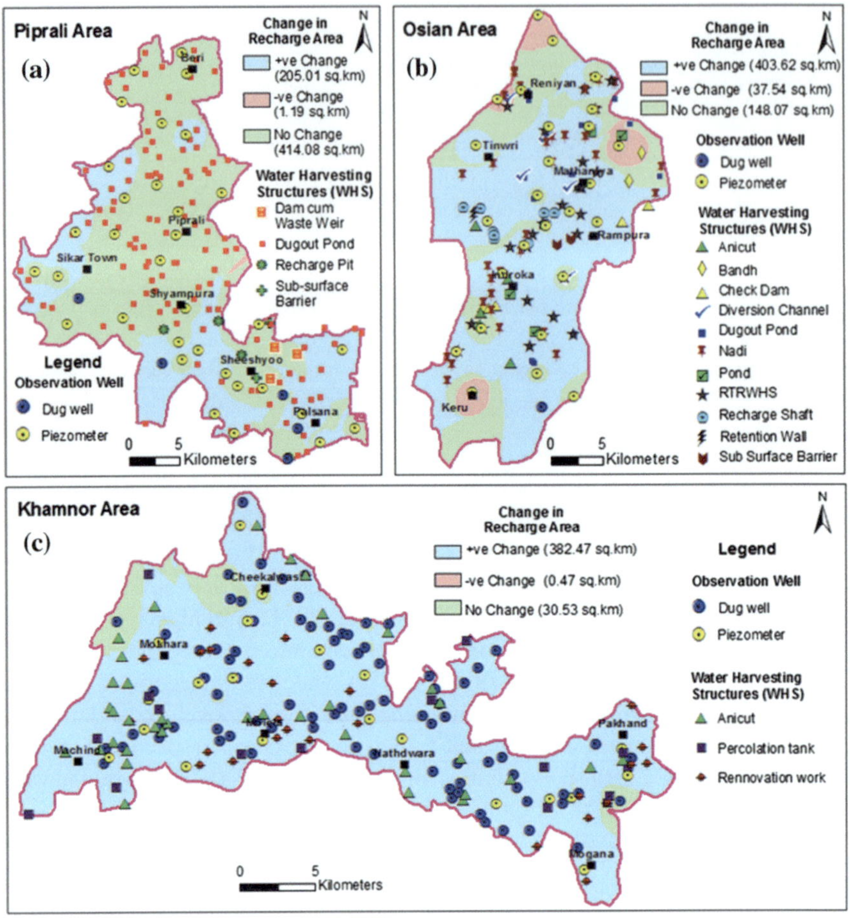

Fig. 7 Spatial change of groundwater recharge area between 2008–2009 and 2011–2012 with construction of water harvesting structures in **a** Piprali, **b** Osian, and **c** Khamnor area

1–2 m (75 km^2) in 2011–2012. The spatial map of groundwater draft (Fig. 8b) shows total negative changes in 269 km^2 (50%) area, positive changes in 62 km^2 (10%) area, and no change in 258 km^2 (4%) area, respectively. Thus, spatial maps due to change in draft area in Piprali and Osian show groundwater draft exceeds recharge between 2008–09 and 2011–12. It has been observed in khamnor area (Table 3c) that the draft area of 178 km^2 (43%) in 2008–09, under fall category of <2 m, has been improved by reducing draft area as 55 km^2 (13%) in 2011–2012. The Khamnor area thus shows mixed response with positive changes in 104 km^2 area (25%) and negative changes in 212 km^2 area (51%) with no change in 97 km^2 area (Fig. 8c).

Table 3 Change matrix for groundwater draft area with falling groundwater level before (2008–09) and after (2011–12) the construction of water harvesting structures in 2009–10

(a) Piprali area

Draft category		Groundwater draft area in 2011–12 (km²)					
		<1 m	1–2 m	2–4 m	>4 m	Total	Area (%)
Groundwater draft area in 2008–09 (km²)	<1 m	**3.67**	Nil	Nil	Nil	3.67	0.59
	1–2 m	17.11	**Nil**	11.99	Nil	29.10	4.69
					1.28		93.04
	>4 m	Nil	6.84	3.58	**Nil**	10.42	1.68
	Total	455.57	125.74	37.69	1.28		100.00
	Area (%)	73.45	20.27	6.08	0.21	100.00	

(b) Osian area

Draft category		Groundwater draft area in 2011–12 (km²)					
		<1 m	1–2 m	2–4 m	>4 m	Total	Area (%)
Groundwater draft area in 2008–09 (km²)	<1 m	**45.31**	75.23	18.72	6.20	145.47	24.69
	1– 2m	3.63	**75.70**	82.53	21.14	183.01	31.06
	2–4 m	20.03	38.01	**121.95**	65.71	245.70	41.70
	>4 m	Nil	Nil	Nil	**15.06**	15.06	2.56
	Total	68.97	188.95	223.21	108.11	589.24	100.00
	Area (%)	11.70	32.07	37.88	18.35	100.00	

(c) Khamnor area

Draft Category		Groundwater draft area in 2011–12 (km²)							
		Draft <2 m	Draft 2–3 m	Draft 3–5 m	Draft 5–7 m	Draft 7–10 m	Draft >10 m	Total	Area (%)
Groundwater draft area in 2008–09 (km²)	Draft <2 m	**34.48**	24.83	71.84	23.45	20.76	2.26	177.63	42.96
	Draft 2–3 m	6.70	**7.79**	21.69	20.60	6.40	0.10	63.28	15.30
	Draft 3–5 m	13.45	9.65	**29.91**	16.11	2.00	Nil	71.13	17.20
	Draft 5–7 m	0.36	4.88	23.94	**17.76**	1.71	Nil	48.64	11.76
	Draft 7–10 m	Nil	1.83	12.57	22.95	**6.99**	0.13	44.47	10.76
	Draft >10 m	Nil	0.24	2.10	3.62	1.77	**0.58**	8.32	2.01
	Total	54.98	49.22	162.05	104.50	39.63	3.09	413.47	100.00
	Area (%)	13.30	11.90	39.19	25.27	9.58	0.75	100.00	

Note The bold values indicate no change in groundwater draft area between 2008–2009 and 2011–2012. The values above and below the bold values indicate negative and positive changes in groundwater draft area, respectively

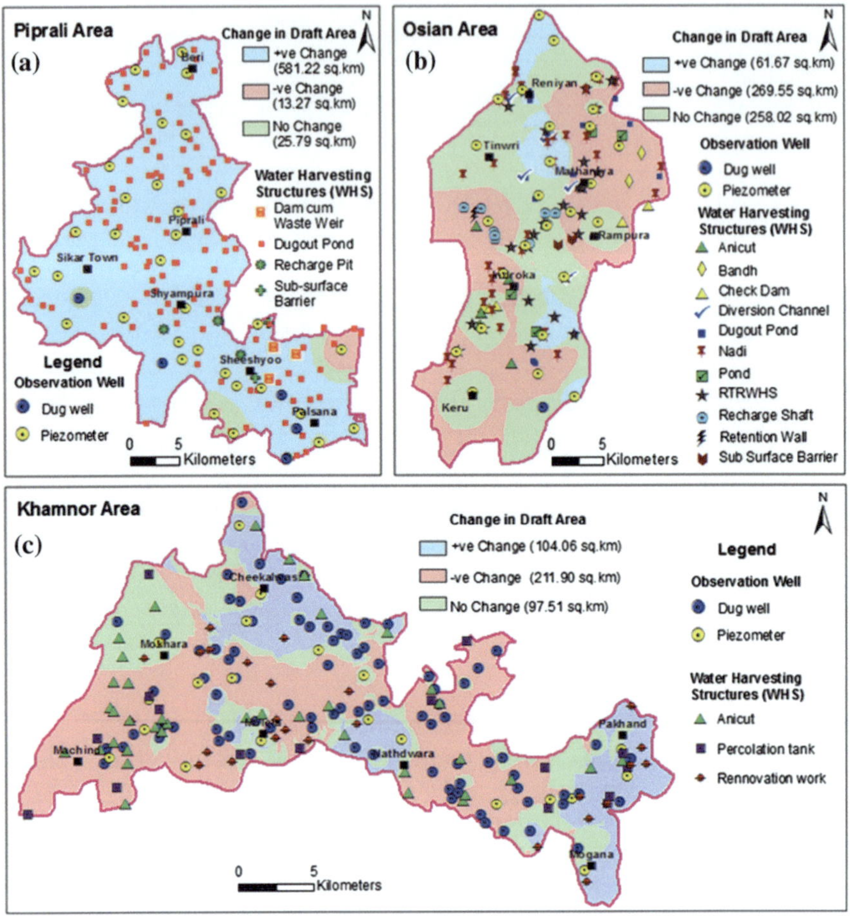

Fig. 8 Spatial change of groundwater draft area between 2008–2009 and 2011–2012 with construction of water harvesting structures in **a** Piprali, **b** Osian, and **c** Khamnor area

5 Comparison of Groundwater Behavior in Different Aquifer System

Salient features of the hydrogeological characteristics due to groundwater recharge and irrigation draft in different aquifer systems are given in Table 4. In comparison with Osian and Khamnor, farmers in Piprali area mainly grow wheat and mustard with sprinkler irrigation practices. Density of groundwater observation wells is found to be the highest (1.20 wells/village) in Khamnor area and least (0.56 wells/village) in Piprali area. In 2011, rainfall intensity in Khamnor area (786 mm) shows more than double than that of Osian area (350 mm). Average groundwater level in gneissic terrain in Khamnor area is much shallow (9.12 m) than the Osian

Table 4 Comparison of different parameters in Piprali, Osian and Khamnor Area, Rajasthan

S. No.	Parameters	Piprali	Osian	Khamnor
1	Study area	620.28 km^2	589.24 km^2	413.47 km^2
2	Villages and gram panchayats	70 Villages of 32 GP	33 Villages of 15 GP	88 villages in 23 GP
3	Major aquifer	Older alluvium/aeolian sand	Jodhpur sandstone	Gneisses
4	Rainfall intensity (2011)	503 mm	350 mm	786 mm
5	Cropping pattern	Bajra, wheat, mustard	Bajra, wheat, vegetables	Maize, wheat
6	No. of observation wells	39 OW	26 OW	106 OW
7	Average. ground water level	Deep (52.26 m)	Very deep (74.97 m)	Shallow (9.12 m)
8	GW recharge in 2011–12 (fluctuation in m)	2.65–0.82 m	5.2–1.28 m	0.16–28.75 m
9	GW draft in 2011–12 (fluctuation in m)	up to −4.39 m	0.7–7.1 m	−12.73 to 1.45 m
10	Rain water harvesting structures	Total = 129 (121 DOP, 3 DWW, 3 RP, 2 SSB)	Total = 118 (36 Nadi, 29 RTRWHS, 19 DOP, 5 DC, 10RS, 5 Anicut, 2 SSB, 1 RTW, 4 Pond, 2 Bandh	Total = 81 (38 Anicut, 16 PT, 13 RTRWHS, 14 RW)
11	Irrigation practices	Sprinkler irrigation	Flood irrigation	Mostly flood irrigation
12	GW declination/inclination and futuristic scenario	Declination rate at 1.50 m/year. Deepest well (130 m) may completely dry by 2068	Declination rate is very fast (1.93 m/year). Deepest well (193 m) may completely dry by 2048	Increase rate at 1.24 m/year. Increasing in water level from 2009–10 to 2011–12 is attributed to increase in rainfall
13	Impact of WHS on GW recharge (2008–09 to 2011–12)	Marginal improvement in GW recharge (positive impact), as reflected by 33.05% increase in recharge area with no change in 66.75% area	Significant improvement in GW recharge (positive impact) as reflected by 68.50% increase in recharge area with no change in 23.87% area	Significant improvement in GW recharge (positive impact) as reflected by 92.5% increase in recharge area with no change in 7.38% area
14	Impact of WHS on GW draft (2008–09 to 2011–12)	Significant positive impact due to WHS, as reflected by reduction in GW draft area (93.7%)	Mixed response by WHS due to increase in GW draft area by 49.74% with no change in 43.79% area	Mixed response by WHS due to decrease (51.25%) and increase (25.16%) in GW draft area

GP Gram Panchayat; *GW* groundwater; *OW* observation well; *WHS* water harvesting structure; *DOP* dug out pond; *DWW* dam cum waste weir; *RP* recharge pit; *RTRWHS* roof top rain water harvesting structure; *PT* percolation tank; *SSB* sub surface barrier; *CD* check dam; *RS* recharge shaft; *RTW* retention wall; *RW* renovation work; *DC* diversion channel

where groundwater level is very deep (74.97 m). During 2011–2012, groundwater fluctuations due to recharge vary from −0.90 to 2.65 m in Piprali, −1.28 to 5.2 m in Osian, and 0.16 to 28.75 m in Khamnor area, respectively. During the same year, groundwater fluctuations due to irrigation draft vary from −4.39 to −0.02 m in Piprali, −7.1 to 0.7 m in Osian and −12.73 to 1.45 m in Khamnor area, respectively. Thus, Khamnor area shows both highest groundwater recharge and irrigation draft during 2011–2012. It has been estimated that declination in groundwater level of Osian is faster (1.93 m/year) than that of Piprali area (1.5 m/year). To augment groundwater recharge and remedies in irrigation draft, several rain water harvesting structures have been constructed in 2009–2010 viz. 129 in Piprali, 118 in Osian, and 81 in Khamnor study area. As a result, a positive impact due to higher groundwater recharge has been observed in all the three areas irrespective of the aquifer conditions. However, Osian and Khamnor areas show negative impact due to increase in draft area, even after construction of WHS.

6 Conclusions

Detailed hydrogeological study show unique behavior due to groundwater recharge and irrigation draft in different aquifer systems in Rajasthan. Sandstone in Osian shows faster rate of groundwater declination (1.93 m/year) due to higher porosity and permeability than that of **alluvium** in Piprali (1.5 m/year). The predicted groundwater situation is going to be worse by drying up of all the wells in these two areas due to continuous rate of declination in the next half century. Higher rainfall with favorable aquifer material in Khamnor area causes groundwater inclination due to higher rate of infiltration in gneisses. Although groundwater recharge is commensurate with the rainfall pattern in shallow aquifer, the groundwater draft behaves differently which depends on irrigation intensity for agricultural crop production. A gap between the **recharg**e and draft for irrigation has been envisaged which has been caused a deterioration of groundwater scenario in the region. Proper management practices of groundwater and construction of effective water harvesting structures can improve the situation.

Acknowledgements Authors are indebted to the Director of National Remote Sensing Centre, Hyderabad and the Chief General Manager of Regional Centres/NRSC, Hyderabad for their support and help. Kind support and cooperation by the Officers of Sikar, Jodhpur and Rajsamand districts belong, Govt. of Rajasthan in providing groundwater level data is duly acknowledged.

References

Aksever F, Davraz A, Karaguzel R (2015) Groundwater balance estimation and sustainability in the Sandıklı Basin (Afyonkarahisar/Turkey). J Earth Syst Sci 124(4):783–798

Allison GB, Gee GW, Tyler SW (1994) Vadose-zone techniques for estimating ground water recharge in arid and semi-arid regions. Soil Sci Soc Am J 58:6–14

Asano T, Cotruvo JA (2004) Groundwater recharge with reclaimed municipal wastewater: health and regulatory considerations. Water Res 38(8):1941–1951

Aswathanarayana U (2001) Water resources management and the environment. A.A. Balkema Publisher, Netherlanders 422p

Bhadra BK, Pathak S, Sharma JR (2013) Impact of industrial effluents on groundwater around Pali City, Rajasthan using field and satellite data. J Geol Soc India 82(6):675–691

Bhadra BK, Sanjay Kumar, Paliwal R (2016a) GIS-based assessment of non-equilibrium pattern between groundwater recharge and irrigation draft in a semi-arid region of Rajasthan, India. J Arid Land 8(2):184–196. doi:10.1007/s40333-015-0059-3

Bhadra BK, Sanjay K, Rakesh P, Jeyaseelan AT (2016b) Modeling of groundwater draft based on satellite derived crop acreage estimation over arid region of Northwest India. Hydrogeol J 24(7)

Bhuiyan C, Singh RP, Flügel WA (2009) Modelling of ground water recharge-potential in the hard-rock Aravalli terrain, India: a GIS approach. Env Earth Sci 59(4):929–938

Chai H, Cheng W, Zhou C, Chen X, Ma X, Zhao S (2011) Analysis and comparison of spatial interpolation methods for temperature data in Xinjiang Uygur Autonomous Region, China. Nat Sci 3(12):999–1010

Chatterjee R, Purohit RR (2009) Estimation of replenishable groundwater resources of India and their status of utilization. Curr Sci 96(12):1581–1591

Crosbie RS, Binning P, Kalma JD (2005) A time series approach to inferring groundwater recharge using the water table fluctuation method. Water Resour Res 41(1):W01008. doi:10.1029/2004WR003077

Das S (2008) Hydrogeological research in India. J Geol Soc India, 589 p (Golden Jubilee Volume)

GSI (1980) Geological quadrangle map of Sikar, Sheet no. 45 M (1:250,000). Geol Surv India. http://www.portal.gsi.gov.in

GSI (1985) Geological quadrangle map of Jodhpur, Sheet no. 45F (1: 250,000). Geol Surv India. http://www.portal.gsi.gov.in

GSI (1993) Geological quadrangle map of Shergarh, Sheet no. 45B (1:250,000). Geol Surv India. http://www.portal.gsi.gov.in

GSI (1997) District resource map of Rajsamand district. Geol Surv India. http://www.portal.gsi.gov.in

Gundogdu KS, Guney I (2007) Spatial analyses of groundwater levels using universal Kriging. J Earth Syst Sci 116(1):49–55

Gundogdu A, Yazicigil H, Kirmizitas H (2000) Comparison of groundwater recharge estimation techniques: a case study from the Kucuk Menderes river basin in Turkey. In: Proceedings of symposium 'integrated water resources management' in California, IAHS Publ. no. 272, pp 417–423

Gupta RP (1991) Remote sensing geology. Springer, Germany, p 356

Harou JJ, Lund JR (2008) Ending groundwater overdraft in hydrologic-economic systems. Hydrogeol J 16(6):1039–1055

Healy RW, Cook PG (2002) Using groundwater levels to estimate recharge. Hydrogeol J 10:91–109

Jahan CS, Mazumder QH, Islam ATMM, Adham MI (2010) Impact of irrigation in Barind area, NW Bangladesh—an evaluation based on the meteorological parameters and fluctuation trend in groundwater table. J Geol Soc India 76:134–142

Kamińska A, Grzywna A (2014) Comparison of deterministic interpolation methods for the estimation of groundwater level. J Ecol Eng 15(4):55–60

Karanth KR (1987) Ground water assessment, development and management. Tata Mcgraw-Hill Publ. Co., Ltd., New Delhi, p 720

Khan HH, Nabi Aadil, Perrin Jerome, Ahmed Shakeel (2008) Remote sensing and GIS in estimating groundwater balance in semi-arid regions. In: 3rd International Conference water resources and arid environments, pp 1–9

Kumar V, Remadevi (2006) Kriging of groundwater levels—a case study. J Spatial Hydrol 6 (1):81–92

Lakshmamma ND, Balasubramanian A, Siddalingamurthy S, Sumithra S (2015) Estimation of groundwater recharge studies in Gundal watershed, Gundlupet Taluk, Chamarajanagar District, Karnataka, India using remote sensing and GIS. Int J Curr Eng Tech 5(3):2138–2148

Marechal JC, Dewandel B, Ahmed S, Galeazzi L, Zaidi FK (2006) Combined estimation of specific yield and natural recharge in semi-arid groundwater basin with irrigated agriculture. J Hydrol 329:281–293

MoWR Report (2009) Ground water resource estimation methodology. In: Report of the Ground Water Resource Estimation Committee, Ministry of Water Resources Government of India, New Delhi, 113 p

Oliver MA (1990) Kriging: a method of interpolation for geographical information systems. Int. J Geog Infor Syst 4:313–332

Pareek H (1984) Pre-quarternary geology and mineral resources of Northwestern Rajasthan. Memoir Geol Surv India 115:1–99

Radhakrishna BP (2003) Groundwater Recharge. J Geol Soc India 62:135–138

Ravikumar G, Shahidhar T, Krishnaveni M, Karunakaran K (2005) GIS based groundwater quantity assessment model. Int J Civil Environ Eng 1(2):21–30

Reddy AGS (2012) Water level variations in fractured, semi-confined aquifers of Anantpur district, Southern India. J Geol Soc India 80:111–118

Saha D, Aqrawal AK (2006) Determination of specific yielding water balance approach—come from torle odhe watershed in the Trap, province, Maharashtra stats, India. Hydrogeol J 14:625–635

Saha D, Dhar YR, Vittale SS (2009) Delineation of groundwater development polointed zones in parts of magind Ganga alluavid plain in South Bihar, Eastern India. Environ Mom Assam. doi:10.1007/&10661-009-0937-2

Scanlon BR, Keese KE, Flint AL, Flint LE, Gaye CB, Edmundsm WM, Ian SI (2006) Global synthesis of groundwater recharge in semiarid and arid regions. Hydrol Process 20:3335–3370

Scott CA, Shah T (2004) Groundwater overdraft reduction through agricultural energy policy: insights from India and Mexico. Int J Water Res Dev 20(2):149–164

Shahid S, Wang X, Rahman MM, Hasan R, Harun SB, Shamsudin S (2015) Spatial assessment of groundwater over-exploitation in northwestern districts of Bangladesh. J Geol Soc India 85 (4):463–470

Sharif M, Ashok KR (2011) Impact of groundwater over-draft on farm income and efficiency in crop production. Agri Eco Res Rev 24(2):291–300

Shiferaw B, Ratna RV, Wani SP (2008) Watershed externalities, shifting cropping patterns and groundwater depletion in Indian semi-arid villages: the effect of alternative water pricing policies. Ecol Econ 67:327–340

Skidmore A (2010) Environmental modeling with GIS and remote sensing. CRC Press, Taylor and Francis Group, Boca Raton, London

Sophocleous MA (1991) Combining the soil water balance and water-level fluctuation methods to estimate natural ground water recharge: practical aspects. J Hydrol 124:229–241

SRSAC (1999) Groundwater atlas of Rajasthan. State Remote Sensing Application Centre (SRSAC), Jodhpur, Government of Rajasthan, pp 1–551

Sukhija BS, Nagabhushanam P, Reddy DV (1996) Groundwater recharge in semi-arid regions of India: an overview of results obtained using tracers. Hydrogeol J 4(3):50–71. doi:10.1007/s100400050089

Surinaidu L, Bacon CGD, Pavelic P (2012) Agricultural groundwater management in the Upper Bhima Basin, India: current status and future scenarios. Hydrol Earth Syst Sci Discuss 9:10657–10686

Evaluation of Water Level Behavior in Coal-Mining Area, Adjacent Township, and District Areas of Jharkhand State, India

Prasoon Kumar Singh, Ashwani Kumar Tiwari and Poornima Verma

Abstract The present paper deals with groundwater level behavior in the coal-mining areas and also in the adjacent township areas of the Jharkhand State. Thickness of saturated zone and the water level behavior depend on spatiotemporal and hydrogeological factors such as soil type, geomorphology, geological formations, rain fall pattern and distribution, and groundwater exploitation. The groundwater regime largely follows the topographical features in the area. In the mining areas, i.e., the East Bokaro, Jharia, and West Bokaro coalfield the water level ranges from 1.64 to 11.20, 3.05 to 11.10, and 1.8 to 16.12 mbgl, respectively, during pre-monsoon period while during post-monsoon it ranges from 1.36 to 10.80, 1.4 to 7.5, and 0.3 to 10.7 mbgl, respectively. The average seasonal water table fluctuation in the area is 2.60, 0.68, and 3.2 mbgl, respectively. Whereas in the adjacent township areas, the average water level during pre-monsoon is 6.99, 6.29, and 5.21 mbgl, respectively, while during post-monsoon the value are 2.46, 6.29, and 3.40 mbgl, respectively. It has been found that mining and industrial activities in some specific locations are impacting both shallow as well as deep aquifer indicating that sustainable groundwater management is urgently needed incorporating geological complexities and aquifer depositions in the area. Artificial recharge and protection of impacted aquifers are the key components of the management plan.

Keywords Groundwater · Hydrogeology · Water level fluctuations Groundwater management

---

P.K. Singh (✉) · A.K. Tiwari · P. Verma
Department of Environmental Science and Engineering, Indian Institute of Technology (ISM), Dhanbad 826004, Jharkhand, India
e-mail: pks0506@yahoo.co.in

D. Saha et al. (eds.), *Clean and Sustainable Groundwater in India*,
Springer Hydrogeology, https://doi.org/10.1007/978-981-10-4552-3_17

1 Introduction

Water is the elixir of life and is crucial for sustainable development. Earlier, it was considered to be infinites or at least fully renewable natural resource. However, in the last two decades, a pressure is felt in water resource which is extremely important for a country's development due to increasing demand (Tabari et al. 2012; Gupta et al. 2016; Mukherjee and Shekhar 2014; Kunar et al. 2011). This natural resources imparts an important role for sustainability in livelihood in different countries of the globe including this country. This resource is dynamic as it is related to yearly monsoon cycle (Mukherjee et al. 2015). Geologically, its availability is limited in hard rock areas in comparison with the soft rock terrain (Srivastava et al. 2012). India spans over 32.87 million km^2 about 65% of which is occupied by consolidated formation such as granites, basalt and other volcanism, gneisses, other metamorphic rocks, and consolidated sedimentary rocks (Mukherjee 2016). A major part of the hard rock terrain particularly in the peninsular states is remaining drought prone and thus groundwater dependence to more (Maggirwar and Umrikar 2011; Ray et al. 2014). Extreme use of groundwater results in over exploitation and lowering of water level. This is such a situation when rich peasantry slowly dominate and poor population suffer from deprivation; the reason is that the rich farmers can offered deeper wells and the poor farmers face the dry shallow aquifers. Postel and Hichman (1999) have observed this kind of stress urged the need of technological intervention. The rate of depletion of groundwater level and deterioration of groundwater quality are of immediate concern in major groundwater depending urban areas (Saha et al. 2007, 2010; Singh et al. 2011; Sharma et al. 2014; Krishan et al. 2014, 2016; Guhe and Mukherjee 2012; Chandra et al. 2015; Tiwari and Singh 2014; Tiwari et al. 2015, 2016). The water level changes with time are influenced by various factors such as rainfall and formation characteristics. The groundwater storage and flow is controlled by various factors such as topography, rock type, tectonic activity of the area, secondary porosities, weathered zone, lineaments, and land use (Bhuiyan 2010). The objective of present study is to apply the groundwater level behavior spatially and temporally in coal-mining and surrounding areas of Jharkhand State, India, by using statistical and GIS techniques. The impact of mining activities is to be studied so that a sustainable groundwater management can be developed.

2 Materials and Methods

The groundwater fluctuation was determined for specific areas of the Jharkhand State. Jharkhand State consists of industrial and mining areas. The groundwater was monitored for the pre- and post-monsoon seasons in year 2013. Depth to water level has been recorded using a sensor-based water level recorder for pre- and post-monsoon periods. The groundwater fluctuation was represented by spatial distribution using ARC GIS 10.2 software.

3 Background of the Study Area

3.1 Mining Areas

The Jharkhand State is a blessed with immense mineral potential. The state is occupied by 80,356.20 million metric tons of the Gondwana coal deposits. The present study includes the East Bokaro, West Bokaro Jharia coalfield. Administratively, they are located in Bokaro and Dhanbad district (Fig. 1). The East Bokaro coalfield, one of the major repositories of medium-coking, metallurgical coal in Peninsular Gondwana Basins in India occupies an area of about 237 km^2. This coalfield contains a number of coal seams in Barakar as well as Raniganj Formation and holds a continuous deposition from Talchir to Supra-Panchet (Mahadeva) Formations (Fig. 2). The West Bokaro coalfield area (207 km^2) consists of metamorphic, Barakar, Barren Measures, Mahadeva, Panchet, Raniganj, and Talchir Formation. The Barakar Formation covers the major part of the coalfield and comprises as coarse to fine-grained sandstone, pebbly conglomerates, gritty sandstones, gray shales, carbonaceous shales, fire clays, and

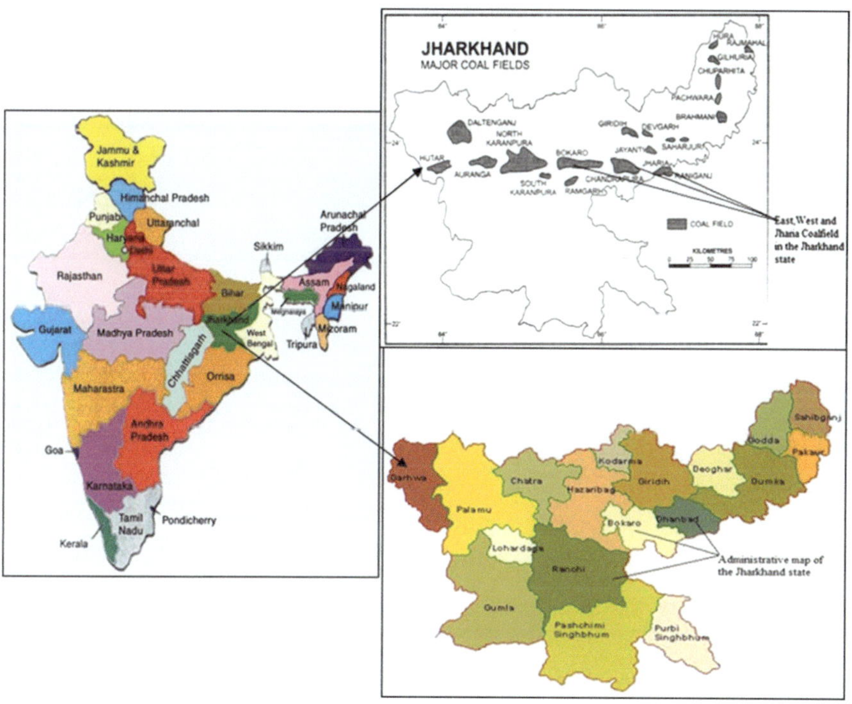

Fig. 1 Administrative areas (Dhanbad, Bokaro and Ranchi district) and coalfield areas (E Bokaro, W Bokaro, and Jharia) shown in Jharkhand State map

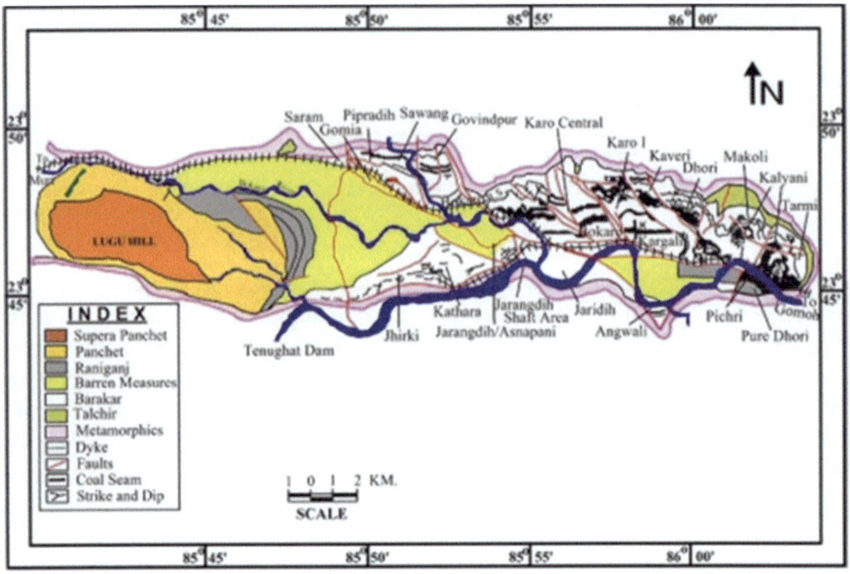

Fig. 2 Geology map of the East Bokaro coalfield

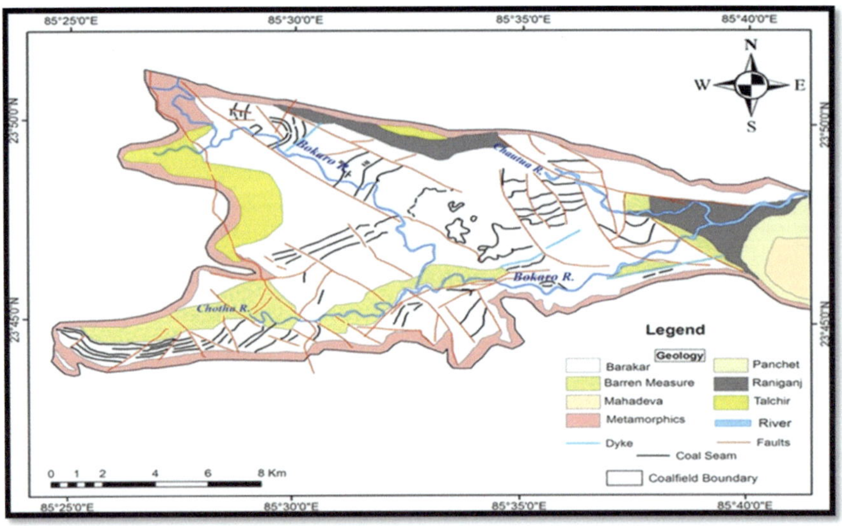

Fig. 3 Geology map of the West Bokaro coalfield

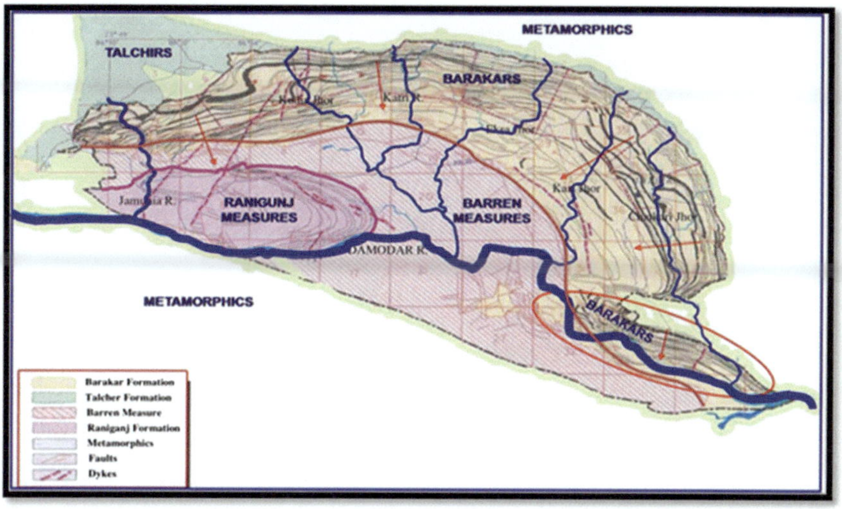

Fig. 4 Geology map of the Jharia coalfield

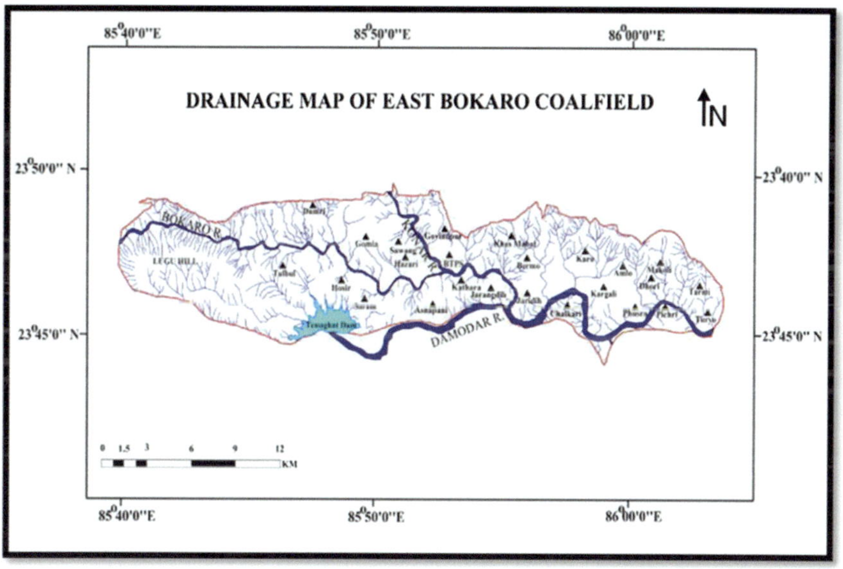

Fig. 5 Drainage map of the East Bokaro coalfield

coal seams found in the West Bokaro coalfield (Fig. 3). The Jharia coalfield consists of area of about 450 km^2 belongs to the lower Gondwana Group containing Talchir, Barakar, Barren Measures, and Raniganj Formation. It is a sickle-shaped coalfield occurring in the form of a basin truncated with a major boundary fault on the southern flank (Fig. 4).

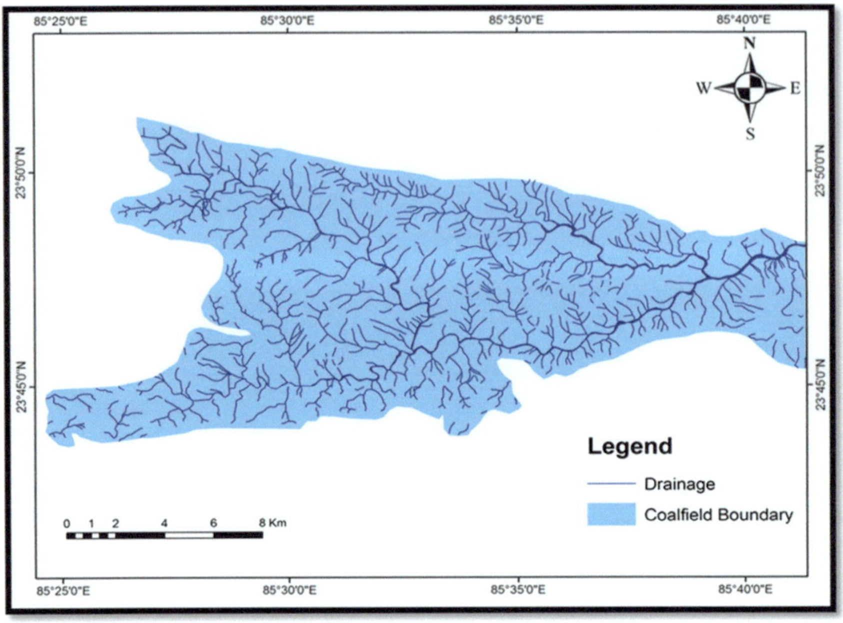

Fig. 6 Drainage map of the West Bokaro coalfield

The Damodar River is the most prominent river in the study area which flows from West to East. Drainage pattern of the East Bokaro coalfield region reflects the characteristic of surface and subsurface geology (Fig. 5). The West Bokaro coalfield falls in Bokaro River sub-basin, where the river flows the west to the east. The Chutua River and Chotha River are the main tributaries of the Bokaro River which drain the northern hilly terrain and southern region of the coalfield, respectively (Fig. 6). In the Jharia coalfield, the general flow direction of tributaries of the Damodar River is from north east to south east and ultimately joining to trunk river Damodar. The average stream length in the region varies from 15 to 110 km with average basin area from 10 to 150 km^2 which shows in the Fig. 7.

Beyond the coal-mining area, geological formations of the Dhanbad district are composed of crystalline rocks of Precambrian. Ranchi area is underlain by the Chotanagpur, granite, and gneissic complex suit of rocks of Proterozoic age. These are mainly meta-sedimentary rocks intercalated with granitic rocks and the rocks of the chotanagpur pleateau around Ranchi. Kishoreganj and Ranchi Pahari consist of Khondalites. The major parts of the Bokaro district are comprised with granite and granite gneiss rock of Archean age and quartzites, mica schists, and phyllites are also found. The oldest rock of the area is unclassified meta-sedimentaries, which comprise quartzite and quartz schists. Three-fourths of the area is occupied by rocks of Chotanagpur, granite, gneiss, coal, shale, and sandstone deposits are found in parts of Bermo and Gomia blocks of the Bokaro district.

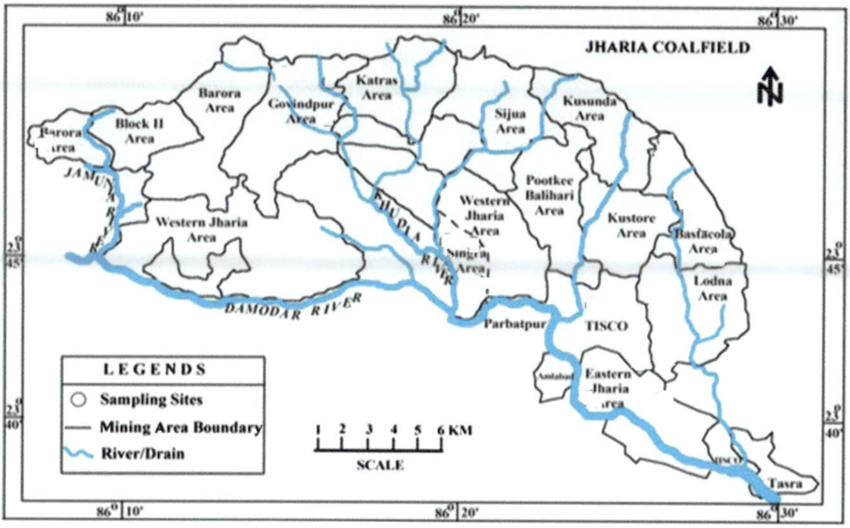

Fig. 7 Drainage map of the Jharia coalfield

3.2 *Climate and Rainfall*

The meteorological parameters depict the weather condition and the climate of the area as dry humid and sub-tropical. It is characterized by hot and dry summer from March to May, rainy season from June to October, and cold winter spreading from November to February. The study area receives rain both during the southwest and northeast monsoons. The rainfall pattern of the study area was observed in decreasing trend in the mining areas. The average annual rainfall of the mining area and the industrial areas is shown in the Tables 1 and 2.

3.3 *Drainage Pattern in the Coal-Mining District of the Jharkhand State*

The drainage of the district belongs to Damodar sub-basin. All the rivers that originate or flow through the district have an easterly or south easterly directions. The Damodar River is the most important river with an easterly course for about 125 km. Streams such as Jamunia, Katri, and Pusai are originating from northern hills of Parasnath and Tundi areas. These are flowing from N–S to NNW–SSE and meeting the Damodar River. In Ranchi district, the drainage system is the

Table 1 Average annual rainfall in the mining areas of Jharkhand State

Year	Annual rainfall (mm) East Bokaro coalfield	Annual rainfall (mm) West Bokaro coalfield	Annual rainfall (mm) Jharia Coalfield
2000	1428.6	1174	1390
2001	2543.8	1623	2094
2002	1737.6	1004	2311
2003	2123.6	1488	1731
2004	732.8	1548	1370
2005	927	1300	1678.5
2006	1362.2	1990	1516
2007	1683.9	1898	1823
2008	1450.4	1576	921
2009	1095.1	1390	1150
2010	1057.3	1016	1099
2011	1834	1834	1400
2012	1498.02	1481	1490

Table 2 Average annual rainfall in the coal-mining adjacent township areas of the Jharkhand State

Year	Annual rainfall (mm) Dhanbad district	Annual rainfall (mm) Ranchi district	Annual rainfall (mm) Bokaro district
2000	938.99	1189	938.1
2001	954.63	1253	1148.4
2002	1154.45	1282	997.5
2003	1752.56	1580	992.3
2004	1348.28	1200	767
2005	1726.8	879	1201.9
2006	1551.1	1468	1458.3
2007	1574.8	1378	1375.3
2008	1193.5	1375	926.4
2009	1145	1128	706.4
2010	879.9	806	1385.7
2011	1316	1578	1096.5
2012	1140.4	1342	1411.1

homogenous distribution of granitic rocks in the west has led the formation of dendritic drainage pattern. All these rivulets acts as tributaries to the Subarnarekha, which has easterly trend up to Hatia, and then, it takes north-easterly trend. Bokaro district shows a major drainage system. The major tributary of the Damodar are Konar and Jamuniya. The minor tributaries of the Damodar River are Isri, Gobai, Tasharkhan, Kadwa, Khanju, etc. The drainage system is mainly confined to weak zones, viz. joints and fractures.

4 Results and Discussion

4.1 Groundwater Level Fluctuation Monitoring in Coal-Mining District of Jharkhand State

During May, the deepest water level (pre-monsoon) was collected while another set of water level was collected in December considered as post-monsoon. The water level data was plotted in respect of depth from surface to get the water level depth map representing the shallow aquifer.

Sixty-five dug wells were measured from Dhanbad district from selected wells. The water level varies between 2.1–11.6 mbgl in the pre-monsoon and 1.3–7.5 mbgl during the post-monsoon. The average water table of the Dhanbad district during the pre- and post-monsoon is 5.21 and 3.40 mbgl, respectively (Table 3). The overall average water table fluctuation for the area is found to be ranged 0.1–4.0 mbgl, and the annual average water level fluctuation of the area is 0.68 mbgl. Depth to water level map (Fig. 8) reveals higher WLF up to 4 mbgl is observed in the southwest, west, and in patches in northern parts. The combined effect of terrain, soil, geomorphology, structural features, and mining impacts the water levels and its fluctuation (Chandra et al. 2015). The lower fluctuation is reported in the north, east, and southeastern parts area. The reason for this is combined impact of hydrogeology of the area impacted by two prominent reservoirs Maithon and Panchet. They are located on Rivers Barakar and Damodar flowing along the northern and southern boundaries. Thus, the west, southwest, and some northern parts experiences acute scarcity owing to less groundwater availability; on the other hand, the eastern region has plenty of resource available below surface.

Totally, 69 dug wells have been monitored seasonally (pre- and post-monsoon season) in Ranchi urban area for study area in 2013. Depth of well in study area varies from 17.66 to 3.32 mbgl and its diameter range from 6.46 to 1 m. Depth to water level registered wide variation throughout Ranchi urban area. The Maximum water level has been recorded as 13 mbgl in front of office of Prabhat khaber, whereas minimum water table has been recorded as 2 mbgl at Kanke and right side Kanke Dam during the pre-monsoon period (Table 3). However, the maximum water level has been recorded as 7 mbgl at Basant Bihar and minimum water table has been recorded as 0.2 mbgl at Kanke during the post-monsoon season. Water table fluctuations in 2013 vary between 0.07 and 9.98 mbgl (Table 3). The maximum water level fluctuation is 9.97 at Hetu location and minimum water level fluctuation as 0.07 at Bariatu has been recorded in the Ranchi urban area (Table 3). Water level has been found low at some location like Kanke because higher elevation and consumption of water are also at this location and higher water level has been found higher at some site due to lower elevation and excessive use of water for different purpose such as domestic, agricultural, and other purposes. During all the season, the well is continuously recharge by the surface water which is being discharge mainly from the Kanke Dam, Dhurwa Dam, and Rukka Dam and rivers.

Table 3 Water level fluctuation in the pre- and post-monsoon seasons in the coal-mining adjacent township areas of the Jharkhand State

	Dhanbad district			Bokaro district			Ranchi district		
	Pre-monsoon	Post-monsoon	WLF	Pre-monsoon	Post-monsoon	WLF	Pre-monsoon	Post-monsoon	WLF
Min.	2.1	1.3	0.21	1.92	1.66	0.1	12.5	8.47	8.76
Max.	11.6	7.5	4.9	11.95	7.4	6.1	1.47	0.35	0.07
Mean	5.21	3.40	1.79	6.29	3.92	2.55	6.77	3.48	3.34

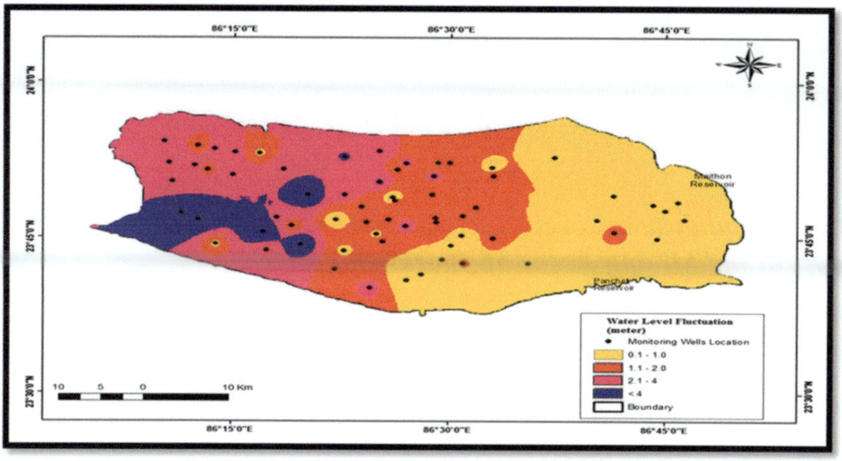

Fig. 8 Contour map of WLF of the Dhanbad district in year 2013

In Bokaro district, 30 dug wells were monitored seasonally in the pre- and post-monsoon season; it observed that water level varies from 1.92 to 11.95 mbgl in the pre-monsoon season with the average water level of 6.29 mbgl while, in the post-monsoon season it ranged from 1.66 to 7.4 mbgl with the average water level of 3.92 mbgl, respectively (Table 3). The overall average water table fluctuation for the area is found to be ranged 0.2–6.1 mbgl, and the annual average water level fluctuation of the area is 2.55 mbgl. In the area, the contour map (Fig. 9) reveals

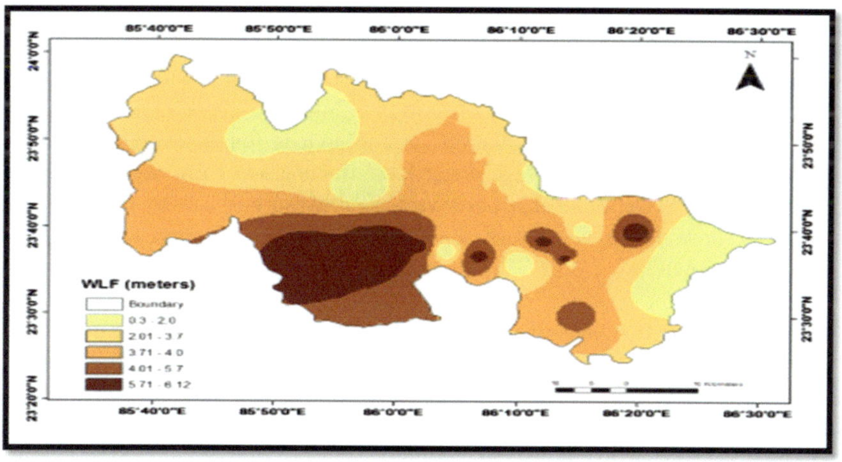

Fig. 9 Contour map of WLF of the Bokaro district in year 2013

that larger WLF, up to 6.1 m, has been observed in Peterwar block in the southwest, western part (Chandankiyari block), and also patches in the central part. Spatially, some parts in the study region are sensitive to WLF due to formation characteristics and various anthropogenic interventions.

4.2 Groundwater Level Fluctuation Monitoring in the Mining Areas of Jharkhand State

In total, 32 wells were established for monitoring in East Bokaro coalfield to study the water levels during the pre-monsoon and post-monsoon seasons. In the study area, the seasonal variation in water level is provided in Table 4. Water table in thirty-two monitored wells in study area varies between 1.64–11.2 mbgl in the pre-monsoon and 1.36–10.8 mbgl during the post-monsoon, respectively (Table 4). The average water table in East Bokaro coalfield during the pre-monsoon and post-monsoon is 4.92 and 4.24 mbgl, respectively. The average fluctuation of water levels is found to be ranged 0.15–2.87 mbgl, and the annual average water level fluctuation of the area is 0.68 mbgl (Fig. 10). The drainage pattern and water level monitoring indicate in general the groundwater flow direction toward the north and northeast in the study area. The groundwater regime largely follows the topographic features. Being a metamorphic terrain with prominent shear zone and other structural features, aquifer systems are largely fractured controlled within the study area comprising the stratiform ore bodies.

In West Bokaro coalfield area, totally 33 wells were measured during the pre- and post-monsoon seasons. In the study area, the seasonal variation in water level is provided in Table 4. Water table in thirty-three monitored wells in study area varies between 1.8–16.12 mbgl in the pre-monsoon and 0.3–10.7 mbgl during the post-monsoon, respectively (Table 4). The average water table in West Bokaro coalfield during the pre- and post-monsoon is 6.2 and 3.0 mbgl, respectively. The overall average water table fluctuation for the area is found to be ranged 1.5–5.42 mbgl, and the annual average water level fluctuation of the area is 3.2 mbgl (Fig. 11). It has been found that northwest, southwest, and central region of the study area show higher water level fluctuation, whereas northeast and southeast region have shown lower WLF. The elevation, slope, and geology of the study area are more important factors that affect the groundwater fluctuation and are also affected by mining activities and related industries (Tiwari et al. 2015).

A total forty-five (45 Nos.) shallow wells were selected for the monitoring of groundwater level from different mining areas of Jharia coalfield. The depth to water level shows in the pre-monsoon is 3.05–11.10 mbgl, and in the post-monsoon, it shows 1.4–7.5 mbgl. The water level fluctuation shows is 1.29–6.9 mbgl (Fig. 12). The maximum water level fluctuation observed in Bastacolla

Table 4 Water level fluctuation in the pre- and post-monsoon seasons in the coal-mining areas of Jharkhand State

	East Bokaro coalfield			West Bokaro coalfield			Jharia coalfield		
	Pre-monsoon	Post-monsoon	WLF	Pre-monsoon	Post-monsoon	WLF	Pre-monsoon	Post-monsoon	WLF
Min.	1.64	1.36	0.15	1.8	0.3	1.5	3.05	1.4	1.29
Max.	11.20	10.80	2.87	16.12	10.7	5.42	11.10	7.5	6.9
Mean	4.92	4.24	0.68	6.2	3.0	3.2	4.12	3.10	2.60

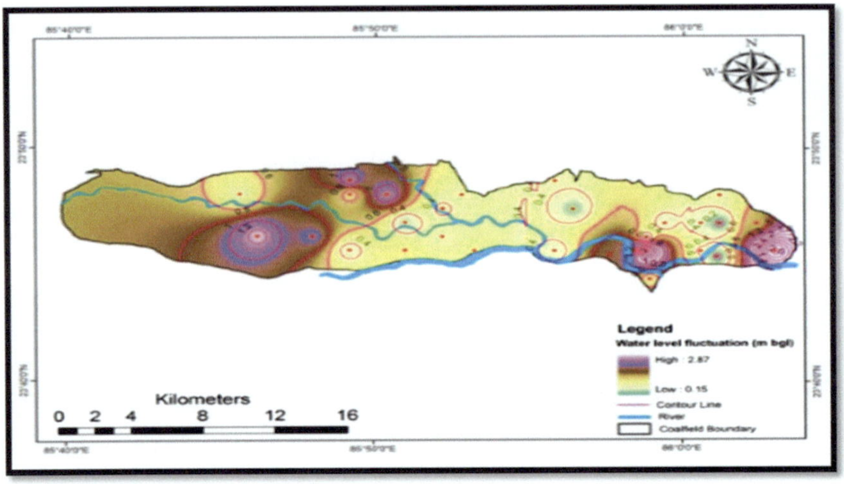

Fig. 10 Contour map of WLF of the East Bokaro coalfield in year 2013

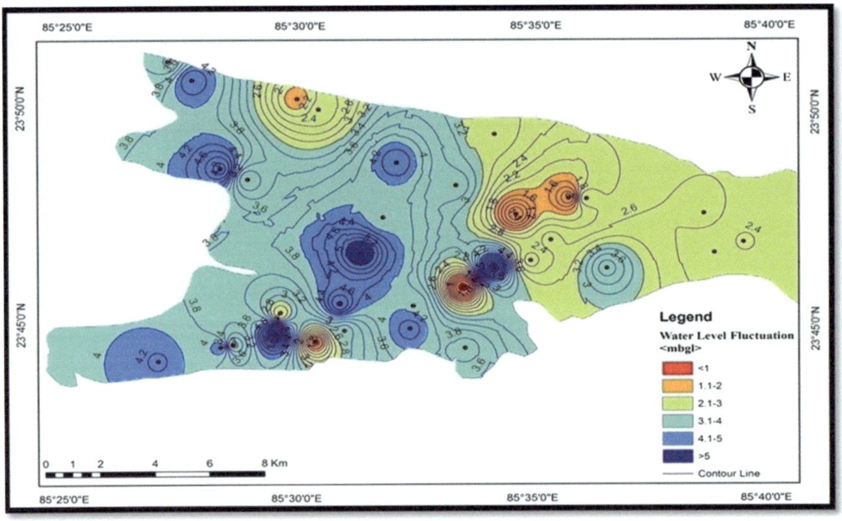

Fig. 11 Contour map of WLF of the West Bokaro coalfield in year 2013

area and Londna area due to some lithological features which affect groundwater fluctuation. The good recharge of groundwater shows in the eastern Jharia area which shows less groundwater fluctuation. The Damodar River is the main sources of recharge of the groundwater. The groundwater is recharged through by cracks, fissure, and planes discontinuity. Finally, it can be stated that groundwater availability is limited in south, west, and patches in the northern part.

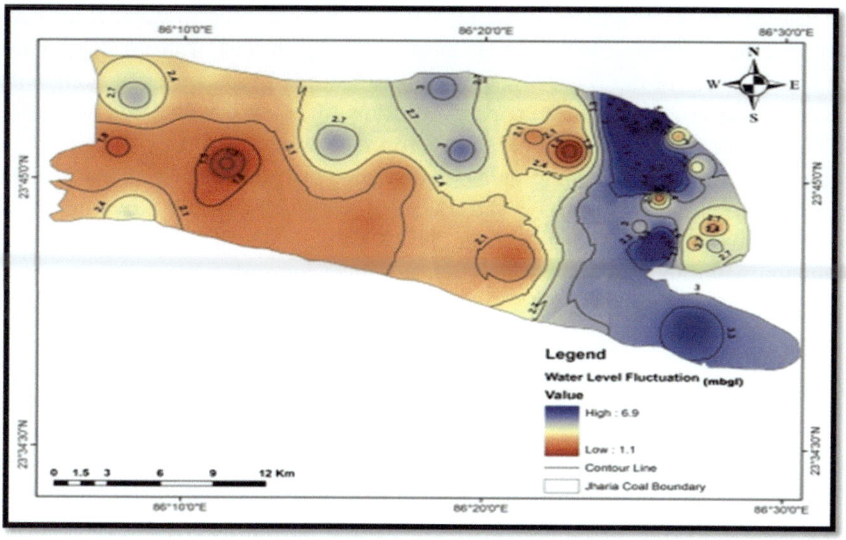

Fig. 12 Contour map of WLF of the Jharia coalfield in year 2013

5 Statistical Analysis of Water Level Fluctuation

The *t*-test is a statistical hypothesis test in which the test statistic follows a Student's *t*-distribution under the null hypothesis. It can be used to determine if two sets of data are significantly different from each other, and is most commonly applied when the test statistic would follow a normal distribution if the value of a scaling term in

Table 5 Paired *t*-tests of the water level fluctuation behavior in the mining and in the Coal-mining adjacent township areas of the Jharkhand State

Locations	*N*	Mean	Std. deviation	Std. error mean	*t*	df	Mean difference	95% confidence interval of the difference		
								Lower limit	Upper limit	
Bokaro district	30	2.51	1.77	0.32	7.77	29	2.51	1.85	3.17	Near coal-mining areas
Ranchi district	69	4.57	2.44	0.29	15.53	68	4.57	3.99	5.16	
Dhanbad district	64	1.73	1.11	0.13	12.41	63	1.73	1.45	2.01	
West Bokaro coalfield	33	3.26	1.26	0.22	14.82	32	3.26	2.81	3.71	In coal-mining areas
East Bokaro coalfield	32	0.67	0.63	0.11	6.03	31	0.67	0.44	0.90	
Jharia coalfield	45	2.60	1.05	0.15	16.61	44	2.60	2.28	2.92	

Note Where *N* is the number of observed well locations

the test statistic were known. A paired sample *t*-test was used to water level fluctuation data, comparing the different locations for each area. Table 5 depicts the mean differences of the water level fluctuation data for all the six areas along with the corresponding *t* values, degrees of freedoms (df), and *p* values for two-tailed paired sample *t*-tests. Water level fluctuation data were statistically higher in the mining areas than near-mining areas, and it shows significant variations in their water level fluctuation (Tables 3 and 4). So, it is also statistically prove that high water level fluctuation in these areas due to huge mining activities and overexploitation with lithology, formation characteristics, and soil.

6 Conclusion

In the area of geology, preciously the hydrogeology governs the storage and flow of groundwater. Groundwater occurs under unconfined state within the weathered zone. They are observed atop the crystalline as well as sedimentary formations. On the other hand in the deeper fractured rocks, it occurs under semi-confined to locally confined conditions. The water level fluctuation behavior during the pre-monsoon and post-monsoon seasons varies from place to place in the study areas. Spatial analysis through GIS made a comparative analysis reveals that various parts in the study area are sensitive to WLF because of various reasons. They are formation characteristics, topography soil, geological structures, etc. Further, water level behavior is greatly affected by mining activities and huge extraction for irrigation and domestic uses. The mining and related activities have marginally affected WLF of the study region.

The large amount of groundwater is being pumped out to sustain the mining operation; there is a need to restore the water potential in and around mining areas. As this quantity of water is being allowed to drain without any planned conservation strategies, there is a critical loss of mine water. Following approaches on water harvesting methods are recommended to be implemented to conserve water in mining rural areas. Storing of pumped out mine water in suitable places. Rainwater harvesting in open areas and rooftop rainwater harvesting should be done in the study areas.

Acknowledgements The authors are grateful to Professor D.C. Panigarhi, Director, Indian School of Mines for providing research facilities and also thankful to Head of the Department of ESE for his kind support. We also thank to laboratory members for their constant support and encouragement.

References

Bhuiyan C (2010) Hydrogeological factors: their association and relationship with seasonal water-table fluctuation in the composite hardrock Aravalli terrain, India. J Environ Earth Sci 60:733–748

Chandra S, Singh PK, Tiwari AK, Panigrahy B, Kumar A (2015) Evaluation of hydrogeological factor and their relationship with seasonal water table fluctuation in Dhanbad district, Jharkhand, India. ISH J Hydraul Eng 21(2):193–206

Guhey R, Mukherjee A (2012) Impact analysis of rain water harvesting and artificial recharge on karstic chandi limestone aquifer of Boriyakhurd micro watershed Raipur, Chhattisgarh. Int J Earth Sci Eng 5(2):276–281

Gupta A, Kunar S, Shekhar S, Mukherjee A (2016) Water and food security of the Central Indian Craton. IWRA India J 5(1):12–14

Krishan G, Rao MS, Loyal RS, Lohani AK, Tuli NK, Takshi KS, Kumar CP, Semwal P, Kumar S (2014) Groundwater level analyses of Punjab, India: a quantitative approach. Octa J Environ Res 2(3)

Krishan G, Lohani AK, Rao MS, Kumar CP, Takshi KS (2016) Groundwater fluctuation and trend in Amritsar, Punjab, India. In: Geostatistical and Geospatial Approaches for the Characterization of Natural Resources in the Environment, pp 183–187

Kunar S, Gupta A, Mukherjee A (2011) Rainwater harvesting and artificial recharge in urban areas of Raipur city: a case study from karstic terrain of central India. National Ground water Congress, New Delhi, CGWB, pp T4-1-5

Maggirwar BC, Umrikar BN (2011) Influence of various factors on the fluctuation of groundwater level in hard rock terrain and its importance in the assessment of groundwater. J Geol Min Res 3(11):305–317

Mukherjee A (2016) Automated generic categorization of hard rock aquifers: an innovative approach, Indian groundwater. 7: 26–35

Mukherjee A, Shekhar S (2014) Aquifer based water security of cities of Eastern India. Int J Eng Dev Res 2(1):73–81

Mukherjee A, Gupta A, Ray RK, Tewari D (2015 On line) Aquifer response to recharge—discharge phenomenon: inference from well hydrographs for genetic classification. J Appl Water Sci doi:10.1007/s13201-015-0293

Postel GA, Hitchman Matthew H (1999) A climatology of Rossby wave breaking along the subtropical tropopause. J Atmos Sci 56(3):359–373

Ray RK, Mukherjee A, Mukherjee R (2014) Estimation of specific yields of individual litho-units in a terrain with multiple litho-units: a water balance approach. J Geol Soc India 84(2):221–226

Saha D, Upadhyay S, Dhar YR, Singh R (2007) The aquifer system and evaluation of its hydraulic parameters in parts of South Ganga Plain, Bihar. J Geol Soc India 69(5):1031–1041

Saha D, Dhar YR, Vittala SS (2010) Delineation of groundwater development potential zones in parts of marginal Ganga Alluvial Plain in South Bihar, Eastern India. Environ Monit Assess 165(1–4):179–191

Sharma M, Rao MS, Rathore DS, Krishan G (2014) An integrated approach to augment the depleting ground water resource in bist-doab, region of Punjab, India. Int J Earth Sci Eng 7:27–38

Singh AK, Mahato MK, Neogi B, Mondal GC, Singh TB (2011) Hydrogeochemistry, elemental flux, and quality assessment of mine water in the Pootkee-Balihari Mining Area, Jharia Coalfield, India. Min Water Environ 30(3):197–207

Srivastava, Hari Mohan, Choi Junesang (2012) Zeta and q-zeta functions and associated series and integrals. Elsevier

Tabari H, Abghari H, Hosseinzadeh Talaee P (2012) Temporal trends and spatial characteristics of drought and rainfall in arid and semiarid regions of Iran. Hydrol Process 26(22):3351–3361

Tiwari AK, Singh AK (2014) Hydrogeochemical investigation and groundwater quality assessment of Pratapgarh district, UttarPradesh. J Geol Soc India 83(3):329–343

Tiwari AK, Singh PK, Chandra S, Ghosh A (2015) Assessment of groundwater level fluctuation by using remote sensing and GIS in West Bokaro coalfield, Jharkhand, India. ISH J Hydraul Eng 22(1):59–67

Tiwari AK, De Maio M, Singh PK, Singh AK (2016) Hydrogeochemical characterization and groundwater quality assessment in a coal mining area, India. Arabian J Geosci 9(3):1–17

Modelling the Potential Impact of Small-Scale Pumping Near Future Water Supply Wells in a Stressed Aquifer in South Western Bengal Basin on Groundwater Flow

P.K. Sikdar and Paulami Sahu

Abstract This paper highlights the importance and effectiveness of groundwater modelling to water managers to protect stressed aquifer systems and to address conflicting interests and requirements in their decision making process. The study area is located in south western Bengal Basin which is a part of the lower deltaic plain of the River Ganga and lies on the eastern side of the River Hugli, a distributory of River Ganga. The aquifer is sandwiched between two aquitards and is semi-confined in nature. Steady-state model results indicate that groundwater pumping in a proposed housing complex in Kolkata city located just outside the eastern margin of East Kolkata Wetlands (EKW) far outweigh the topographic and geologic controls of the unperturbed groundwater flow system and shifts recharge areas towards the pumping centres. A small part of the groundwater is also recharged from far areas in the south and southwest. Drawdowns due to present and future pumping in the project area extend over a large area which results in interference effect in wells outside the project area. Therefore, part of the city wells will experience an additional fall in the piezometric head. The simulation results indicate that the aquifer is already stressed and therefore it may not be wise to abstract groundwater further. The required water for the housing complex may be obtained from other sources. If other sources of water are not available then as the last resort groundwater may be developed. In such case further groundwater development by wells of high yield around the project area should be avoided. It is also imperative that steps should be taken to artificially recharge the aquifer by roof top rainwater harvesting.

Keywords Bengal Basin · Kolkata city · Housing complex · Groundwater flow modelling · Piezometric head and drawdown · Recharge areas

---

P.K. Sikdar (✉)
Department of Environment Management, Indian Institute of Social Welfare and Business Management, Kolkata 700073, India
e-mail: p_sikdar@hotmail.com

P. Sahu
School of Environment and Sustainable Development, Central University of Gujarat, Gandhinagar 382030, Gujarat, India

D. Saha et al. (eds.), *Clean and Sustainable Groundwater in India*,
Springer Hydrogeology, https://doi.org/10.1007/978-981-10-4552-3_18

1 Introduction

Surface water and groundwater is abundant in Bengal Basin of India and Bangladesh where heavy rainfall occurs for about 90–120 days every year. The surface water is commonly contaminated with microbial pathogens and anthropogenic pollutants (Michael and Voss 2009). Therefore, to provide drinking and irrigation water governments of the two countries resorted to extensive groundwater development post 1970 (Bhattacharya et al. 2004) for more than 150 million inhabitants and an estimated 10 million wells have been constructed. At present majority of the groundwater abstracted for drinking and irrigation purposes comes from depths <150 m. Much of these wells are concentrated in the cities and towns dotted all over the Bengal Basin, the megacity Kolkata being one of them.

Kolkata has experienced much development and growth from its birth in 1757 as a center of trade, commerce and industry. The growth was rather haphazard and unplanned. As a result, settlement was concentrated in the core area of the city which was, in course of time, oversaturated with population growth. Economic opportunities have attracted people from neighbouring states during the last 40–50 years. The population growth has been further increased due to migration of refugees from erstwhile East Bengal (now Bangladesh) after the partition of Bengal. Apart from this, there are millions of floating populations who crowd the vast industrial and commercial area daily, commuting from its hinterland.

In the early days of Kolkata, affluent citizens drew water from masonry dug wells tapping the shallow groundwater and the common people from ponds, tanks and streams that were often contaminated by both bacteria and saline water. Filtered water supply started since 1870 by setting up a water works at Palta about 30 km north of the city by tapping and purifying the water of River Hugli. As the demand for water rose sharply with the population growth, water works were set up at Garden Reach in 1982, Jorabagan in 2011 and Dhapa in 2014. The surface water supply was unable to cater the needs of growing population and hence heavy-duty tube wells were installed to supplement the surface water supply. The trend of increasing groundwater development in Kolkata can easily be understood if we look into the pattern of groundwater withdrawal from the middle of 1950s till date. In 1956 the groundwater withdrawal from the Kolkata Municipal Corporation (KMC) tube wells only was very modest to the tune of only 55 million liters per day (mld) while in 1989 it increased sharply to about 182 mld. At present about 305 mld is abstracted from KMC tube wells. Apart from this there are unaccounted number of heavy duty tube wells installed by private industries, housing estates and high rise apartments. There are also several thousands of small diameter tubewells owned by private household as well as by municipal agencies for public use. As a result of this huge groundwater development over the last sixty years the piezometric surface in Kolkata has lowered significantly and at places by over 10 m (Sahu et al. 2016). Similar lowering of water level has been reported from the urban areas located on the Gangetic Plains and heavily dependent on aquifers for domestic water supply (Mukherjee and Shekhar 2014; Saha et al. 2013).

At present, Kolkata is experiencing a mushroom growth of housing complexes. These complexes have their own motor-fitted wells tapping the stressed aquifer to supply water to the residents of the housing complexes. One such housing complex is coming up in the eastern fringe of East Kolkata Wetlands (EKW). A detailed hydrogeological study (SWRE 2011) in the project area has recommended that the total groundwater withdrawal should not exceed 1071 m^3/day and the maximum withdrawal from a single tubewell should not exceed 65 m^3/h (SWRE 2011). Following the recommendation, the housing complex authority constructed a tube-well with pumping capacity of 50 m^3/h to cater the constructional, drinking and domestic purposes. The study also recommended that the pump should operate for 6 h a day. The project also envisages two more tubewells with pumping capacity of 65 m^3/h for 6 h a day to meet the demand of water for the residents. But the aquifer underlying the project area is already stressed and more groundwater abstraction may increase this stress further which would impact groundwater recharge, storage and discharge. Apart from this, excess groundwater withdrawal may lead to change in groundwater flow pattern, depreciation of groundwater quality, enhance the possibility of land subsidence and increase the cost of pumping. Therefore, it is necessary to understand the quality and quantity of groundwater below the project area.

Study Area

The study area (Fig. 1) is situated in the eastern part of Kolkata Municipal Corporation area and is bounded by North latitude 22° 30′ 09″ to 22° 32′ 07″ and East longitudes 88° 23′ 18″ to 88° 25′ 14″ and extends from Garfa Bus stand in the southwest to Chowbagha High School in the northeast and Hossainpur in the southeast to Majdurpara VIP Bridge in the northwest. A housing complex is being built in the eastern fringe of East Kolkata Wetlands (EKW) (Fig. 1).

The area is a part of the lower deltaic plain of the River Ganga and lies on the eastern side of the River Hugli, a distributory of River Ganga. A hot summer with nor'westers, a rainy season from June to October, a pleasant winter and a short spring characterizes the climate of the area. The highest temperature is 42 °C and the lowest is 10 °C. The average annual rainfall is about 1650 mm, out of which 80% takes place during the rainy season.

The sub-surface geology of the area consists of a sequence of clay, silty clay, sand and sand mixed with occasional gravel of Quaternary age deposited by the River Hugli. The aquifer is a confined one with an upper 40 m thick confining aquitard of clay and silty clay with peaty intercalations (Sikdar 2000; Sahu and Sikdar 2011). The vertical continuity of this layer is broken by the presence of isolated thin layers of fine sand at some locations. At places the aquitard has been scoured by past channels with the deposition of fine sands theses channels were carrying. This confining bed is underlain by a 300 m aquifer comprising sands with occasional gravels and thin clay lenses. The well-screens are placed in the aquifer at depths ranging between 80 and 160 m (Sikdar 2000). Below the aquifer a thick sequence of clay has been encountered in boreholes which are of Pliocene age (Chaterji et al. 1959; Sikdar 2000).

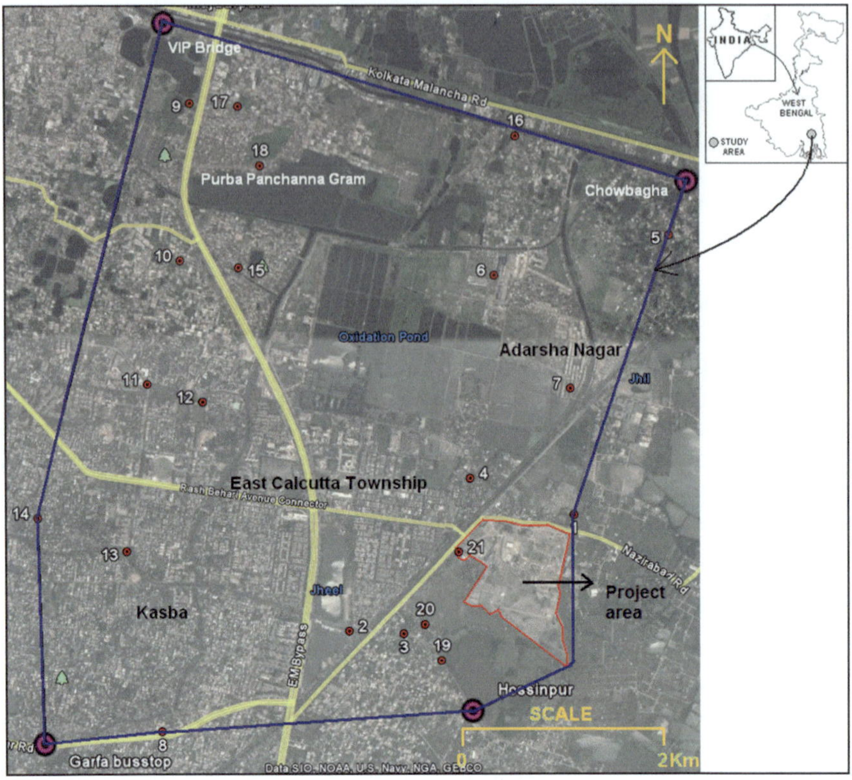

Fig. 1 Location of the study area *Red dots* indicate water level monitoring stations

Lithological logs of two boreholes drilled in the project area (SWRE 2011) indicate the presence of an upper clay/sandy clay aquitard up to 14 m depth followed by fine to medium sand (Fig. 2a, b). This aquitard holds groundwater in the aquifer below under high pressure.

During the post-monsoon period of 1956 the piezometric surface beneath the city had a gentle southerly slope. The piezometric heads were about 0.5–1 m above the mean sea level in the northern part of the city and they progressively fell below the mean sea level towards the south (Chaterji et al. 1964). Heavy groundwater pumping resulted in the formation of a groundwater trough in the south-central part of the city in the pre-monsoon period of 1985 with the core of the trough defined by the −10.5 m contour (Sikdar 2001). This trough persisted in the post-monsoon period of 1985 with the piezometric surface rising by about 2 m. Similar nature of the piezometric surface was observed in the post-monsoon period of 1993, in the pre-monsoon period of 1994 (Sikdar 2001), in 2005 (Sahu and Sikdar 2008; Sikdar and Sahu 2009) and at present (Sahu et al. 2016). The changes in the configuration of the piezometric surface and consequent groundwater flow direction are due to excess groundwater withdrawal of >70 million litres per day in this area (Sikdar

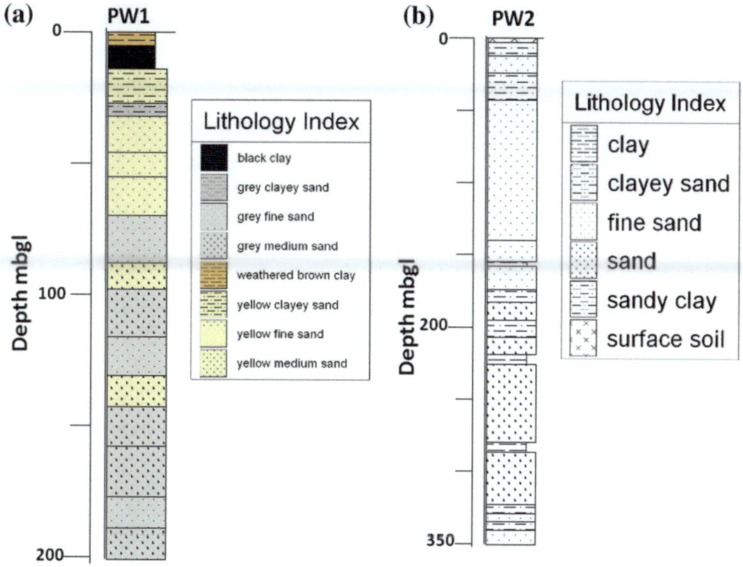

Fig. 2 Lithological log of boreholes in the project area (*Source* SWRE 2011)

2000). The average transmissivity, storage coefficient of the aquifer below the project area and radius of influence of the pumping well are 1491 m^2/day, 6.4×10^{-3} and 682 m respectively (SWRE 2011). The average storage coefficient of the aquifer below the project area suggests that the aquifer is in a semi-confined condition.

2 Objectives

The objectives of the present study are to explain the principle hydrogeological controls on groundwater flow system and then to understand the consequences of small-scale pumping of groundwater below the housing complex on the regional and local groundwater flow patterns by numerical modelling using MODFLOW.

The proposed focus of groundwater modelling arose from the development that has led to a number of groundwater supply and quality concerns, which include: (i) effect of current and future heavy groundwater withdrawal on the head and drawdown in and around the project area, (ii) effect of future heavy groundwater withdrawal on local and regional groundwater flow patterns, and (iii) possible contamination of groundwater of aquifer in the near future.

Therefore, the secondary objectives of the modelling effort are (i) determining the head distribution pattern under past, current and future-development scenarios in the project area, (ii) determining the drawdown under current and future-development scenarios, and (iii) delineating the recharge areas of the present and proposed wells under pre, past (pre-project), current and future-development scenarios.

3 Methods

Field work was carried out in 2012 to locate the tubewells fitted with motor-pump and hand pump to determine the total discharge of groundwater within the study area and to establish network stations (wells fitted with hand pump) for measuring water level and determining arsenic (As) concentration in well water. Twenty one network stations, were selected and their geographic position was recorded by a hand-held GPS and referenced to WGS 84 as datum. Water levels were measured to establish hydraulic heads and the directions of groundwater flow. Water level of the network stations were measured by using a hydro-tape. The surface topography was generated from the elevation data of Shuttle Radar Topography Mission (SRTM) with a spatial and altitude resolution of 90 and 1 m respectively (EROS 2002). The arsenic (As) concentrations of groundwater of the selected eighteen network stations were measured by a Wagtech Digital Arsenator in the field.

3.1 Groundwater Flow Calculation

The piezometric surface elevation map was used to calculate the quantity of groundwater flowing through the project area using the Darcy's equation:

$$Q = TiL$$

where,

Q Rate of flow of groundwater
T Transmissivity
i Hydraulic gradient
L Length of the cross-section

3.2 Groundwater Flow Modeling and Particle Tracking

In a flat deltaic terrain it is hard to evaluate groundwater flow systems on a local-scale because of the lack of appropriate boundary conditions. Therefore, to understand the flow behaviour in such a case it becomes essential to consider the flow system on a regional-scale with proper boundary conditions at the edges of the model. The flow system in the south Bengal Basin is difficult to conceptualize even without any pumping, and hence in this modelling exercise the parameters having major controls on the flow behaviour have been included.

The conceptual model of hydrostratigraphy is generalized based on observed lithological logs (Deshmukh et al. 1973), Vertical Electric Sounding (VES) data (Saha et al. 2002, 2006), and the geology of the area (Sikdar 2000; Sikdar and Sahu 2009). The sediments have been categorized as sand (aquifer) and silty clay/clayey silt (aquitard). The upper aquitard occurs up to ~30 m below ground surface, and the lower one starts from 340 m depth and continues down to the basement (Sahu et al. 2013), which occurs at variable depths. The composition of the unit conceptually classified as aquifer is not homogenous, because layers of silty clay and clayey silt of various dimensions are present. Most of these layers occur as small lenses and cannot be related among boreholes. But there are some aquitards with large spatial extent, but these do not create a laterally-extensive stack of aquifers that are fully separated hydraulically. The units of low hydraulic conductivity (K) are important because their spatial distribution reduces the ease of vertical groundwater flow. The top low-K unit is also not persistent. In places, it is incised by past channels and grades into silty clay and fine sand making easy passage for surface water to infiltrate to the shallow and deep portions of the aquifer. Sikdar and Sahu (2009) had reported high tritium in groundwater at a depth of 22 m (5.5 TU) and 92 m (4.5 TU) indicating mixing of waters of various depths.

Following the approach of Michael and Voss (2009) and Sahu et al. (2013), for the purpose of the present model development, the south Bengal Basin sediments in the model area are considered to be a single large, homogeneous and anisotropic system down to a consistent low-permeability unit, the Precambrian granite or shale or a thick Pliocene clay basement.

In the present study, USGS MODFLOW software (McDonald et al. 2000) with preprocessor Modflow GUI (Winston 2000 and ArgusONE™) was used to develop a three-dimensional numerical model of the south Bengal Basin aquifer system. To simulate groundwater flow through porous media, MODFLOW uses finite-difference method. Modeled flowpaths were tracked in the forward direction from locations of possible groundwater recharge area using MODPATH (Pollock 1994). The MODFLOW-GUI pre-processor (Winston 2000) based on ArgusONE™ commercial software was used to create the models.

The topography of the area being very flat it is difficult to assign proper boundaries. Therefore, the model of Sahu et al. (2013) is used as the base model for the present work. The model area extends to the boundaries of the aquifer system, which, because of the layered but connected nature, results in a very large model area. The model area extends from the River Ganges and River Padma in the north to the Bay of Bengal on the south and from the outcrop of Precambrian crystalline bedrock of the Chotanagpur Plateau on the west, to the River Meghna on the east in Bangladesh (Fig. 3). The three-dimensional (3D) model consists of thirty-two vertical layers of finite-difference cells with lateral cell sizes that vary from 500 m × 500 m to 5000 m × 5000 m, with a total of 930,240 cells.

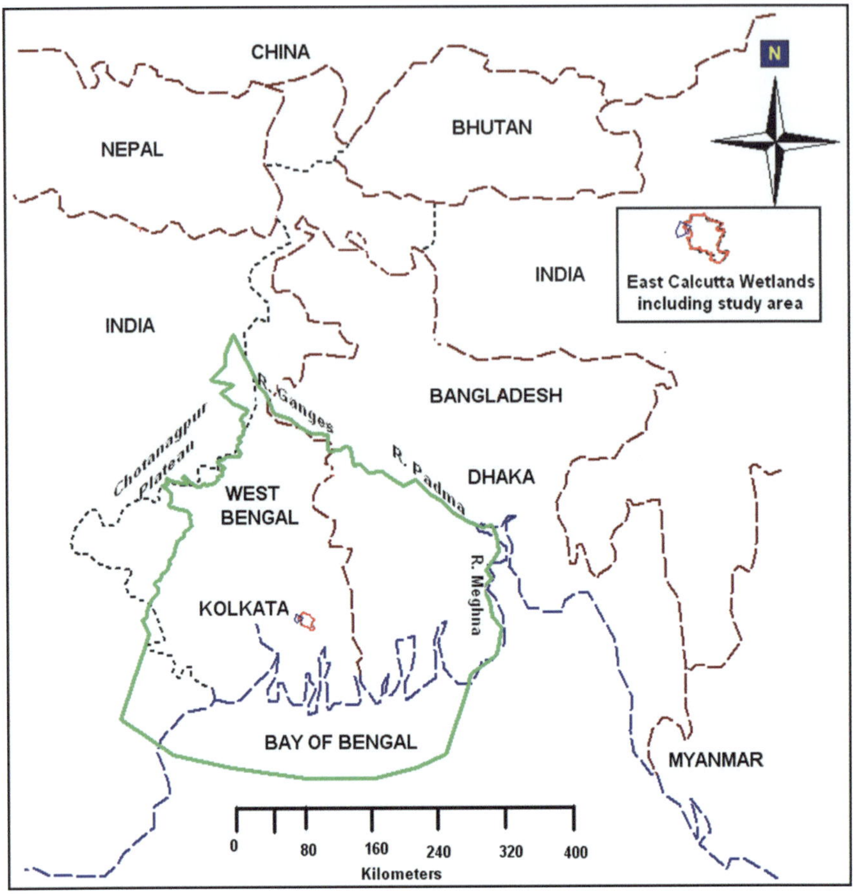

Fig. 3 Location of the model area. *Green line* indicates model boundary, *red line* indicates East Calcutta Wetlands (ECW) boundary, *small blue continuous line* indicates study area boundary, *blue dashed line* indicates coastline, *brown line* indicates international boundary and *black dashed line* indicates state boundary

MODPATH (Pollock 1994) particles were placed in the three wells within the project area at a depth interval of 123.5–153.5 m (strainer position of the wells) below ground surface. The particles follow advective flowpaths, that is, movement of the contaminants at the same speed as the average groundwater flow as predicted by MODPATH. The particles are tracked backward from pumping zones of the project area in different simulations to determine their recharge locations.

4 Results and Discussion

4.1 Groundwater Abstraction

In the study area groundwater is abstracted for drinking, domestic and industrial purposes (Table 1). For drinking/domestic purpose, groundwater is abstracted from wells fitted with motor-pumps and hand-pumps. The motor-pump fitted wells have been installed by the Kolkata Municipal Corporation (KMC) as well as by the housing and commercial complexes. The total groundwater abstraction by these wells is 11,143 m^3/day, which is about 89.4% of the total groundwater abstraction (Table 1). There are also about 96 hand-pump fitted wells catering to the drinking/domestic needs of the residents of the area. Out of these 96 wells, 33 wells are owned by the KMC and the rest are owned by individual households. The total discharge from hand-pump fitted wells is 457 m^3/day (assumptions: discharge of each well is 0.68 m^3/day, and pumping hours per day is 7). For industrial use groundwater is being abstracted through wells fitted motor-pump constructed within the Kasba Industrial Estate. The industrial use of groundwater is only about 6.9% (Table 1).

4.2 Field Piezometric Head

Water level below ground level of the study area ranges from 6.10 to 20.28 m and the piezometric head ranges from −1.8 to −16.5 m mean sea level (Table 2). The piezometric surface contour map is shown in Fig. 4. A perusal of this figure reveals three groundwater mounds around wells SW-5, SW-7 and SW-13 and three groundwater troughs around wells SW-11, SW-18 and SW-21. The groundwater

Table 1 Summary of groundwater abstraction for drinking/domestic and industrial purpose in the study area

Sl. No.	Purpose	Type of wells	Owner	Total abstraction (m^3/day)	% of total abstraction
1	Industrial	Motor-pump fitted	Kasba industrial estate	859	6.9
2	Drinking/domestic	Motor-pump fitted	Kolkata municipal corporation	6722	53.9
3	Drinking/domestic	Motor-pump fitted	Commercial and housing complexes	4421	35.5
4	Drinking/domestic	Hand-pump fitted	Kolkata municipal corporation and individual household	457	3.7
Total				12,459	100

Table 2 Piezometric surface data of the study area

Sl. No.	Network stations	Piezometric surface			
		Total depth (m)	Reduced level (m)	Below ground level (m)	Mean sea level
1	SW-1	85.34	3.0	13.28	−10.28
2	SW-2	85.34	4.0	14.48	−10.48
3	SW-3	91.44	4.0	14.40	−10.40
4	SW-4	85.34	5.1	11.65	−6.55
5	SW-5	85.34	4.8	6.60	−1.80
6	SW-6	91.44	2.7	12.10	−9.40
7	SW-7	85.34	1.1	6.10	−5.00
8	SW-8	85.34	2.1	9.76	−7.66
9	SW-9	91.44	3.8	15.72	−11.92
10	SW-10	85.34	5.8	15.08	−9.28
11	SW-11	91.44	2.2	15.70	−13.50
12	SW-12	91.44	5.7	16.21	−10.51
13	SW-13	85.34	5.0	11.50	−6.50
14	SW-14	30.48	6.3	15.60	−9.30
15	SW-15	85.34	4.4	12.88	−8.48
16	SW-16	85.34	3.6	15.45	−11.85
17	SW-17	91.44	4.9	15.43	−10.53
18	SW-18	79.25	0.7	13.70	−13.00
19	SW-19	91.44	3.4	17.70	−14.30
20	SW-20	91.44	4.3	14.95	−10.65
21	SW-21	91.44	3.7	20.28	−16.58

flow into the aquifer of the project area is controlled by the trough formed around well SW-21 and groundwater flows into this trough from all directions of which the majority being from nearby areas. A part of the flow comes from a distant location in the southwest (Fig. 4).

4.3 Groundwater Flow Rate Across the Project Area

The average hydraulic gradient in and around the project area is 9×10^{-3} (Table 3) and the average transmissivity of the aquifer is 1491 m^2/day (SWRE 2011). Using Darcy's equation, the groundwater flow across the maximum width AB (638.4 m) of the project area (Fig. 4) is 8567 m^3/day. Therefore, the anticipated future groundwater abstraction, that is, 780 m^3/day is only 9.1% of the total groundwater flowing through the aquifer below the project area. But it should be remembered

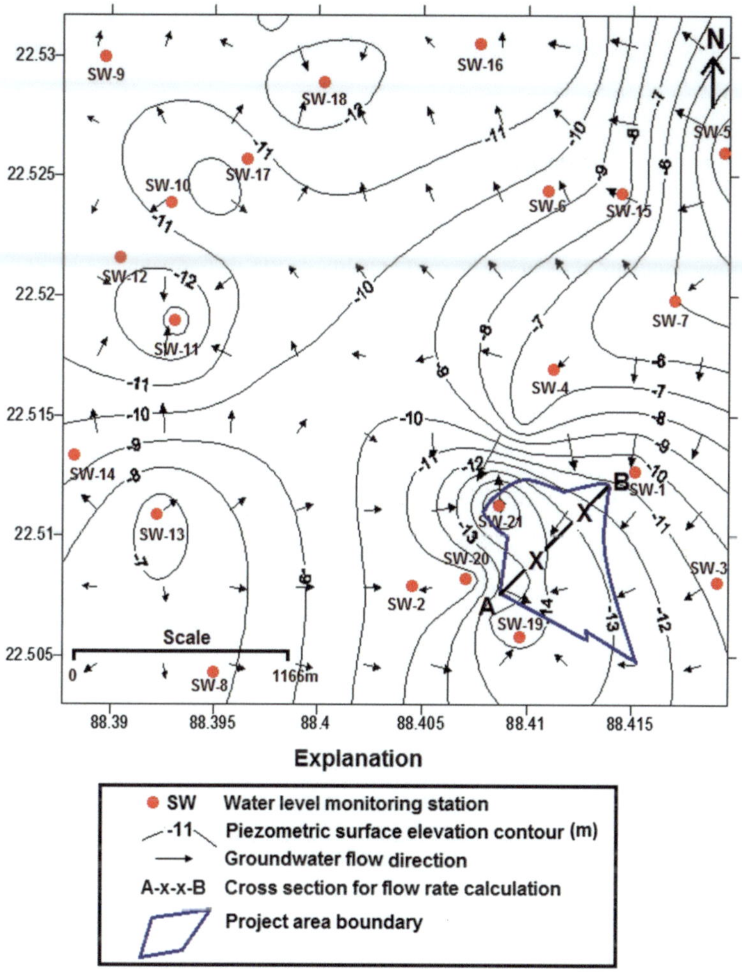

Fig. 4 Piezometric surface elevation contour map of the study area showing network stations and groundwater flow directions

Table 3 Hydraulic gradient in and around project area

Sl. No.	Head loss (h_L) (m)	Length over which head loss taken place (L) (m)	Hydraulic gradient (i)
1	1.0	45.6	0.02
2	1.0	68.4	0.01
3	1.0	319.2	0.003
4	1.0	273.6	0.003
5	1.0	91.2	0.01
Average			0.009

that the calculated total groundwater flow is occurring in an already stressed aquifer, and any further groundwater withdrawal should be done with caution.

4.4 Arsenic Concentration

The As concentration in groundwater in and around the study area is very low, the maximum concentration being 2 μg/L (Table 4). Therefore, the threat of arsenic contamination in the groundwater of project area is low.

4.5 Model Validation

The root mean square (RMS) error of model runs ranges from 0.67 to 31.02 m. The minimum RMS error (Table 5) among all the tested combinations was obtained for the combination, horizontal hydraulic conductivity (Kh) = 2.47×10^{-4} m/s, vertical hydraulic conductivity (Kv) = 1.65×10^{-7} m/s, and anisotropy (Kh/Kv) = 1500. For the sake of simplicity and acknowledging the approximate nature of the model, the 'base case' hydrogeologic parameters are set to: Kh = 2.4×10^{-4} m/s, Kv = 1.6×10^{-7} m/s (and therefore, Kh/Kv = 1500). The same value of large-scale anisotropy had been previously estimated for the aquifer

Table 4 Concentration of arsenic in groundwater in the study area

Sl. No.	Network station	As concentration (μg/L)
1	SW-1	1
2	SW-2	1
3	SW-3	2
4	SW-4	1
5	SW-5	2
6	SW-6	0
7	SW-7	0
8	SW-8	0
9	SW-9	1
10	SW-10	1
11	SW-11	1
12	SW-12	2
13	SW-13	0
14	SW-14	0
15	SW-15	1
16	SW-16	0
17	SW-17	0
18	SW-18	1

Table 5 Results of model fit with 'base case' parameter values

Sl. No.	Network station	Observed head at field with respect to MSL (m)	Computed hydraulic head with respect to MSL (m)	Residual (m)	(Residual)^2
		(A)	(B)	(A − B)	$(A - B)^2$
1	SW-1	−10.28	−10.12	−0.16	0.0256
2	SW-2	−10.48	−10.51	0.03	0.0009
3	SW-3	−10.40	−10.02	−0.38	0.1444
4	SW-4	−6.55	−6.17	−0.38	0.1444
5	SW-5	−1.80	−2.55	0.75	0.5625
6	SW-6	−9.40	−8.95	−0.45	0.2025
7	SW-7	−5.00	−5.42	0.42	0.1764
8	SW-8	−7.66	−7.81	0.15	0.0225
9	SW-9	−11.92	−11.75	−0.17	0.0289
10	SW-10	−9.28	−9.16	−0.12	0.0144
11	SW-11	−13.50	−14.03	0.53	0.2809
12	SW-12	−10.51	−10.27	−0.24	0.0576
13	SW-13	−6.50	−6.81	0.31	0.0961
14	SW-14	−9.30	−9.11	−0.19	0.0361
15	SW-15	−8.48	−8.12	−0.36	0.1296
16	SW-16	−11.85	−11.52	−0.33	0.1089
17	SW-17	−10.53	−10.22	−0.31	0.0961
18	SW-18	−13.00	−12.81	−0.19	0.0361
19	SW-19	−14.3	−13.23	−1.07	1.1449
20	SW-20	−10.6	−11.31	0.71	0.5041
21	SW-21	−16.5	−14.11	−2.39	5.7121
Sum					9.5250
Mean					0.4536
R.M.S. Error					0.6735

system (Sahu et al. 2013). The plots of the simulated head data using 'base case' hydrogeologic parameters and field head data closely follow the 1:1 line indicating a good agreement between the simulated heads and the observed heads (Fig. 5).

4.6 *Model Hydraulic Head and Drawdown*

The model generated head distribution pattern of the piezometric heads for the three development scenarios is more or less similar. Therefore, the head distribution of the current-development condition with only one tubewell functioning in the project area (Fig. 6) is given in this paper. The heads simulated under the three development scenarios show a large groundwater trough in the east-central part of the city.

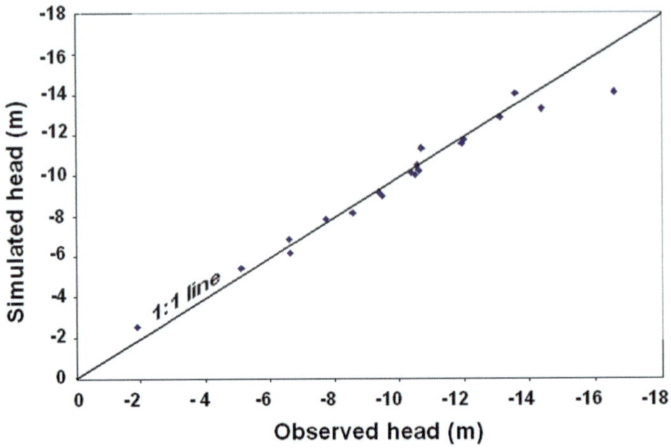

Fig. 5 Scatter plot showing the model simulated head versus field observed head

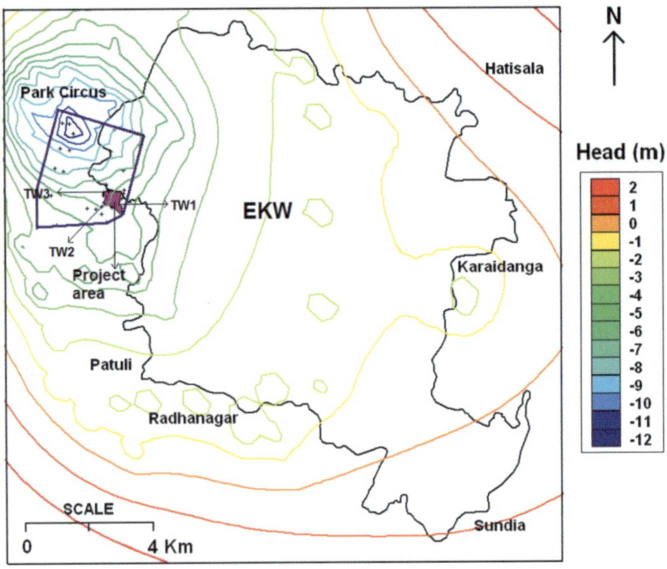

Fig. 6 Piezometric head distribution of the area in and around the study area at a depth range of 100–140 m under current-development condition

The piezometric surface contours are closed in nature and the groundwater trough which had formed in the east-central Kolkata in the mid-eighties still persists today. The core of the trough is located in Park Circus area and groundwater from all direction flows towards this trough. The deepest part of the trough in past (without wells in the project area), current (with one pumping well in the project area) and future development (with three pumping wells in the project area) scenarios is 12.48, 12.50 and 12.54 m below the mean sea level respectively. Therefore, increasing the pumping rate in the project area may lead to a fall in the piezometric head in the worst affected part of Kolkata city.

The simulated current drawdown with respect to the past-development condition (Fig. 7) varies between 0.02 and 0.18 m, the maximum being observed in the existing tubewell (TW1) in the project area (Table 6). The simulated future drawdown (with respect to the current-development condition) ranges from 0.03 to 0.32 m (Fig. 8), the maximum drawdown being observed in TW1 (Table 6). The two proposed tubewells, TW2 and TW3, in the project area show drawdown of 0.26 and 0.28 m respectively. The effect of pumping will be felt outside the project area (Figs. 7 and 8) which may result in interference effect in wells. Therefore, part of

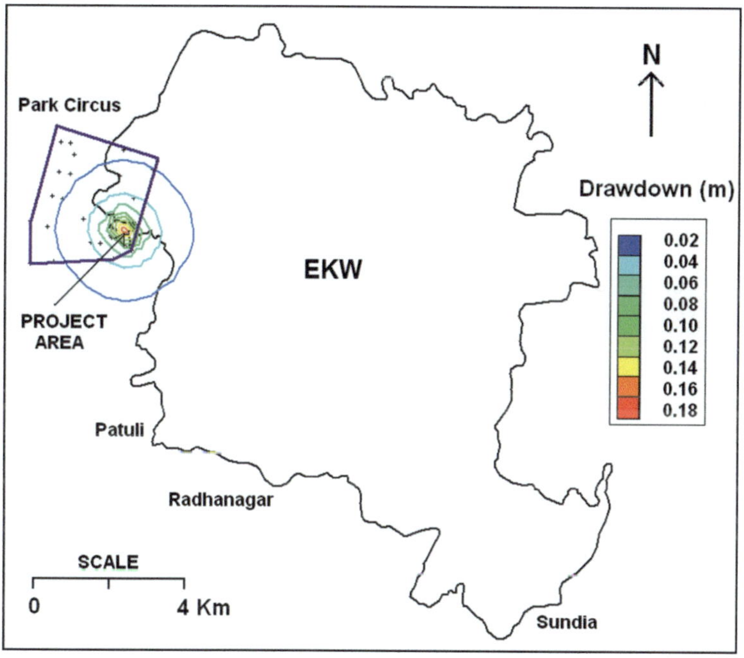

Fig. 7 Spatial distributions of drawdown (depth: 100–140 m) in and around the study area under current-development condition

Table 6 Model drawdown in present and proposed tubewells within the project area under different development scenarios

Development scenarios	Drawdown (m)		
	TW1	TW2	TW3
Past-development	–	–	–
Current-development	0.18	–	–
Future-development	0.32	0.26	0.28

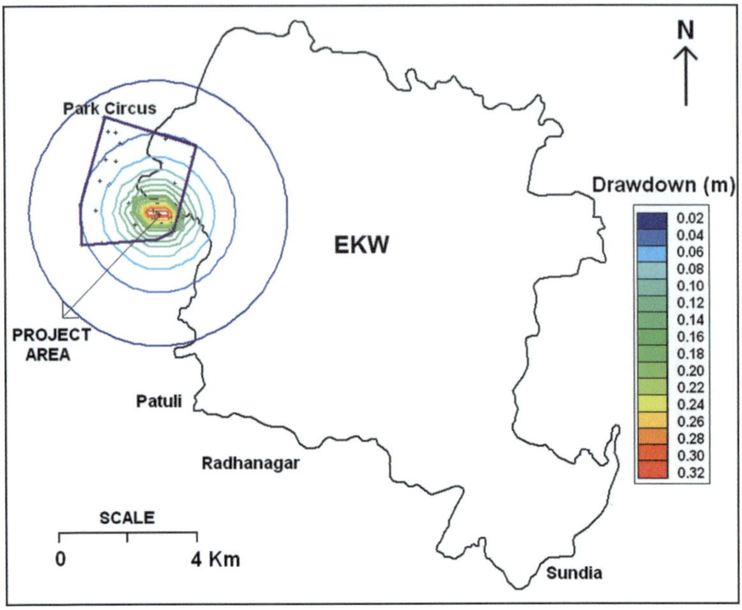

Fig. 8 Spatial distributions of drawdown (depth: 100–140 m) in and around the study area under future-development condition

the city wells may experience an additional fall in the piezometric head with pumping in the project area.

4.7 Recharge Areas

Modeled flowpaths are tracked in the backward direction from possible recharge areas using MODPATH (Pollock 1994). In the system unperturbed by pumping (pre-development condition), groundwater flow flows follows the topography of the land surface. Under this condition all the simulated pathlines of a hypothetical well in the project area originate from Bankra area in Howrah district (Fig. 9; Table 7). Introduction of pumping in Kolkata and surrounding areas (a region with low natural topographic gradients) have altered the flow paths. Therefore, before the first well was constructed in the project area, that is, during past-development or pre-project scenario, the groundwater flowpaths completely change and the hypothetical well in the project area is recharged by groundwater flowing mainly from EKW and areas around the project area outside EKW (Fig. 10; Table 7). Small amount of water also flows from south (Amtala) and south-west (Joka) (Table 7).

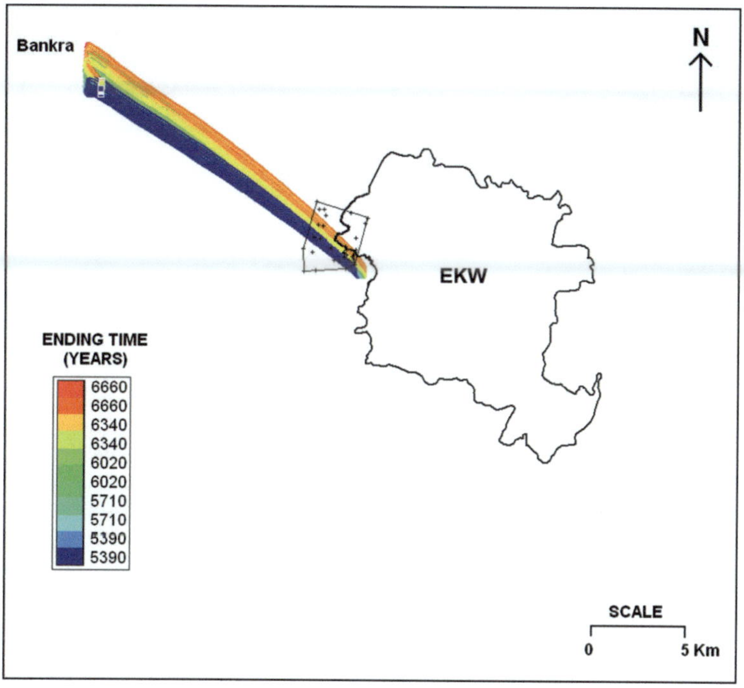

Fig. 9 Recharge areas of the wells of the proposed housing complex groundwater shown by recharge points of backwards-tracked particle pathlines under pre-development conditions

Table 7 Percentage of groundwater recharge from different locations in the present and future wells in the housing complex under different pumping scenarios

Pumping scenario	Groundwater recharge (%)				
	Bankra (Howrah)	Within EKW	In and around project area outside EKW	Near Amtala	Near Joka
Pre-development	100	0	0	0	0
Past-development/pre-project	0	65	29	0.5	5.5
Current pumping in project area (300 m^3/day)	0	45	34	5.5	15.5
Future pumping in project area (1071 m^3/day)	0	33	45	3	19

Increasing the pumping rate, the recharge areas remain more or less similar, but more groundwater flows from areas in and around the project area and less from EKW (Figs. 11 and 12; Table 7). More water will also flow from the southwest in

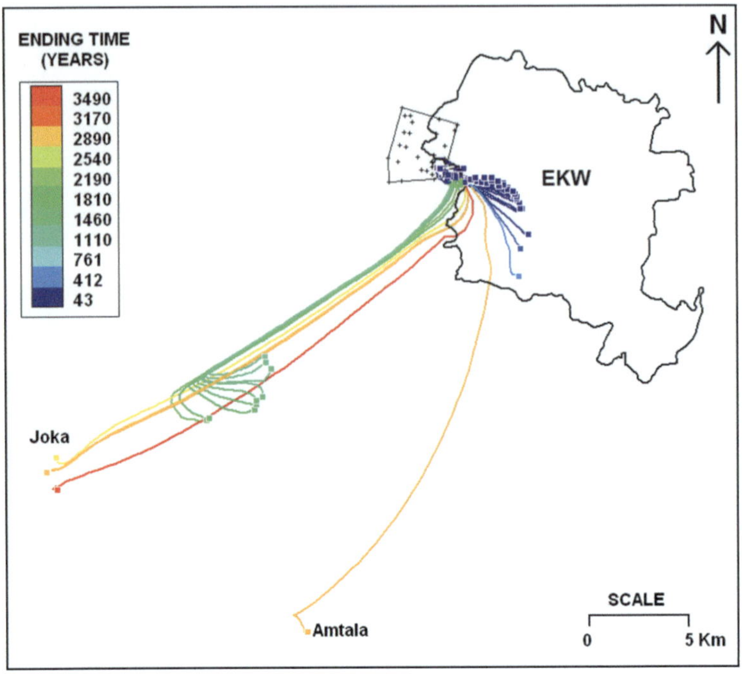

Fig. 10 Recharge areas of the wells of the proposed housing complex groundwater shown by recharge points of backwards-tracked particle pathlines under pre-project conditions

between Amtala and Joka. In the recharge area groundwater is As-free. Therefore, from quality point of view pumping in the project area may not have an adverse impact in the near future.

With introduction of pumping in the project area the horizontal path length (hpl) of the groundwater flow path varies between 0.4 and 24.4 km with an average length of 5.9 km (Table 8). The travel time of water varies between 55 and 3930 years with an average of 745 years (Table 8). If the existing tubewell in the housing complex was not present then the average hpl will decrease by 200 m but the average travel time remains the same (Table 8). In the future when two new wells will be operational, with each discharging at a rate of 65 m^3/h, most of the waters may be recharged from areas in and around the project area outside EKW (Fig. 11; Table 7). The mean hpl may increase by 200 m compared to the current-development condition but the average travel time of water may be drastically reduced to 391 years (Table 8).

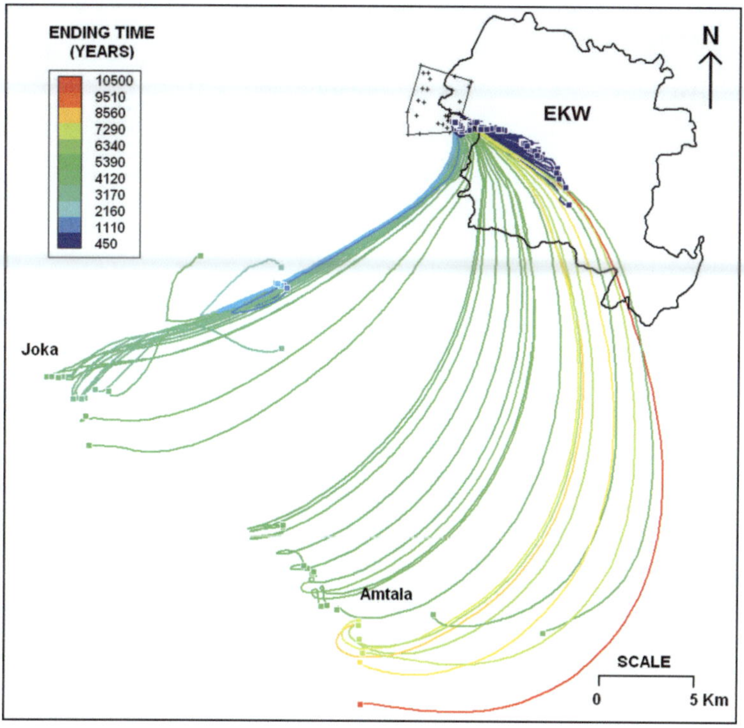

Fig. 11 Recharge areas of wells of the proposed housing complex with groundwater shown by recharge points of backwards-tracked particle pathlines under current-development conditions

4.8 Implications for Mitigation

The study reveals that the effect of a small-scale pumping in the Kolkata city aquifer will be felt outside the zone of pumping. The piezometric head may fall in the deepest part of the groundwater trough and hence a part of the city wells may experience an additional fall in the piezometric head. Therefore, it is better not to stress the aquifer further by groundwater development. The required water for the project area may be obtained from other sources. If other sources of water are not available then as the last resort groundwater may be developed. But after the wells are operational further groundwater development using wells of high yield in and around the project area should be avoided. Under such circumstances, it is absolutely necessary to artificially recharge the stressed aquifer by rooftop rainwater harvesting.

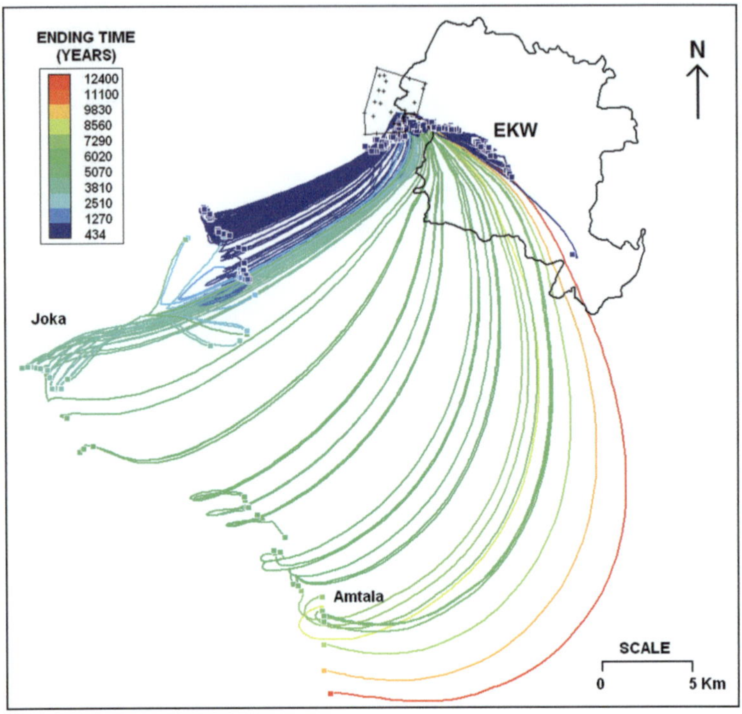

Fig. 12 Recharge areas of wells of the proposed housing complex with groundwater shown by recharge points of backwards-tracked particle pathlines under future-development conditions

Table 8 Horizontal path length (hpl) and travel time of water from screen depth of project area wells to their recharge locations under different pumping scenarios

Well field area	Min. hpl (km)	Max. hpl (km)	Mean hpl (km)	Min. ravel time (Year)	Max. travel time (Year)	Mean travel time (Year)
Past-development/ pre-project	0.4	24.4	5.9	56.4	3490	745
Current pumping in project area	0.4	24.5	6.1	55.5	10,500	745
Future pumping in project area	0.4	24.7	6.3	3.7	12,400	391

5 Conclusion

This paper highlights the importance and effectiveness of groundwater modelling to water managers to protect stressed aquifer systems and to address conflicting interests and requirements in their decision making process. The results of this modelling exercise can be used to communicate key messages to public and decision-makers during environmental impact assessment process, formulate policies and regulations of groundwater system, issue permits to develop new wells and well fields, develop enforcement actions and design monitoring and data collection systems. The paper also brings out the need to undertake a basin-wise national groundwater modelling programme for sustainable development and management of India's groundwater.

Acknowledgements The authors thank Dr. S.P. Sinha Ray, Emeritus President, Centre for Ground Water Studies for providing relevant data required to carry out the research work. Prof. A. Sarkar Director, Indian Institute of Social Welfare and Business Management (IISWBM) is also thaked for providing the infrastructure for the research work. PS conveys thanks to Prof. S.A. Bari, Vice Chancellor, Central University of Gujarat and Prof. M.H. Fulekar, Dean, School of Environment and Sustainable Development, Central University of Gujarat.

References

Bhattacharya P, Welch AH, Ahmed KM, Jacks G, Naidu R (2004) Arsenic in 1593 groundwater of sedimentary aquifers. Appl Geochem 19:163–167

Chaterji GC, Deshmukh DS, Sinha SC, Niyogi BN et al (1959) Some recent observations on the depth of sediment in the Ganga valley below Varanasi, Ganga valley. A symposium. ONGC, Dehra Dun

Chaterji GC, Biswas AB, Basu S, Niyogi BN et al (1964) Geology and groundwater resources of the greater Calcutta metropolitan area, West Bengal. Bull Geol Surv India B21:1–150

Deshmukh DS, Prasad KN, Niyogi BN, Biswas AB, Sinha BPC, Chatterjee GC et al (1973) Geology and groundwater resources of the alluvial areas of West Bengal. Bull Geol Surv India Ser B 34

EROS (2002) Shuttle radar topography mission (SRTM) elevation data set. National Aeronautics and Space Administration (NASA), German Aerospace Center (DLR), Italian Space Agency (ASI), From: the National Center for Earth Resources Observations and Science, Sioux Falls, U.S. Geological Survey

McDonald MG, Harbaugh AW, Banta ER, Hill MC (2000) MODFLOW-2000, the U.S. Geological Survey modular ground-water model-user guide to modularization concepts and the ground water flow process. U.S. Geological Survey Open-File Report 00-92. U.S. Geological Survey, Reston

Michael HA, Voss CI (2009) Estimation of regional-scale groundwater flow properties in the Bengal Basin of India and Bangladesh. J Hydrol 17(6):1329–1346

Mukherjee A, Shekhar S (2014) Aquifer based water security of cities of Eastern India. Int J Eng Dev Res **2**(1):73–81. ISSN: 2321-9939

Pollock DW (1994) User's guide for MODPATH/MODPATH-PLOT, Version 3: A particle tracking post-processing package for MODFLOW, the U. S. Geological Survey finite-difference ground-water flow model. USGS Open-File Report, pp 94–464

Saha D, Dwivedi SN, Singh RK (2013) Aquifer system response to intensive pumping in urban areas of the Gangetic plains, India: the case study of Patna Urban area. Environ Earth Sci. doi:10.1007/s12665-013-2577-7

Saha DK, Choudhury K, Chakraborty PK et al (2006) Geophysical survey for delineation of subsurface saline water zones in Southeastern part of Kolkata megacity. J Geol Soc India 68:885–892

Saha DK, Choudhury K, Murty BSR et al (2002) Geophysical studies for subsurface saline water zones in the south eastern part of Calcutta megacity. Geol Surv India Spec Publ 75:262–271

Sahu P, Sikdar PK (2008) Hydrochemical framework of the aquifer in and around East Calcutta Wetlands, West Bengal, India. Environ Geol 55(4):823–835

Sahu P, Sikdar PK (2011) Groundwater potential zoning of a peri-urban wetland of south Bengal Basin, India. Environ Monit Assess 174:119–134

Sahu P, Sikdar PK, Chakraborty S (2016) Geochemical evolution of groundwater of southern Bengal basin: the example of Rajarhat and adjoining areas, West Bengal, India. J Earth Syst Sci 125(1):129–145

Sahu P, Michael HA, Voss CI, Sikdar PK (2013) Impacts on groundwater recharge areas of megacity pumping: analysis of potential contamination of Kolkata, India, water supply. J Hydrol Sci 58(6):1340–1360

Sikdar PK (2000) Geology of the quaternary aquifers of the twin city of Calcutta—Howrah. J Geol Soc India 56(8):169–186

Sikdar PK, Sarkar SS, Palchoudhury S (2001) Geochemical evolution of groundwater in quaternary aquifers of Calcutta and Howrah, India. J Asian Earth Sci 19(5):579–594

Sikdar PK, Sahu P (2009) Understanding wetland sub-surface hydrology using geologic and isotopic signatures. J Hydrol Earth Syst Sci 13:1313–1323

SWRE (School of Water Resources Engineering) (2011) Technical study for finding different hydrological parameters including chemical quality of groundwater to ascertain the optimum yield from groundwater source in the locality at Mouza Madurdaha 782, Anandapur, Kolkata —107. Technical report. SWRE, Jadavpur University

Winston RB (2000) Graphical user interface for MODFLOW, Version 4: U.S. Geological Survey Open-File Report 00-315. U.S. Geological Survey, Virginia

Hydrodynamics of Groundwater Flow in the Arsenic-Affected Areas of the Gangetic West Bengal, India

Abhijit Mukherjee

Abstract Hydrostratigraphy and groundwater flow in a ~21,000 km^2 area of the arsenic-contaminated districts of West Bengal [Murshidabad, Nadia, North 24 Parganas and South 24 Parganas (including Calcutta)], India, have been delineated. Based on 143 lithologs, a regional hydrostratigraphic model has been developed to a depth of 300 m below mean sea level. Lithologic interpolations that were done manually and by computer models (resolution: 1000 m × 1000 m × 2 m) demonstrate a near-continuous, unconfined to semi-confined aquifer dominated by sand underlain by a thick clay aquitard. The aquifer thickens toward basin center in Bangladesh, toward south and east. Toward the Bay of Bengal, at the southern boundary, several heterogeneous aquitard layers of clay sub-divide the primary aquifer, into several deeper, regionally-discontinuous but laterally-connected, confined to semi-confined aquifers. Eight 21-layer regional groundwater models are based on observed topography and hydrostratigraphy. Groundwater flows were simulated in the pre-monsoon, monsoon and post-monsoon seasons with presence (2001) and absence (pre-1970s) of irrigation pumping and projected pumping for pre-monsoon seasons of 2011 and 2021. Modeling results indicate topographically dominated, seasonally variable, continuos regional-scale groundwater flow, which have been largely distorted by pumping. Groundwater flow is predominantly in the upper ~100 m of the aquifer but occurs to a depth of ~200 m.

Keywords West Bengal · Hydrostratigraphy · Numerical modeling Groundwater flow

---

A. Mukherjee (✉)
Department of Geology and Geophysics, Indian Institute of Technology (IIT), Kharagpur, Kharagpur 721302, India
e-mail: abhijit@gg.iitkgp.ernet.in; amukh2@gmail.com

D. Saha et al. (eds.), *Clean and Sustainable Groundwater in India*,
Springer Hydrogeology, https://doi.org/10.1007/978-981-10-4552-3_19

1 Introduction

More than 120 million (2% of global population) lives in ~200,000 km^2 in eastern parts of India (West Bengal state) and most of Bangladesh. This area known as the Bengal basin (BB) or Ganges–Brahmaputra–Meghna river basin is considered to be the world's largest fluvio-deltaic system (Coleman 1981; Alam et al. 2003). The rivers and surface water bodies have been indiscriminately used as sewerage and industrial waste discharge by the residents of the Bengal basin, thereby making the surface water unsafe. Further, intense irrigation of groundwater for high-yielding dry-season rice (Boro) (Harvey et al. 2005) largely increased the water-demand in the area. During the early 1970s, this resulted to the change of water-usage policy to groundwater from surface water in both West Bengal and Bangladesh, resulting to installation of millions of wells that range from low discharge hand-pumped wells to motor-driven, high-discharge wells (Smith et al. 2000; Kinniburgh and Smedley 2001; Harvey et al. 2005; Horneman et al. 2004). However, unfortunately, wide areas of the groundwater of BB were found to be polluted by elevated concentrations of carcinogenic arsenic (As) (WHO 1993). Since then (1990), numerous studies by governmental/non-governmental agencies and researchers tried to understand the causes and effects of this contamination. These studies have led to a detailed understanding of the water–sediment chemistry and proposition of several (often conflicting) hypotheses on the occurrence and fate of arsenic in groundwater on both sides of the Indo–Bangladesh border (e.g. Saha 1991; AIP/PHED 1991, 1995; Mallick and Rajagopal 1995; Das et al. 1995, 1996; Bhattacharya et al. 1997; CGWB 1997; Mandal et al. 1998; Nickson et al. 1998, 2000; Ray 1999; Acharyya et al. 1999, 2000; Kinniburgh and Smedley 2001; McArthur et al. 2001, 2004; Ravenscroft et al. 2001, in press; Harvey 2002; Ghosh and Mukherjee 2002; JICA 2002; Smedley and Kinniburgh 2002; Harvey et al. 2005; Dowling et al. 2003; Stuben et al. 2003; Horneman et al. 2004; Mukherjee et al. 2007; Mukherjee and Fryar 2004).

In West Bengal, the regional hydrologic framework and detailed hydrochemistry remain largely unrevealed, prior to our studies (Mukherjee and Fryar 2004). Thus, numerical simulations of regional-scale, seasonal groundwater flows through observed hydrostratigraphic framework were undertaken (see Mukherjee et al. 2007 for details). The objectives for the groundwater flow modeling were to:

(i) understand the groundwater flow pattern at a regional scale in the western Bengal basin;
(ii) understand the interactions between shallow and deep groundwater flow and hence the possibility of cross contamination of arsenic across the vertical extent of the aquifers;
(iii) calculate the relative magnitude of flow in the shallower and deeper parts of the aquifers;

(iv) analyze the effects of pumping at present and projected future rates on the flow patterns and predict the future status of the groundwater resource for the study area.

2 Study Area

Extent: The study area is part of the Gangetic fluvial delta–plain, surrounded in the north by the River Ganges, in the west by the River Bhagirathi-Hooghly, in the east by rivers Jalangi, Ichamati and several tidal rivers to the Bay of Bengal in the south (Fig. 1). This includes the four arsenic-enriched districts, parts of Murshidabad, Nadia, North and South 24 Parganas (including Kolkata or Calcutta) (~21,000 km^2). About 50 million people reside in this area, which includes the city of Kolkata (formerly Calcutta) and several other important towns. The rationale for selecting this area is that it can be considered to be a hydrologically closed system for purpose of groundwater flow modeling, considering the major hydrologic boundaries on all four sides (Mukherjee et al. 2009).

This study uses a spherical coordinate system with datum WGS84. For discretization, UTM 1983, Zone 45 (central meridian 87, reference latitude 0, scale factor of 0.9996, spheroid GRS 80, false easting 500,000, false northing 0) has been used. The origin coordinates for modeling were fixed at northing 2,380,000, easting 600,000.

3 Modeling of Hydrodynamics

Conceptual model: Before planning the details of the model design, a conceptual model was conceived. Being a tropical fluvio–deltaic system with monsoonal rainfall, groundwater flow in the western Bengal basin is complicated to conceptualize, even without considering extensive anthropogenic impacts. Therefore, in this modeling effort, only the parameters that were thought to have significant control on the groundwater flow were incorporated. The model design is based on the observed hydrostratigraphy, including an upper aquifer and a basal confining unit. Recharge is primarily meteoric, while discharge occurs both through outflow to surface water bodies and pumping for irrigation. Specified head boundaries were applied for the Ganges, Bhagirathi-Hooghly and Jalangi/Ichamati rivers, and a constant head boundary to the Bay of Bengal.

Discretization and design: The 3-D groundwater flow in the study area is modeled by MODFLOW (McDonald and Harbaugh 1988). Groundwater flow in heterogeneous, anisotropic medium can be expressed by the following equation:

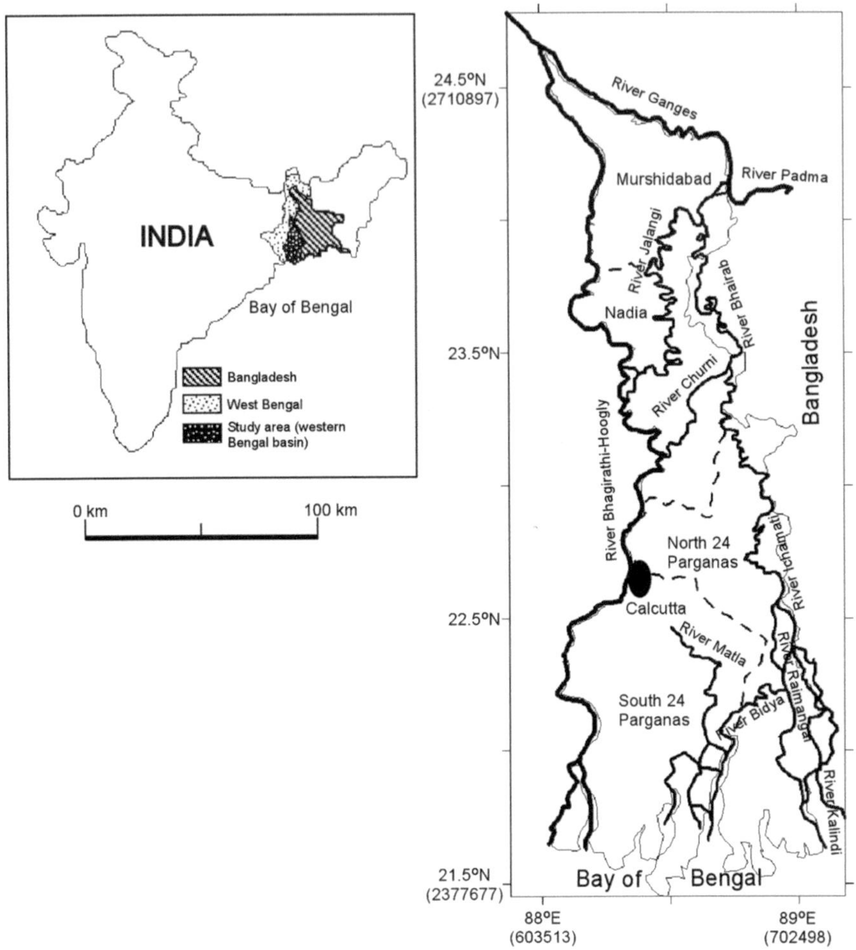

Fig. 1 Location map of the study area showing the districts and important rivers (district boundaries are in *dashed lines*). UTM values (in *parentheses*) are listed beside respective latitude and longitude values

$$\frac{\partial}{\partial x}\left(K_{xx}\frac{\partial h}{\partial x}\right)+\frac{\partial}{\partial y}\left(K_{yy}\frac{\partial h}{\partial y}\right)+\frac{\partial}{\partial z}\left(K_{zz}\frac{\partial h}{\partial z}\right)-W=Ss\frac{\partial h}{\partial t} \quad (1)$$

where

K_{xx}, K_{yy} and K_{zz} are hydraulic conductivity values in x, y and z directions, which are assumed to be parallel to the major axes of K (L/T)
h is the hydraulic head (L)

W is volumetric flux per unit volume and represents sources and/or sinks of water ($T - 1$)
Ss is the specific storage of the porous material ($L - 1$)
t is time (*T*)

Groundwater Vistas® (GV) v. 4 (Environmental Simulations International, Reinshold, PA, USA) was used as a pre-processor (for design and discretization) and post-processor (for viewing results and mass balance). The modeled study area stretches for 335 km from north to south, and 110 km from east to west. The horizontal discretization of the model was in a finite difference grid network of 335 (x) × 110 (y) (36,850 horizontal grid cells) with unit grid cell dimensions of 1 km–1 km. The model thickness varies from 330 to 300 m, depending on topography, with 21 layers (total of 773,850 3-D grid cells). Except for the top layer, which has variable thickness due to topographic features, each layer is 15 m in thickness. In order to make the model realistic, the hydraulic framework of the different layers was based on the lithologic modeling of the study area in Mukherjee et al. (2007) (Fig. 3). The finite difference discretization was based on a finite difference, hydrostratigraphic solid model of dimension 1000 m × 1000 m × 15 m (Mukherjee et al. 2007). Each of the sediment types was assigned an index value (*G*) to the *K* value for the aquifers and aquitards. Then, the lithologic layers from *RW* were imported to *GV* via Surfer as grid layers with nodal values of hydraulic conductivity ($K_x = K_y$ and $K_z = 1/10$ of K_x). The basal layer was designed with no-flow boundary, which reflects the hydrostratigraphy. The initial values of aquifer parameters were taken from the literature. CGWB (1994a, b, c, d) and SWID (1998) estimated that transmissivity (*T*) values (in m^2/d) vary from 3300 to 7000 in the north (Murshidabad), from 500 to 8800 in the south (North and South 24 Parganas), with 0.3 as a mean storativity (*S*). The value of porosity (*n*) was assumed to be 0.2 (Harvey 2002; JICA 2002). The final aquifer parameters were estimated by sensitivity analyses.

The surface topography was defined through the NASA SRTM-90 DEM, which was imported into GV for the surface elevation/upper bound of the topmost layer (Fig. 2). According to the conceptual model, ArcGIS® (ESRI, Redlands, CA, USA) shape files were developed for each river and then imported into GV as specified head boundary river objects. To resemble reality, all of the rivers shown in Fig. 1 have been included. The rivers were divided into 11 reaches according to geometric similarities along each reach. For each of the stream reaches, the starting and ending cells were specified with attributes like channel width, stream stage, channel bed elevation and *K* of the bed sediments. Depending on the reach length and surrounding topography, *GV* automatically interpolated the attribute values for the cells lying between the starting and ending cells of each reach. The width of each of the reach was specified according to measurement from the base map. In Bangladesh, the stage of the rivers varies from approximately 5 to 13 m above the streambed (Dowling et al. 2003). For this model, the stage values for each reach were selected from field experience and the topographic values specified for the cells in the vicinity, assuming that the stage of the river is near the land surface. The

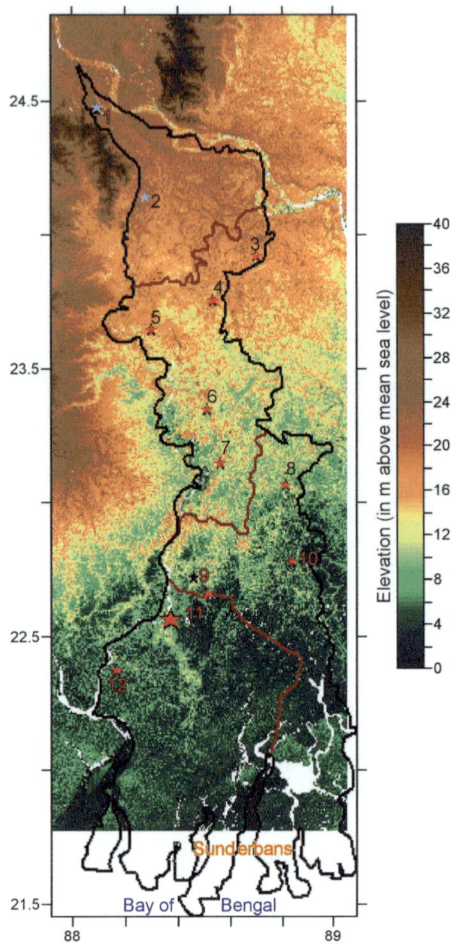

Fig. 2 Elevation map of the study area prepared from SRTM-90 DEM along with locations of important towns (*red* and *blue* stars): *#1* Jangipur, *#2* Behrampur, *#3* Karimpur, *#4* Tehatta, *#5* Debagram, *#6* KrishnaNagar, *#7* Ranaghat, *#8* Bongaon, *#9* Barasat, *#10* Basirhat, *#11* Calcutta (Kolkata), *#12* Diamond Harbor

other attributes were specified according to estimation, field experience and local information. Constant head boundaries with a value of 0 m representing the Bay of Bengal were assigned to the southernmost rows of the top four layers of the model. The constant head boundaries for each layer were guided by the original bathymetric data (Survey of India 1971). The extreme southern boundary was marked at the southernmost tip of the study area (the active delta front) (Fig. 3).

Seasonality and Recharge: The study area has distinct wet (monsoon) and dry seasons. On average (1901–1970), the seasonal rainfall can be considered as pre-monsoon (January–May) 16.23% of annual rainfall, monsoon (June–October) 82.21% and post-monsoon (November–December) 1.57%. Increased pumping lowers the water level in the dry seasons. The aquifer is mostly full during the monsoon season as an effect of increased precipitation and lower rates of pumping. Hence any excess potential recharge is rejected (Kinniburgh and Smedley 2001),

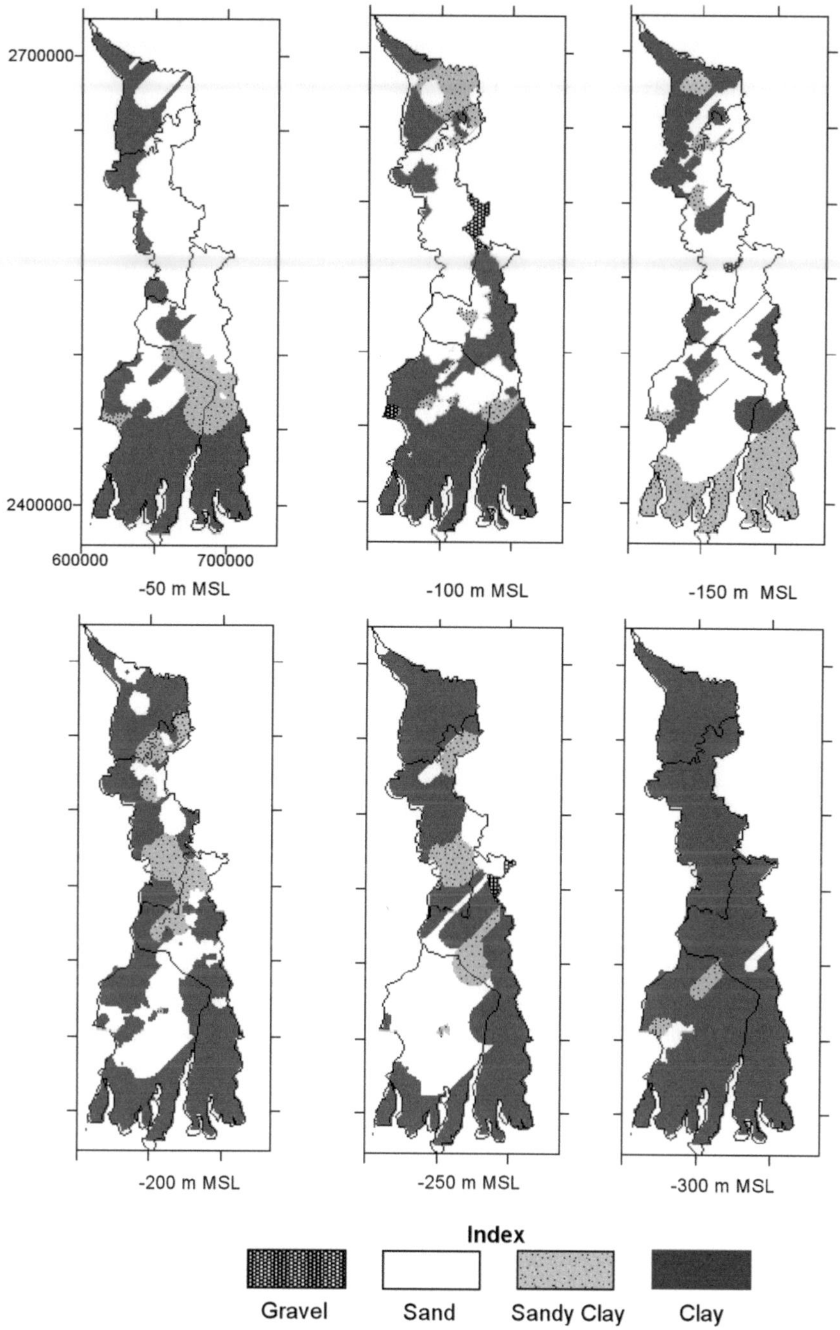

Fig. 3 Modeled plan maps of the study area showing the subsurface distribution of lithologic units at definite depths

thereby causing regular, widespread runoff and flooding. The division of seasons for this study was done in the context of irrigation schedules. The seasons have different daily recharge rates and pumping rates for irrigation.

In West Bengal, rice cultivation is dominated by boro (mainly dependent on groundwater irrigation) during the pre-monsoon (January to April/May), followed by kharif (mainly rainwater fed) during the monsoon (July–October/November). Rabi (post-monsoon) cultivation is also mostly groundwater fed. The high-yielding boro cultivation caught momentum with the onset of groundwater irrigation pumping in the early 1970s. In Bangladesh, the cultivation increased from less than 0.05% of the total area in the early 1970s to more than 20% (more than 45% of total cultivable area) around 2000 (Hossain et al. 2003; Harvey et al. 2005). Boro requires from about 1 m (Harvey et al. 2005) to about 3 m (AIP/PHED 1995) of annual irrigation, while kharif requires only about 0.3 m (S.P. SinhaRoy, Arsenic Task Force, Government of West Bengal, personal communication, 2004). In spite of having some localities with an overdraft problem (where the groundwater is not replenished annually), groundwater levels at the beginning of each irrigation season tend to be similar from year to year, which suggests that the groundwater levels are controlled by local river stages (Kinniburgh and Smedley 2001). This observation and the lack of detailed storativity data for the different hydrologic units led to construction of three separate models for different annual irrigation seasons, instead of a single transient model with three seasons (stress periods). A single transient model would have been preferable, but the disproportionate precipitation (and hence recharge) during the monsoon adversely affects the water levels during other seasons of a transient model, resulting in poorly constrained values of the volume of water in storage. Moreover, calibration of the pre-monsoon model with pumping (at the beginning of the irrigation season) indicates a reasonable fit, as noted below, supporting the rationale of separate seasonal models.

The exact recharge values in the study area were not available. Previous studies have estimated 0.3–5.5 mm/d (Kinniburgh and Smedley 2001) and 1.6 mm/d (annual average) (Basu et al. 2001; Dowling et al. 2003) for the Bangladesh part of the basin. For the West Bengal part, SWID (1998) estimated that recharge is up to 15% of total precipitation. Because recharge rate is one of the primary controlling factors for groundwater flow in a low-topographic-gradient study area, a detailed approach for estimating the potential recharge (PR) was developed. The term "potential recharge" is used in place of "recharge" because the rainwater available for recharge will also runoff, and there is no record or simple method available for estimating either the recharge or runoff in an area crowded by numerous ponds, oxbow lakes, small streams and larger rivers. Harvey et al. (2005) noted that the pond/lake bottoms and streambeds can provide significant areas for infiltration of recharge, and irrigation water applied to cultivated fields can also re-infiltrate, resulting in a very complex cycle of recharge and discharge. Actual recharge values can be obtained by the difference in groundwater levels between wet and dry seasons. However, no such data were available except for general values obtained for some small localities. Moreover, the daily oscillation of water levels due to irrigation pumping can be more than 0.2 m, thus presenting a very difficult scenario for estimating recharge and groundwater flow directions (Harvey et al. 2005).

Seasonal mean rainfall data, as shown in Table 1, were calculated for 25 locations within and near the study area based on multiple years' data obtained from National Climatic Data Center (http://www.ncdc.noaa.gov/oa/pub/data/ghcn/v2/ghcnftp.html). Table 2 shows the monthly mean temperature data for Calcutta from 1811 to 2005. Based on these data, potential evapotranspiration (PET) values were estimated for each location according to Malmstrom (1969):

$$\text{PET (mm/month)} = 40.9 \times e^{*}(T) \quad (2)$$

Table 1 Distribution of rainfall and calculated potential recharge (PR) data from global historic climatology network (GHCN v.2) stations in and around the study area

Location	Data coverage		Pre-monsoon (Jan–May)		Monsoon (Jun–Oct)		Post-monsoon (Nov–Dec)	
	From	To	Rain (mm/d)	PR (mm/d)	Rain (mm/d)	PR (mm/d)	Rain (mm/d)	PR (mm/d)
Behrampur	1901	1970	1.40	0.14	7.57	3.30	0.31	0.00
Arambagh	1901	1968	1.51	0.13	7.24	3.10	0.39	0.00
Bongaon	1901	1967	2.09	0.31	8.34	3.96	0.35	0.00
KrishnaNagar	1901	1970	1.77	0.21	7.55	3.26	0.38	0.00
Ranaghat	1901	1967	1.73	0.20	6.80	2.73	0.28	0.00
Kalyanganj	1906	1954	1.31	0.12	6.98	2.86	0.30	0.00
Haringhata	1908	1966	1.35	0.10	6.43	2.41	0.30	0.00
Calcutta	1829	2005	1.67	0.19	8.92	4.42	0.47	0.01
Budge Budge	1901	1970	1.64	0.16	9.27	4.74	0.39	0.00
Dum Dum	1947	1980	1.47	0.14	8.58	4.13	0.43	0.01
Barrackporc	1901	1965	1.52	0.14	8.01	3.74	0.49	0.01
Barasat	1901	1959	1.62	0.16	8.11	3.75	0.39	0.00
Basirhat	1901	1970	1.67	0.20	8.54	4.16	0.47	0.01
Diamond Harbor	1901	1967	1.46	0.13	8.93	4.49	0.37	0.00
Gosaba	1921	1956	1.39	0.15	9.41	4.86	0.64	0.02
Sagar Island	1975	1984	1.46	0.11	9.23	4.78	0.31	0.00
Potkabari	1901	1956	1.68	0.24	6.88	2.78	0.39	0.01
Hooghly	1901	1962	1.83	0.22	7.52	3.30	0.40	0.00
Kalna	1901	1963	1.61	0.16	6.78	2.73	0.33	0.00
Howrah	1901	1948	1.59	0.15	8.32	3.95	0.35	0.00
Serampore	1901	1962	1.73	0.20	8.24	3.88	0.39	0.00
Azimganj	1901	1961	1.41	0.14	7.26	3.09	0.25	0.00
Boinchee	1932	1962	1.63	0.17	7.92	3.63	0.42	0.01
Katwa	1901	1966	1.44	0.13	7.37	3.17	0.32	0.00
Ulubeia	1901	1968	1.83	0.22	8.90	4.47	0.38	0.00

Table 2 Monthly temperature (1811–2005) and precipitation (1829–2005) data for Calcutta along with calculated evapotranspiration (ET) and potential recharge from GHCN v.2

Month	Temperature (°C)			Precipitation (mm/month)			ET (mm/month)	PR (mm/month)
	Mean	Max	Min	Mean	Max	Min		
January	19.68	22.70	15.00	11.53	73.70	0.00	11.45	0.08
February	22.65	25.40	18.80	24.25	202.20	0.00	23.70	0.54
March	27.52	30.20	23.60	32.61	159.30	0.00	31.87	0.74
April	30.34	33.10	26.80	52.59	196.20	0.00	50.43	2.17
May	30.80	33.20	28.40	130.64	434.60	2.00	106.19	24.45
June	30.01	32.20	27.70	292.36	946.00	40.60	149.71	142.66
July	29.10	30.80	27.20	333.55	879.00	109.00	148.14	185.41
August	28.98	31.00	27.70	337.26	683.50	122.40	147.69	189.57
September	29.01	30.60	27.70	264.65	944.70	56.10	139.73	124.92
October	27.87	31.20	26.50	136.23	474.00	0.00	102.02	34.21
November	24.00	27.10	21.80	22.37	238.50	0.00	22.01	0.36
December	20.14	23.40	17.80	6.47	146.00	0.00	6.46	0.01

where

$$e^{*} = 0.611 \times \exp\left(\frac{17.3 \times T}{T + 237.3}\right)$$

T is average monthly temperature (°C).

Evapotranspiration (ET) values were calculated (Pike 1964) for each of these 25 locations as:

$$\text{ET (mm/month)} = \frac{R}{\left[1 + \left(\frac{R}{\text{PET}}\right)^2\right]^{1/2}}) \tag{3}$$

where

R average precipitation (mm/month)
PET potential evapotranspiration (mm/month)

Potential recharge (PR) was then calculated as:

$$\text{PR (mm/month)} = R\text{–ET} \tag{4}$$

The monthly PR values were aggregated according to the previously described seasons. These values were then used to form 2-D daily PR grids of the study area by kriging in Surfer, where

$$\text{daily PR (mm/day)} = \frac{R_s}{d_s} \tag{5}$$

Where

R_s cumulative mean monthly rainfall for that season (in mm)
d_s number of days in that season

The daily PR grid files were then imported into GV to form the different zones of recharge in the top layer of each model.

Groundwater abstraction and simulation scenarios:

The number of deep tube wells (DTW) (MID 2001) and public/community water supply wells (PDTW) and their estimated usage were used for the calculation of groundwater draft. These values were used to determine the abstraction for across the modeled area, following the equation:

$$Q - \left(\frac{N \times r \times t}{A}\right) \times a \tag{6}$$

where

Q the assigned total pumpage (outflow) value to each grid cell for a specified time [$L3$]
A area of each district within the model boundary [$L2$]
N number of pumps
r pumpage rate [$L3/T$]
t hours of operation per day [T]
a area of each grid cell

Model scenarios were developed based on the seasonality, pumping regime and predictivity:

(i) with pumping: pre-monsoon (PREM-p), monsoon (M-p) and post-monsoon (POSTM-p), based on numbers of DTWs and PDTWs in 2001;
(ii) without pumping: pre-monsoon (PREM), monsoon (M) and post-monsoon (POSTM), without any pumping (pre-1970s conditions);
(iii) projected pumping: projected pumping rates for 2011 (2011-p) and 2021 (2021-p)

Calibration and sensitivity analyses: Ideally, the models should have been calibrated for all three seasons for steady state simulations without any pumping. However, because water level data were only available for the study area for pre-monsoon in presence of pumping, the PREM-p model was calibrated in steady state against known head values from head values observed from more than 120 locations across the study area. Sensitivity analyses were performed on this calibrated model (PREM-p) by sequentially varying the $K_x{-}_y$ of sand and clay, S and n values.

The model was found to be most sensitive to the K*x* values of sand and clay. Using the selected values from the analyses, the final model was recalibrated and found to have good agreement with present water level and hydraulic head contours. The differences between the observed and computed values were probably a cumulative manifestation of a specified head boundary in the vicinity, discrepancy in topographic elevations, exclusion of STW and uncertain aquifer parameters. The finalized values obtained from the sensitivity analyses were then used for defining the aquifer properties of the other seven models (PREM, M, POSTM, M-p, POSTM-p, 2011-p and 2021-p) (Tables 3 and 4).

Simplifications and limitations: As discussed earlier, the complexities of the western Bengal aquifers are so great that it is very difficult to construct a groundwater model that reflects local-scale variations of aquifer properties. Hence, based on the similarity to other data from the study area or previous studies, numerous generalizations and pre-assumptions were involved in order to construct the model. As a consequence, there are several limitations. Thus, the modeling exercises for this study should be considered as a guideline or template on which

Table 3 Results of sensitivity analyses on PREM-p for uncertain aquifer parameter values

	RSS	Comment
Sand (K_x) m/d		
25	78.9	Formation of large dry cells
37.5	34.4	Used model parameter value
50	21.9	Formation of small dry cells in Nd and NP
70	21.7	Formation of large dry cells all over the study area
Sand (K_y) m/d		
25	34.4	Used model parameter value
37.5	31.2	Formation of dry cells near CCU and SP
15	39.5	Formation of dry cells near CCU and Md
Clay (K_{x-y}) m/d		
0.1	34.4	Used model parameter value
0.5	47.7	Not much effect, long time to converge
1	49.1	Not much effect, long time to converge
0.01	42.2	Distorted contours and flow toward River Ganges
S		
0.03	34.4	Used model parameter value
0.015	34.4	No effect
0.06	34.4	No effect
0.3	34.4	No effect
n		
0.2	34.4	Used model parameter value
0.1	34.4	No effect
0.4	34.4	No effect

Bold initial values, *italicized* used values, *RSS* residual sum of squares
Md Murshidabad, *Nd* Nadia, *NP* North 24 Parganas, *SP* South 24 Parganas, *CCU* Calcutta

Table 4 Results of model calibration with the best-fitted parameter values

Calibration locations	Latitude	Longitude	Layer	Observed (hydraulic head in m above MSL)	Computed (hydraulic head in m above MSL)	Residual
Raghunathganj	24.5	88.08	4	22	21.32	0.68
Lalgola	24.42	88.25	4	20.7	20.40	0.30
Bhagawangola	24.34	88.32	4	19.9	18.38	1.52
Raninagar-II	24.25	88.6	4	16.5	15.27	1.23
Jiaganj	24.23	88.27	4	17.2	18.55	−1.35
Lalbagh	24.18	88.27	4	18.6	17.52	1.08
Raninagar-I	24.15	88.47	4	14.5	11.75	2.75
Behrampur	24.09	88.26	4	17.5	16.34	1.16
Nawda	23.91	88.46	4	14.3	14.09	0.21
Potkabari	23.86	88.52	6	13.5	13.36	0.14
Debagram	23.69	88.33	6	11.3	11.05	0.25
Chapra	23.61	88.56	6	11.9	11.03	0.87
Nakashipara	23.6	88.35	6	12.4	10.98	1.42
Banguin	23.33	88.66	6	8.2	8.24	−0.04
Bongaon	23.05	88.83	7	7.1	6.75	0.35
Amdanga	22.77	88.53	10	4.5	3.64	0.86
Baduria	22.76	88.79	7	7.8	7.19	0.61
DumDum	22.73	88.38	9	3	1.77	1.23
Haroa	22.61	88.68	12	4	6.50	−2.50
East Calcutta	22.56	88.42	9	0	−0.18	0.18
Rajpur-Sonarpur	22.45	88.43	12	5	4.19	0.81
Baruipur	22.4	88.48	12	5	3.43	1.57
Jalerhat	22.3	88.56	12	3.5	2.74	0.76
Sonakhali	22.3	88.74	12	2.8	3.04	−0.24
Kulpi	22.16	88.29	12	2.2	1.38	0.82
Dakhin Durgapur	22.06	88.39	12	2.2	0.89	1.31

Residual mean: 0.61
Residual standard deviation: 0.97
Sum of squares: 34.35
Absolute residual mean: 0.93
Minimum residual: −2.50
Maximum residual: 2.75

future workers may modify and elaborate, although our results reasonably resemble current conditions. The simplifications are listed below.

(i) Values for potential recharge at each cell are based on the interpolated values from the 25 monitoring stations. Calculation of PET was based on the temperature values that were available for Calcutta only. Moreover, as mentioned earlier, there may be differences between the potential recharge and absolute recharge. Hence the amount of water that is actually percolating into the system is not well constrained.

(ii) In order to make the numerical flow model executable within the present constraints of time and computational resources, a 15-m vertical resolution was used. This leads to a broad generalization for the hydrostratigraphic layers, so smaller scale variations in the lithology may be over- or under-emphasized.

(iii) Precise aquifer property ($K_{x}-_{y}-_{z}$ and S) values for each locality were not obtained. Hence generalized aquifer property values for the different hydrogeologic units were used throughout the model, and the ranges of observed values obtained from the literature were used for sensitivity analyses.

(iv) Most of the specified head values were estimated based on previous literature, field experience, local information and topography, except the stream widths, which were known for each reach.

(v) The smaller water bodies (such as small streams, ponds and lakes), which are present throughout the modeled area, were not built into the model because they were thought to have insignificant effects on the regional flow.

(vi) No comprehensive data were found for the DTW pumps (location, rate of pumping, hours of operation or screen depth and length). Hence a number of generalizations were undertaken as explained earlier.

(vii) The STWs, which would probably have significant control on the groundwater flow, were not included in the model because of insufficiency of data.

4 Results and Discussion

The model simulation results (Fig. 4a, b) for without pumping (PREM, M and POSTM) indicate that there is a topographically induced, natural regional-scale flow from the northern parts of the study area, toward the southern and southeastern parts. The rivers in the system are in general effluent. The PREM contours indicate lateral flows in the vicinity of the Ganges in the north and northwest and along the Bhagirathi-Hooghly north of Calcutta. Flows also converge in the vicinity of the Jalangi, whereas contours meet the Bhairab and upper part of the Ichamati at high angles. PREM also indicates the development of a few groundwater mounds above the land surface, illustrating aquifer saturation in those localities, with radially outward flow for excess water. Two such prominent localities can be noted near the western part of North 24 Parganas (near Basirhat) and northern Murshidabad (near Jangipur). These locations match with present-day waterlogged areas (CGWB 1994e). The water logging may have been caused by presence of a shallow aquitards and very low hydraulic gradients that intersect topographic lows. The possible result is flow stagnation and groundwater accumulation in these areas. Near the

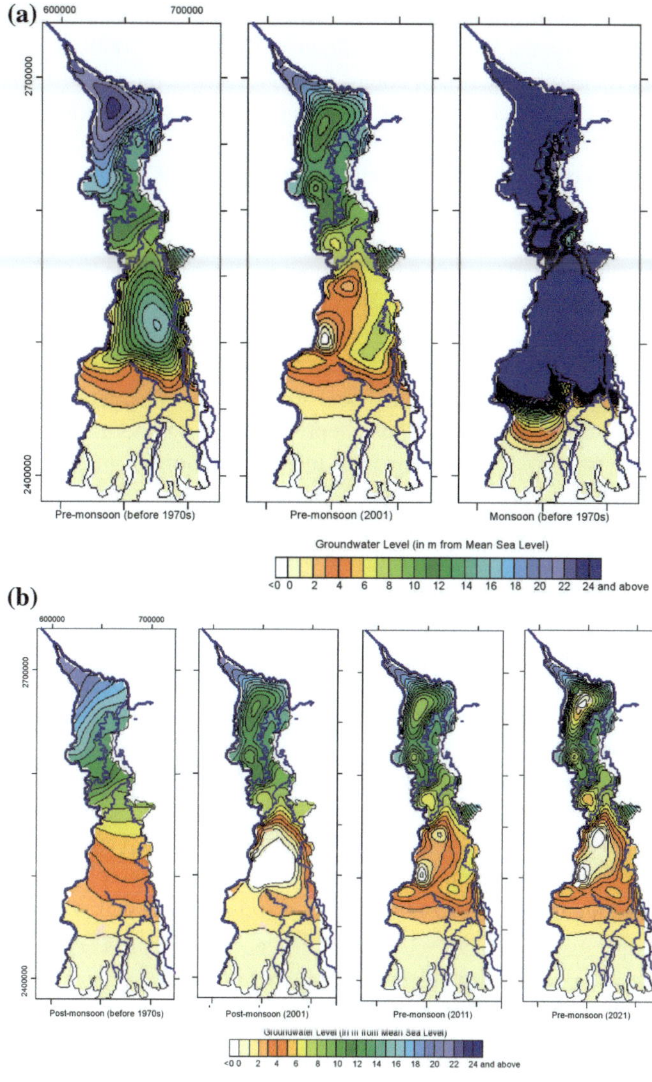

Fig. 4 **a** Modeled groundwater level maps obtained from PREM, PREM-p, M and M-p. The *bold blue lines* indicated the major rivers in the area. **b** Modeled groundwater level maps obtained from POSTM, POSTM-p, 2011-p and 2021-p. The *bold blue lines* indicated the major rivers in the area

active delta front, contours become more broadly spaced, which is consistent with tidal fluctuations, although the models did not include those. The groundwater flow in M is not conspicuously visible on the map. Due to the large amount of water available for recharge during monsoonal rainfall, aquifer saturation and excess water ponding takes place in terms of recurrent wide floods and increased runoff to the streams. These are in agreement with Harvey et al. (2005), who observed in Munshiganj (Bangladesh) that groundwater flow ceases as stream stages rise and hydraulic head becomes uniform in flooded areas. Although no such historic data were available for the study area, a devastating flood in September 2000 affected 18.3 million people (with demise of 1187 people) in 6177 villages of nine districts of West Bengal, including Calcutta (http://www.bapscare.org/services/disaster/2000/2000floods.htm). In Jumpukur (location 46), Nadia, floodwater marks of 2000 were observed during fieldwork at 1.98 m above land surface. Of the 65 blocks within the study area, 75% have flood hazard potential of medium to very high (Sanyal and Lu 2003). The POSTM model demonstrates north-to-south flow with perpendicular to semi-perpendicular intersection of contour lines with the boundary rivers. The "with pumping" models (PREM-p, POSTM-P, 2011-p and 2021-p) suggest that except the southern half of South 24 Parganas, in all other parts of the study area, the natural flow paths have been thoroughly disordered by pumping-induced conning. In vertical sections, local-scale convective flow cells are found to have developed corresponding to these depressions. In southern South 24 Parganas, flow reversal has developed along with increased inflow from the constant head boundary of the Bay of Bengal. This phenomenon can be easily explained by the large saltwater intrusion that has been documented in these areas (CGWB 1994d, e; SWID 1998; Allison et al. 2003). PREM-p demonstrates the effects of abstraction all over the study area, with prominent pumping centers in the blocks of Murshidabad-Jiaganj, Behrampur, Hariharpara and Domkal in central Murshidabad; Kaliganj and Nakashipara in northern Nadia; Shantipur, KrishnaNagar-I and Ranaghat-I in south-central Nadia; Chakdah, Haringhat, Habra-II and Amdanga in southern Nadia; and northern North 24 Parganas and Calcutta. Calcutta (CGWB 1994e; Sikdar et al. 2001) is possibly most affected by these conning, due to the huge amount of pumping for municipal water supply through PWDTs. The groundwater mounds near Jangipur and Basirhat, though still observable, have substantially reduced in size and extent. In the 2011-p and 2021-p models, the cones have become more acute. In 2021-p, some dry cells corresponding to deep cones of depression have formed in central Murshidabad, northern North 24 Parganas, and south-southeast Calcutta. These dry cells imply severe overdraft. In the POSTM-p model, in addition to the effects of the PREM-p models, a huge cone of depression has developed in north and central North 24 Parganas with centers at Calcutta, Rajarhat, Barasat-I and II and Deganga. However, in reality, the cone of depression may not be so large because of the availability of monsoonal water in storage that has not been accounted for in the model design. The effect of pumping is not at all conspicuous in the monsoon model (M-p), indicating that the saturated aquifers are not stressed in spite of pumping. The errors in mass balance range between 0.02% (PREM-p) and −0.06% (POSTM-p) (Fig. 5). It indicates that at present the submarine groundwater discharge (SGD) through the delta front of South 24 Parganas seems to have decreased by 6% annually since the onset of pumping, while the annual saltwater intrusion has increased by 98%. The inflow of river water into the system has increased by 92% in the last three decades, but the rivers are still effluent mostly because of the monsoonal groundwater discharge. The model demonstrates that the rivers have changed from slightly effluent to influent for the post-monsoon season. This estimation is probably not viable

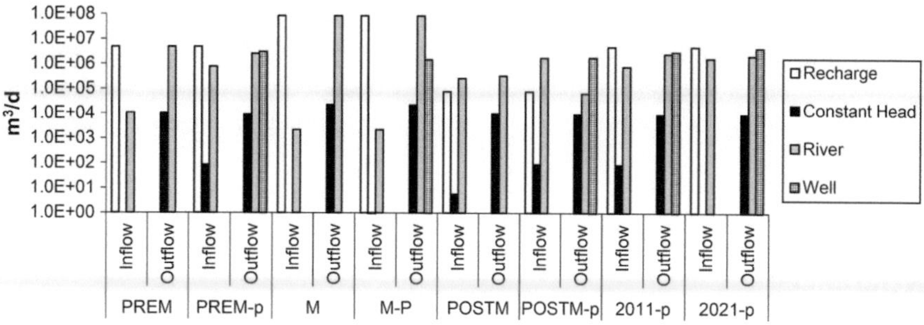

Fig. 5 Plot of mass balance of the eight flow models shown in Fig. 4a, b

because of the monsoonal water in storage, as mentioned above. An annual SGD as about 1.5 ± 0.5 × 1011 m^3 of water for the Bengal delta front (~500 km) was estimated by Dowling et al. (2003). Considering that the delta front in this study area is 110 km, the present estimates for SGD of 5.37 × 106 m^3 are much lower, possibly because the results show that a considerable volume of groundwater upwells from the aquifer >30 m to the rivers. The estimation of abstracted water (only by DTW and PDTW) from the modeled area is about 8 × 108 m^3 annually. On the basis of estimation of resident groundwater (4.52 × 1011 m^3) from the lithologic model, it seems that the total volume of groundwater resource is as yet underutilized. Dowling et al. (2003) suggested an average actual recharge rate of 0. 6 m/year. Using that estimation, the total annual recharge in the study area (recharge rate × rechargeable surface area) would be about 9 × 109 m^3, which shows that the groundwater abstraction by DTW and PDTW is only about 9%. This value is similar to Dowling et al.'s estimate of 12 ± 4%. However, the composite cone of depression around Calcutta (as visible in PREM-p and POSTM-p) is perennial and formed by overdraft pumping, which has superseded any recharge in that area. The cause may be exponential metropolitan growth (which is directly proportional to increases in groundwater demand for drinking and industrial use) and the lack of recharge due to impervious urban cover and the near-surface shallow aquitard (Fig. 3). The groundwater in and around Calcutta has not exhausted, probably because of the recharge that takes place in the northern areas (southern Nadia and northern North 24 Parganas) and subsurface inflow from the Bhagirathi-Hooghly. The bulk of the groundwater flow in the system takes place through layers 2–6 (~15 m–105 mbgl). This indicates that the local flow systems predominate in the system that is in agreement to Harvey et al's (2005) assessment for the Bangladesh part of the basin. Although lesser in extent, the medium-to-regional-scale flow to a depth of about 200 m (layer 13) is not insignificant. Below that depth, to the base of the model, there is minimal recharge, and groundwater has relatively long residence times. These results indicate that the Golden Bengal aquifer is continuously being recharged and there is substantial flow even in the deeper part of the aquifer, apart from the isolated aquifers mentioned above (Mukherjee et al. 2011).

In conclusion, groundwater flow in the study area is dependent on the amount and timing of precipitation and is controlled by the relatively flat topography and locations of major streams, but has been heavily distorted by pumping. This finding is in agreement with the results of 3-D regional modeling in western Bangladesh (JICA 2002).

References

Acharyya SK, Lahiri S, Raymahashay BC, Bhowmik A (1999) Arsenic poisoning in the Ganges delta. Nature 401:545

Acharyya SK, Lahiri S, Raymahashay BC, Bhowmik A (2000) Arsenic toxicity of groundwater in parts of the Bengal basin in India and Bangladesh: the role of quaternary stratigraphy and holocene sea-level fluctuation. Environ Geol 39(10):1127–1137

AIP/PHED (1991) National drinking water mission submission project on arsenic pollution in groundwater in West Bengal. Final report of Steering Committee, Arsenic Investigation Project (AIP) and Public Health Engineering Directorate (PHED), Government of West Bengal, India 58 pp

AIP/PHED (1995) Perspective plan for arsenic affected districts of West Bengal. Arsenic Investigation Project (AIP) and Public Health Engineering Directorate (PHED), Government of West Bengal, India

Alam M, Alam MM, Curray JR, Chowdhury MLR, Gani MR (2003) An overview of the sedimentary geology of the Bengal basin in relation to the regional tectonic framework and basin-fill history. Sed Geol 155:179–208

Allison MA, Khan SR, Goodbred SL, Kuehl SA (2003) Stratigraphic evolution of the late Holocene Ganges-Brahmaputra lower delta plain. Sed Geol 155(3–4):317–342

Basu AR, Jacobsen SB, Poreda RJ, Dowling CB, Aggrawal PK (2001) Large groundwater strontium flux to the oceans from the Bengal Basin and marine strontium isotope record. Science 293(1470–1473):21

Bhattacharya P, Chatterjee D, Jacks G (1997) Occurrence of arsenic-contaminated groundwater in alluvial aquifers from the delta plain, eastern India: options for a safe drinking water supply. Water Res Dev 13:79–92

CGWB (1994a) Hydrogeology and groundwater resource of Murshidabad district, West Bengal. Technical report, series D, Central Ground Water Board (CGWB), Eastern Region, Government of India

CGWB (1994b) Hydrogeology and groundwater resource of Nadia district, West Bengal. Technical report, series D, Central Ground Water Board (CGWB), Eastern Region, Government of India

CGWB (1994c) Hydrogeology and groundwater resource of North 24 Parganas district, West Bengal. Technical report, series D, Central Ground Water Board (CGWB), Eastern Region, Government of India

CGWB (1994d) Hydrogeology and groundwater resource of South 24 Parganas district, West Bengal. Technical report, series D, Central Ground Water Board (CGWB), Eastern Region, Government of India

CGWB (1994e) Hydrogeological atlas of West Bengal, scale 1: 2,000,000. Central Ground Water Board (CGWB), Eastern Region, Government of India

CGWB (1997) High arsenic groundwater in West Bengal. Technical report, series D, Central Ground Water Board (CGWB), Eastern Region, Government of India

Coleman JM (1981) Deltas: processes of deposition and models of exploration, 2nd edn. Burgess, Minneapolis

Das D, Chatterjee A, Mandal BK, Samanta G, Chakraborti D, Chanda B (1995) Arsenic in groundwater in six districts of West Bengal, India: the biggest arsenic calamity in the world. Part 2: Arsenic concentration in drinking water, hair, nails, urine, skin-scale and liver tissue (biopsy) of the affected people. Analyst 120:917–924

Das D, Chatterjee A, Mandal BK, Samanta G, Chakraborti D, Chanda B (1996) Arsenic in groundwater in six districts of West Bengal, India. Environ Geochem Health 18:5–15

Dowling CB, Poreda RJ, Basu AR (2003) The groundwater geochemistry of the Bengal basin: weathering, chemosorption, and trace metal flux to the oceans. Geochim Cosmochim Acta 67 (12):2117–2136

Fergusson J (1863) On recent changes in the delta of the Ganges quarterly. J Geol Soc London 19:321–354

Folk RL, Ward WC (1957) Brazos river bar: a study in the significance of grain size parameters. J Sediment Petrol 22:125–145

Ghosh AR, Mukherjee A (2002) Arsenic contamination of groundwater and human health impacts in Burdwan District, West Bengal, India. Geol Soc Am, Abstr Programs 34(2):107

Ghosh NC, Chakraborty B, Mazumder PK (1999) Groundwater flow mode. Proceedings of workshop on groundwater and its protection with special reference to arsenic contamination, Calcutta, CGWB, India

Goodbred SL, Kuehl SA (2000) The significance of large sediment supply, active tectonism, and eustasy on margin sequence development: late quaternary stratigraphy and evolution of the Ganges-Brahmaputra delta. Sed Geol 133:227–248

Harvey CF (2002) Groundwater flow in the Ganges delta. Science 296:1563

Harvey CF, Swartz CH, Badruzzaman ABM, Keon-Blute N, Yu W, Ali A, Jay J, Beckie R, Niedan V, Brabander D (2005) Groundwater arsenic contamination on the Ganges Delta: biogeochemistry, hydrology, human perturbations, and human suffering on a large scale. Comptes Rendus Geosciences 337(**1**–**2**):285–296

Horneman A, VanGeen A, Kent DV, Mathe PE, Zheng Y, Dhar RK, O'Connel S, Hoque MA, Aziz Z, Shamsudduha M (2004) Decoupling of As and Fe release to Bangladesh groundwater under reducing conditions. Part I: evidence from sediment profiles. Geochim Cosmochim Acta 68(17):3459–3473

Hossain MD, Lewis M, Bose A, Chowdhury A (2003) Rice research: technological progress and impacts on the poor: the Bangladesh case (Summary report), International Food Policy Research Institute (IFPRI)

JICA (2002) The study on the ground water development of deep aquifers for safe drinking water supply to arsenic affected areas in western Bangladesh. Draft final report, book 1–3. Japan International Cooperation Agency, Kokusai Kogyo Co. Ltd. and Mitsui Mineral Development Engineering Co. Ltd

Kinniburgh DG, Smedley PL (eds) (2001) Report of arsenic contamination of Bangladesh. British Geological Survey (BGS) Technical report WC/00/19 Vol. 2 Final report, Keyworth, 255 pp

Mallick S, Rajagopal NR (1995) Groundwater development in the arsenic-affected alluvial belt of West Bengal–some questions. Curr Sci 70:956–958

Malmstrom VH (1969) A new approach to classification of climate. J Geogr 68:351–357

Mandal BK, Chowdhury TR, Samanta G, Mukherjee D, Chanda CR, Saha KC, Chakraborti D (1998) Impact of safe water for drinking on five families for 2 years in West Bengal, India. Sci Total Environ 218:185–201

McArthur JM, Banerjee DM, Hudson-Edwards KA, Mishra R, Purohit R, Ravenscroft P, Cronin A, Howarth RJ, Chatterjee A, Talukder T, Lowry D, Houghton S, Chadha D (2004) Natural organic matter in sedimentary basins and its relation to arsenic in anoxic groundwater: the example of West Bengal and its worldwide implications. Appl Geochem 19:1255–1293

McArthur JM, Ravenscroft P, Safiullah S, Thirlwall MF (2001) Arsenic in groundwater: testing pollution mechanisms for aquifers in Bangladesh. Water Resour Res 37:109–117

McDonald MG, Harbaugh AW (1988) A modular three-dimensional finite-difference groundwater flow model. U.S. Geological survey techniques of water-resources investigations, book 6, A1, 586 pp

MID (2001) Report of 3rd minor irrigation census in West Bengal 2000–2001. Minor Irrigation Department, Government of West Bengal, India

Morgan JP, McIntire WG (1959) Quaternary geology of Bengal Basin, East Pakistan and India. Geol Soc Am Bull 70:319–342

Mukherjee A, Fryar AE (2004) Regional-scale hydrostratigraphy and groundwater chemistry in the western Bengal basin, India. Geolog Soc Am Abstr Programs 36(5):566

Mukherjee A, Fryar AE (2005) A composite approach to characterize the deeper aquifer of the arsenic contaminated western Bengal basin, India. Geol Soc A Abstr Programs 37(7):170

Mukherjee A, Fryar AE, Howell P (2007) Regional hydrostratigraphy and groundwater flow modeling of the arsenic contaminated aquifers of the western Bengal basin, West Bengal, India. Hydrogeol J 15:1397–1418

Mukherjee A, Fryar AE, Thomas WA (2009) Geologic, geomorphic and hydrologic framework and evolution of the Bengal basin, India. J Asian Earth Sci 34(3):227–244

Nickson R, McArthur JM, Ravenscroft P, Burgess WG, Rahman M (1998) Arsenic poisoning of groundwater in Bangladesh. Nature 395:338

Nickson R, McArthur JM, Ravenscroft P, Burgess WG, Ahmed M (2000) Mechanism of arsenic release to groundwater, Bangladesh and West Bengal. Appl Geochem 15:403–411

Pike JG (1964) The estimation of annual runoff from meteorological data in a tropical climate. J Hydrol 2:116–123

Ravenscroft P, McArthur JM, Hoque B (2001) Geochemical and palaeohydrological controls on pollution of groundwater by arsenic. In: Chappell WR, Abernathy CO, Calderon R (eds) Arsenic exposure and health effects IV. Elsevier Science Ltd, Oxford, pp 53–77

Ravenscroft P, Burgess WG, Ahmed KM, Burren M, Perrin, J (in press) Arsenic in groundwater of the Bengal basin, Bangladesh: distribution, field relations, and hydrogeologic setting. Hydrogeol J doi:10.1007/s10040-003-0314-0

Ray AK (1999) Chemistry of arsenic and arsenic minerals relevant to contamination of groundwater and soil from subterranean source. Everyman's Sci 35(1)

Saha AK (1991) Genesis of arsenic in groundwater in parts of West Bengal. Annual volume, Center for Studies on Man and Environment, Calcutta, India

Sanyal J, Lu XX (2003) Application of GIS in flood hazard mapping: a case study of Gangetic West Bengal, India. Proceedings of map Asia conference, GISdevelopment.net, 8 pp

Sikdar PK, Sarkar SS, Palchoudhury S (2001) Geochemical evolution of groundwater in the quaternary aquifer of Calcutta and Howrah India. J Asian Earth Sci 19(5):579–594

Smedley PL, Kinniburgh DG (2002) A review of the source, behaviour and distribution of arsenic in natural waters. Appl Geochem 17:517–568

Smith AH, Lingas EO, Rahman M (2000) Contamination of drinking-water by arsenic in Bangladesh: a public health emergency. Bull World Health Organ 78(9):1093–1103

Stuben D, Berner Z, Chandrashekharam D, Karmakar J (2003) Arsenic enrichment in groundwater of West Bengal, India: geochemical evidence for mobilization of as under reducing conditions. Appl Geochem 18(9):1417–1434

Survey of India (1971) Map of West Bengal, scale 1: 1000,000. Publishing group of Survey of India, Government of India, Calcutta, India

SWID (1998) A comprehensive hydrogeological information of Murshidabad district. Geologic Division No. III, State Water Investigation Directorate (SWID), Government of West Bengal, Calcutta, India, 43 pp

WHO (1993) Guidelines for drinking-water quality. Vol. 1 recommendations, 2nd edn. World Health Organization (WHO), Geneva

Recording High Natural Gamma Radiations in Fluoride-Prone Area of Shivpuri District, Madhya Pradesh

Subhash C. Singh, Rakesh Singh and Parvinder Singh

Abstract The occurrence of fluoride in groundwater is a major issue for aquifer-based drinking water supply. Many parts in state of Madhya Pradesh exhibit elevated fluoride content in groundwater. Present study has been conducted in Shivpuri district, underlain by granite and gneiss with some patches of recent alluvium. Ten boreholes have been drilled, and out of them, seven boreholes have been logged geophysically for natural gamma count. Chemical analyses of groundwater samples from the drilled wells have maximum fluoride content up to 5 mg/L. The natural gamma loggings conducted into the boreholes down to the depth of 200 mbgl have recorded 100–1600 cps for various lithological units encountered into the boreholes. It has been observed that water-rich fractures encountered in some of the boreholes with higher fluoride concentration is also showing higher range of natural gamma counts. Fluorite as a mineral in granites is considered as marker horizons for prospecting of radioactive minerals. Fluorite-bearing rocks releases fluoride in groundwater as well as related to high natural gamma radiation in the study area. The paper address geogenic groundwater quality problem of fluoride, associated with radioactive radiations in parts of Shivpuri district, Madhya Pradesh.

Keywords Fluoride · Natural gamma · Groundwater · Radioactive mineral Granite · Madhya Pradesh

S.C. Singh (✉) · R. Singh · P. Singh
Central Ground Water Board, North Central Region, 'A' Block, 4th Floor, Paryavas Bhawan, Bhopal, India
e-mail: singh_subhash_ch@hotmail.com

R. Singh
e-mail: roksingh20@gmail.com

D. Saha et al. (eds.), *Clean and Sustainable Groundwater in India*,
Springer Hydrogeology, https://doi.org/10.1007/978-981-10-4552-3_20

1 Introduction

Natural occurrence of fluoride in groundwater is a geogenic contamination. Fluoride concentration in groundwater are found to be exceeding the limit of 1.5 mg/L for potable use in many parts of India, particularly in hard rock terrain (WHO 1993; NAS & NAE 1972; EEC 1975; BIS 2003). Fluoride is a strongly electronegative ion commonly found in combination with calcium or sodium, forming compounds that occur naturally in soil rock and water. Drinking water is the major pathway of fluoride ingestion by humans (WHO 1994). In India, elevated concentration of fluoride (>1.5 mg/L) has been reported from many districts (CGWB 1999). Fluorosis, resulting from consumption of water of high fluoride content (>1.5 mg/L), has become endemic in many parts of country (Chaddah 2001; Romani 2001; Saha et al. 2008; Raghuwanshi and Thakur 2004). Fluorosis is a crippling disorder due to effect of excessive fluoride in the body, which affects invariably every organ, tissue, and cells in the body. Fluoride damages the pineal gland, which secretes melatonin hormone in the brain. It also affects the reproductive systems and intelligence (Susheela 2001). The most obvious early effects of fluoride in humans are dental and skeletal fluorosis, which are endemic in areas affected by fluoride contaminations (Chabre 1990; Teotia and Teotia 1984). Besides, high concentration of fluoride in environment affects animals and plants life as well (Yur et al. 2003; Franke 1989). The effects of fluoride in reduction to dental caries are well established, whereas its effect on periodontal tissues is obscure (Vandana and Reddy 2007). Fluoride is also known to cross the cell membranes and to enter soft tissues (Schliching and Reinstein 1999). Fluoride has a unique chemical behaviour towards most of the anions and can be easily replaced even under temperature and pressure condition (Wenzel and Blum 1992). Severe skeletal and dental fluorosis have been reported from the Narwar and Karera blocks of Shivpuri district in state of Madhya Pradesh located in central part of India (Singh and Verma 2000; Singh 2003; Saksena and Narwariya 2012).

Geophysical logging is the practice of making a detailed record (a well log) of the geologic formations penetrated by a borehole. This helps in correlation of visual inspection of drill cut samples (geological logs) brought to the surface and physical measurements made by instruments lowered into the borehole (geophysical logs). Well logging measures various physical properties of subsurface horizon such as electrical conductivity, density and natural gamma. Various type of geophysical logging including natural gamma logging have been described by many researchers (Hallenburg 1987, 1992; Keys 1990; Yearsley and Crowder 1990; Samworth 1992). The objective of the study is to ascertain the fluoride concentration depth wise in different productive zones and establish the relation between natural gamma emission and elevated concentration of fluoride in groundwater particular zone. The natural gamma has been recorded in the borehole. Three naturally occurring radioisotopes have decay chains and modes involving the emission of gamma rays, specifically 40K, 238U and its daughter products, 232Th and its daughter products.

The energy spectrum of these decays is concentrated between 0.2 and 3.0 meV. A natural gamma log records total decay events across the gamma energy spectrum. Results are reported in CPS units (counts per second) (Belknap et al. 1960; Keys 1990).

2 Study Area

The study area forms a part of the Shivpuri district which is situated in the northern most part of Madhya Pradesh. The district is bounded in the north by district Gwalior, in the south by the district Guna, in the east by the district Datia, and Jhansi and in the west by the Morena district and Rajasthan state (Fig. 1). The area forming a part of the Shivpuri district lying between latitude 25° 21′–24° 47′N and longitude 77° 45′–78° 27′E and covering an area of about 1968 km^2. The study area comprising of Narwar and Karera blocks is a part of the Sind sub-basin which is tributary of Chambal River, which further joins the Yamuna River. In general, the study area is undulating with random occurrence of granitic hills. The south-west monsoon is the source of rainfall and spreads in the month of June to September. The normal annual rainfall in the area is about 745 mm.

3 Methodology

During the summer season, providing potable drinking water which is largely groundwater based is a prime concern for the area. Elevated concentration of fluoride (up to 5 mg/L) has aggravated the problem. Atomic Mineral Division and Geological Survey of India, two departments under Govt. of India, are involved in radioactive mineral investigations in the adjoining area. Occurrence of fluorite minerals considered as marker horizon for prospecting of radioactive mineral such as Uranium and Thorium. In the study area, ten exploratory borewells have been drilled down to the depth of 200 m below ground. The rock types were studied from drill cut samples, and discharges were obtained for weathered zone as well as from the fracture zone encountered. The groundwater samples were collected and analyzed for fluoride concentrations following the methodology proposed by APHA in the laboratory of Central Ground Water Board Bhopal. Natural gamma logging has been conducted in seven out of ten exploratory boreholes. The natural gamma counts have recorded using digital Geologger. The occurrences of high fluoride content in aquifers have correlated with corresponding natural gamma counts against the aquifer zone. The objective of study is to work out the occurrence of fluoride in groundwater by integrated geological, geophysical, and geochemical investigation.

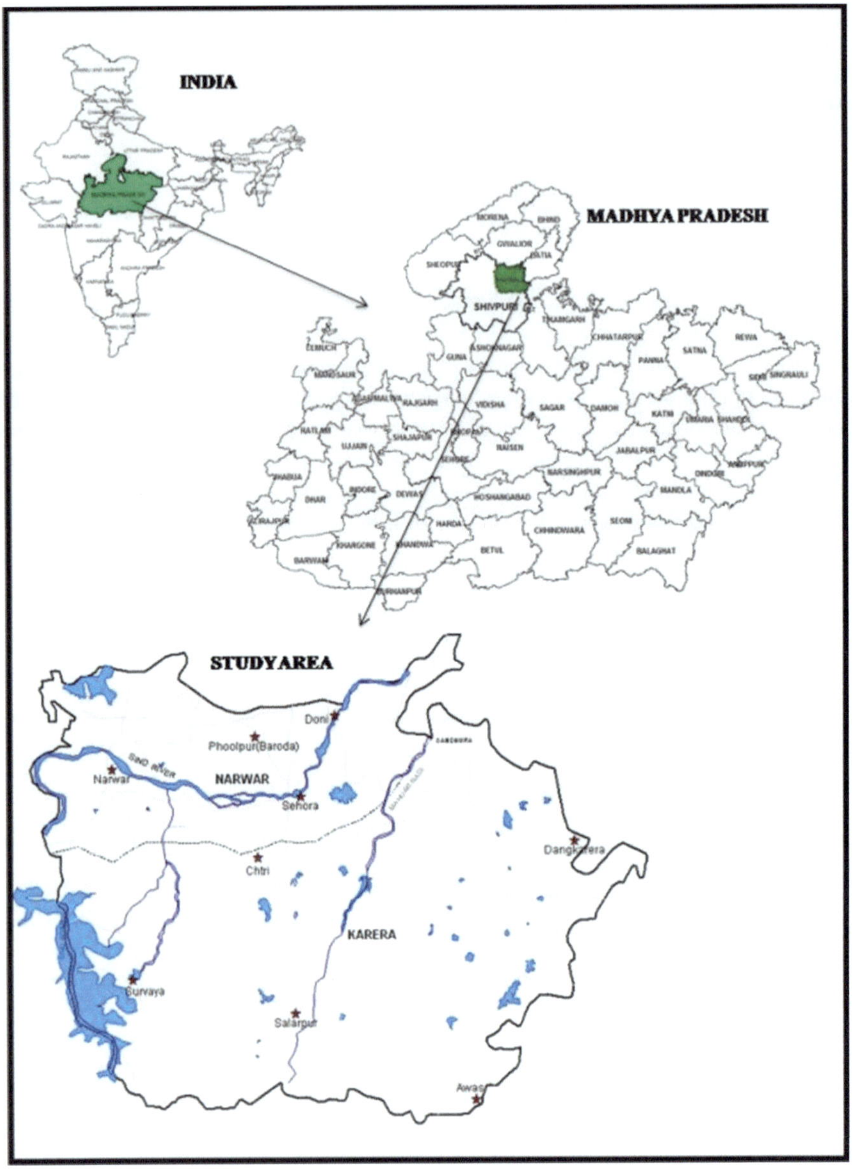

Fig. 1 Location map of study area

4 Hydrogeology

The study area is mainly occupied by Bundelkhand gneissic complex (BCC), Vindhyan Supergroup, Deccan traps with some patches of recent alluvium. The geological succession is presented in Table 1. The BCC constitutes the major part of the area and represented by granites and gneisses. The granite shows great heterogeneity in texture and mineralogical composition. In general, they are leucocratic to mesocratic, phaneric, fine to coarse grained, mainly consist of quartz and orthoclase feldspar. At several places, alternate bands of pink and grey coloured granite are quite distinct. The intensity of weathering in granites varies from place to place. In low laying zones, inter mountain valleys and depressions of the weathered zone is thick and exceeds 30 m. The gneisses are phaneric, medium to coarse grained consisting of pink and grey feldspar, quartz,and ferromagnetism mineral and found to be highly jointed (Shrivastwa et al. 1999; Jain et al. 2001; Sinha et al. 2011). Such jointed hard rocks forming potential aquifers are reported from gneissic–granulitic terrain in Jharkhand and basaltic terrain in Maharashtra (Saha and Agrawal 2006; Saha et al. 2013). Quartz reefs consist of fine-grained pink to greenish cherty material associated with milky quartz occurring as veins. The milky quartz is jointed and shattered and has usually reddish brown stains along the joint planes. Prominent joints are almost vertical trending north-east–south-west, parallel to the strike of the reefs. Another set of joints with horizontal low angle dip is observed. The quartz reefs are comparatively less weathered. In north-east part of area, along the Sind River, 15–25 m thick marginal alluviums observed which is composed of fine to coarse-grained sand, gravel, silt, clay, and kankar of Quaternary age.

The hydrogeological condition of the study area is primarily controlled by the prevailing geological setup of the area. The lineaments and shear zones are functioning as a conduit for groundwater movement and storage. They are the places getting quick recharge. The groundwater in the area lies under unconfined to locally semi-confined condition (Singh and Verma 2000). The top unconfined aquifer consists of weathered granite extend down to the depths of 8–20 mbgl. Seven out of ten boreholes drilled for the study were logged (Table 2). The depths of borewells are ranging from 124 to 201 m. Expect two; remaining five wells are more than 195 m depth. The maximum weathered zone is observed at Doni (23 m), where minimum is observed in Sarsod (5 m). Generally, thin sandy layers are observed

Table 1 Geological succession of study area

Age	Geological formation	Lithology
Recent	Alluvium	Clay fine sand and boulders
Pleistocene	Laterite	–
Upper Cretaceous to Eocene	Deccan traps	Basaltic Lava flows and inter trappeans
Neo-Meso Proterozoic	Upper Vindhyan	Sandstone and shale
Archaeans	Bundelkhand granite	Granites gneisses

Table 2 Details of aquifers characteristics encountered into exploratory boreholes of study area

S. No.	Location	Drilled depth (mbgl)	Aquifer zone depth range (m)	SWL (mbgl)	Natural gamma (cps)	Yield (lps)	Fluoride concentration (ppm)	EC µS at 25°	pH
1	Sihor	195	3–5 ⲭ	[a]	380	Wet	5.6	828	7.16
			5–8.5 ⲭ	3.5	550	0.5	5.6	743	8.41
			109–111	7.32	455	0.80	5.00	778	7.13
2	Baroda	124	2–6 ⲭ	[a]	380	Wet	5.6	[a]	[a]
			6–8 ⲭ	4.6	450	0.5	5.6	543	7.23
			121–123	6.75	530	3.0	6.0	795	7.60
3	Dangkarera	144	3–5 ⲭ	[a]	190	Wet	0.88	1463	7.22
			14–18 ⲭ	6.40	210	0.3	1.20	1576	7.59
			144–146	12.98	400	3.5	4.0	1180	8.15
4	Chitri	200	2–5 ⲭ	[a]	80	Wet	0.56	797	7.93
			5–12 ⲭ	5.8	150	0.3	1.05	478	7.96
			50–55	6.70	500	0.8	1.2	696	7.66
5	Sarsod	201	2–3.5 ⲭ	[a]	220	Wet	0.25	556	7.4
			3.5–7 ⲭ	2.34	205	0.20	0.56	560	7.51
			112–118	3.10	400	1.50	4.00	774	7.74
6	Doni	201	5–7 ⲭ	[a]	270	Wet	2.24	1957	7.22
			14–21 ⲭ	4.0	310	0.50	2.70	1610	7.7
			30–55	6.10	1500	0.80	5.3	1297	6.91
7	Gasarhi	201	4–10 ⲭ	[a]	200	Wet	4.8	841	7.86
			10–16 ⲭ	5.00	250	0.6	3.00	771	7.8
			125–130	NR	550	[a]	[a]	[a]	[a]

ⲭ Sandy layer within the weathered zone
[a]Not available

within the weathered zone have yield capacity ranging from negligible to 3 lps. Within the undulating massive rock, one set of fracture is generally encountered confined between 109 and 146 mbgl (except at Doni, where the fracture encountered between 30 and 55 m depth). Water level in the area ranges between 6.5 and 12.9 mbgl during the month of May (pre monsoon) whereas during November (post-monsoon) the level ranges from 3.3 to 5.4 mbgl. The overall groundwater flow direction is towards north (Fig. 2). The groundwater from the wreathed zone as well as from the fracture is mildly alkaline. The electrical conductivity generally remains within 850 μS/cm. There is not much variation between the fracture and weathered zone in term of EC and pH. In two wells that is Dongkarera and Doni, the mineralization is high, where the EC lies above 1150 μS/cm. On both the site, weathered zone groundwater is higher mineralized than the fracture zone groundwater.

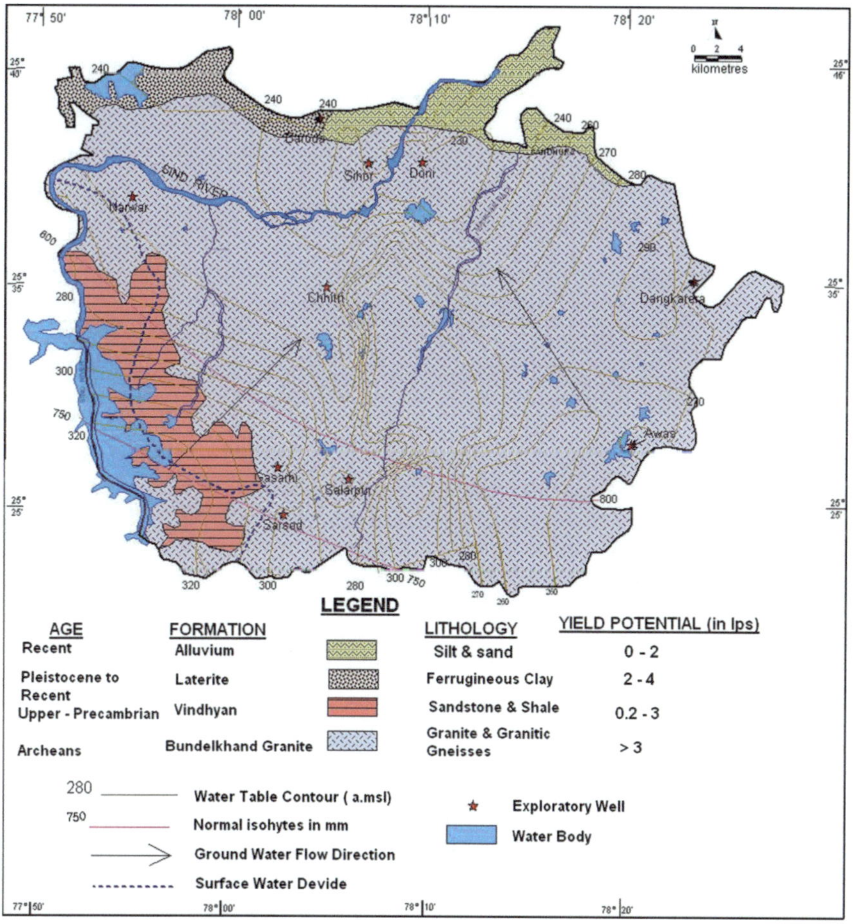

Fig. 2 Hydrogeological map of study area showing the locations of exploratory boreholes

5 Geophysics Borehole Logging

The borehole geophysical logging provides information on the character of the rocks and the fluid penetrated by boreholes. This information is useful for litho-structural interpretation of aquifer systems as well as for evaluation groundwater quality. The measuring sensors, commonly known as probes, are lowered into the borehole to measure continuous changes in physical properties vertically, and the continuous record so obtained is known as a geophysical log. Multiparameter digital geophysical logging have been conducted in exploratory boreholes.

Natural gamma-ray log is a measure of the natural radiation emitted as a result of the disintegration of the radioactive elements such as uranium, thorium and potassium. Examples of litho units which may rich in radioactive elements are feldspar rich granite, etc. If such litho units are present, gamma logs is very useful in deciphering such layers/zones. The interpretations were done on a qualitative

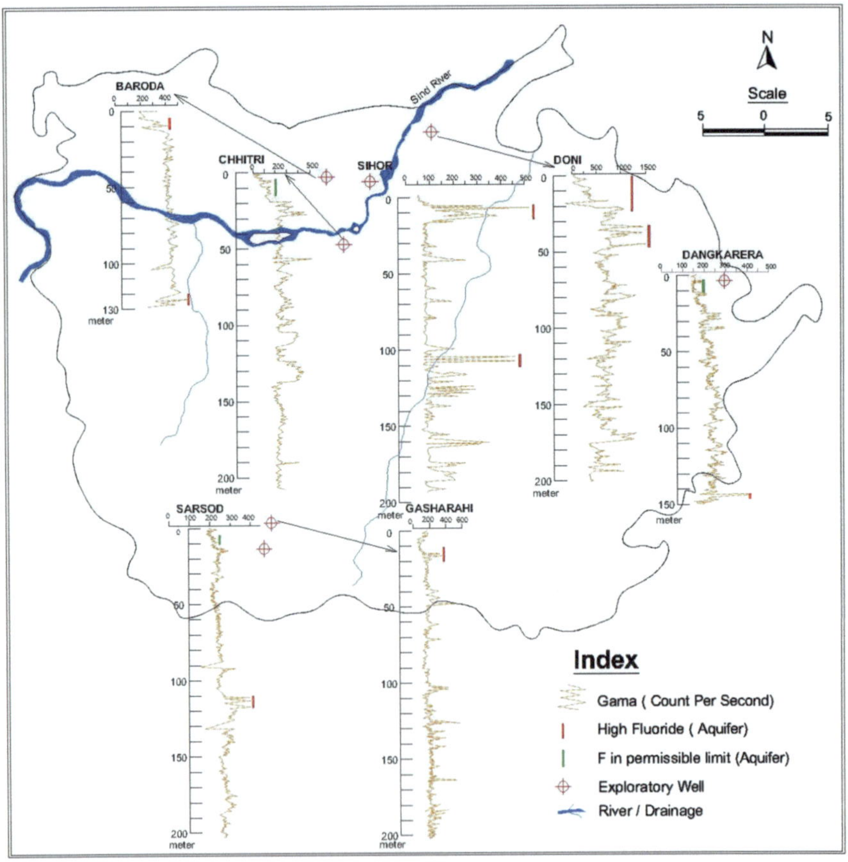

Fig. 3 Natural gamma logs recorded in study area of Shivpuri district, MP

basis in conjunction with the drill cut samples. In study area, geophysically logged boreholes are located at Sihor, Doni, Baroda, Ghasari, Sarsod, Dongkarera and Chhitri (Fig. 3).

6 Results Discussion

Fluoride concentration of groundwater samples collected from each sand layer of the weathered zone as well as from fractures of gamma logged seven borehole are given in Table 2. Except at Chhitri site, everywhere the fluoride concentration exceed the limit of potable water (1.5 mg/L). Highest concentration is observed at Baroda site (6.0 mg/L) from the fracture zone at 121–123 mbgl. Almost everywhere it has been observed that the concentration of fluoride in fracture zone is higher than that of weathered zone. At Dongkarera and Sarsod site, fluoride content in weathered zone remains within permissible limit (<1.5 mg/L), while from the fractures below it grossly exceeds the limit, with 4 mg/L. At Sihor and Baroda sites, the F concentration is high (>5.6 mg/L), but remain same in weathered and fractured zone. The occurrences of high fluoride content in aquifers are correlated with corresponding natural gamma counts of the formation. The natural gamma logs have been analyzed qualitatively in terms of minimum and maximum gamma counts recorded during logging. It is observed that high natural gamma radiations have been recorded against the different aquifer zones having high fluoride concentration. It is known fact that the occurence of fluorite mineral is considered to be a marker horizon in prospecting of radioactive elements such as thorium and uranium. The recording of high gamma values up to 1600 cps in study area may be indicative of presence of radioactive elements. It is notable that the some of the aquifer zones having the fluoride in range of 5 mg/L are having natural gamma counts up to 1500 cps. This observable fact of recording of high gamma values with the high fluoride content in the aquifers may be indicating the presence fluorite mineral in granitic formation encounter into the boreholes.

It has been observed the natural gamma logging have recorded gamma values in range of 100–1500 cps against fracture zones. It is noticed that the aquifers having the fluoride content within the permissable limit (<1.5 mg/M) is showing natural gamma radiation values less than 200 cps, whereas the phreatic aquifer from the marginal alluvium along the Sind River section is having fluoride in aquifer with the natural gamma values in range of 350–500 cps. It is observed that the water-bearing zones from the marginal alluviam and weathered rocks section in downstream of Sind River in north-east part of the study area is having fluoride 2.5–5.5 mg/L with natural gamma values recorded in range 300–450 cps. It is also indicated that in some boreholes in granitic formation maximum natural gamma values in the range of 1000–1600 cps whereas these layers are not having any fracture zone. These observations may be indicative of radioactive elements in lithology. Occurrence of high fluoride in groundwater combined with high natural gamma counts may be the reason for more physical deformaties in local people in comparison with other parts of Madhya Pradesh having much more fluoride values in groundwater (5–18 mg/M).

7 Conclusions

The study area is underlain by Bundelkhand gneiss, representing undulatory topography and facing acute shortage of drinking water due to less availability as well as poor quality of groundwater. The Karera and Narwar blocks of Shivpuri district are affected with severe fluorosis problems. Severe skeletal and dental fluorosis have been seen in the study area. Hydrogeological investigation comprised of monitoring, borehole drilling, analysis of drill cut samples, yield test of individual fracture zones, surface and subsurface geophysical surveys and chemical analysis of groundwater collected from weathered zone and also from individual fracture zone within the depths of 200 mbgl. It is found that the occurrence of fluoride of the order of 5 mg/L or more is prevalent in the area. The natural gamma loggings conducted into the boreholes have recorded 100–1600 cps natural gamma counts for various lithological units. The fracture encountered into few boreholes are having higher range of fluoride and also showing higher range of natural gamma radiations.

Various investigations by different Government agencies for exploration on radioactive minerals in the study area and recording of high natural gamma counts into the exploratory boreholes indicate the presence of radioactive minerals in the area. Occurrence of fluorite in granite forms the marker horizons for prospecting of radioactive minerals. Recording of high gamma counts is indicative for the presence of fluorite which is also responsible for genesis of fluoride in groundwater. Natural gamma logs also demonstrated that a geological formation or sequence encountered into the borehole can be classified in terms of its natural gamma signature. The study reveals that the formation may have radioactive elements in the rocks and may be their radiation effect acting as catalyst for having fluoride concentration in fracture zones. Occurrence of high fluoride in aquifer system combined with high natural gamma counts may be the reason for more physical deformaties in local people in comparison with other parts of Madhya Pradesh those are having much more fluoride value in groundwater (5–18 mg/L). Further study is needed to examine the issues of release of fluoride in groundwater from country rock and presence and role of radioactive minerals in the lithology.

Acknowledgements The authors are thankful to The Chairman, Central Ground Water Board for his kind permission to publish this paper. We are also grateful to Dr. E. Sampath Kumar, Member (SML) and Dr. D. Saha Member (SAM) for his encouragement for writing this paper. Thanks are also due to The Regional Director, CGWB, North Central Region, Bhopal, for providing necessary facility for conducting the study. Close association of Shri H.S. Namdeo, Scientists CGWB during the investigations is thankfully acknowledged.

References

Belknap WB, Dewan JT, Kirkpatrick CV, Mott WE, Perason AJ, Rabson WR (1960) API calibration facility for nuclear logs. American Petroleum Institute, Drilling Production, p 289

BIS (2003) Drinking water specifications, IS: 10500. Buearu of Indian Standards, New Delhi

CGWB (1999) Report on high fluoride in India occurrence, genesis and remedies. Central Groundwater Board, Govt. of India, MOWR, Faridabad, p 29

Chabre M (1990) Aluminofluoride and beryllofluoride complex, new phosphate analogues in enzymology. Trands Biochem Sci 15:6–10

Chaddah DK (2001) Hydrogeological aspects of high fluoride content groundwater in India. In: International workshop on fluoride in drinking water: strategies, management and mitigation, 22–24 Jan, pp 184–195

European Economic Community (1975) Proposal for a council directive, relating to quality of water for human consumption. Off J Eur Commun C214

Franke J (1989) Differences in skeletal responses to fluoride in human and animals. An over review. Fluoride 22(1):10–19

Hallenburg JK (1987) Geophysical logging for mineral and engineering applications. Penn Well Books, Tulsa, Oklahoma

Hallenburg JK (1992) Nonhydrocarbon logging. Log Anal 33(3):259–269

Jain SC et al (2001) Recent find of cauldron structure in Bundelkhand craton. Geol Surv India Spec Publ 64:289–297

Keys WS (1990) Borehole geophysics applied to ground-water investigations. Techniques of Water-Resources Investigations of the United States Geological Survey, Book 2, Chapter E2, Washington

National Academy of Science and National Academy of Engineering (1972) Water quality criteria, Rep. EPA-R3-73-033, NAS & NAE, Washington

Raghuwanshi RS, Thakur GS (2004) Irrigation suitability of surface and sub surface water for agricultural development of the region around choral river basin Indore and Khargone Districts M.P. J Appl Hydrol XVII(2&3):27–33

Romani S (2001) Problem of high fluoride in groundwater and measures for its control international workshop of fluoride in drinking water strategies. Manag Mitig 177–183

Saha D, Agrawal AK (2006) Determination of specific yield using water balance approach—a case study of Torla Odha water shed in Deccan Trap Province, Maharashtra State, India. Hydrogeol J 14:625 635

Saha D, Dhar YR, Sikdar PK (2008) Geochemical evolution of ground water in the pleistocene aquifers of south Ganga Plain, Bihar. J Geol Soc India 71:473–482

Saha D, Dwivedi SN, Roy GK, Reddy DV (2013) Isotope-based investigation on the groundwater flow and recharge mechanism in a hard-rock aquifer system: the case of Ranchi urban area, India. Hydrogeol J. doi:10.1007/s10040-013-0974-3

Saksena DN, Narwaria YS (2012) Incidence of fluoride in groundwater and its potential health effects in ten villages of Karera block in Shivpuri district, Madhya Pradesh, India. Int J Env Sci 3(3):1141

Samworth JR (1992) Quantitative open-hole logging with very small diameter wireline tools. In: SPWLA 33rd annual logging symposium, Paper NN, 24 p

Schliching I, Reinstein J (1999) pH influences fluoride coordination number of the phosphoryl transfer transition state analog. Nat Struct Biol 8:721

Shrivastwa SK et al (1999) Report on specialized thematic mapping of Bundelkhand gneissic complex in parts of Datia and Shivpuri district. GSI Portal

Singh R (2003) Role of hydrogeological studies in fluoride affected area of granitic terrain in Shivpuri district of Madhya Pradesh. J Appl Hydrol

Singh P, Verma SK (2000) Report on hydrogeological framework and development prospect of Shivpuri district, M.P. Unpublished report of CGWB, NCR, Bhopal

Sinha KK, Pandey P, Bhairam CL, Parihar PS (2011) Peperite occurrence and its implications on origin and temporal development of the Proterozoic Dhala Basin, Mohar Area, Shivpuri District, Madhya Pradesh. J Geol Soc India 77:183–189

Susheela AK (2001) A treatise on fluorosis. Fluorosis Research and Rural Development Foundation, New Delhi

Teotia SPS, Teotia M (1984) Endemic fluorosis India: a challenging national health problem. J Assoc Physic India 32:347–352

Vandana KL, Reddy SM (2007) Assessment of periodontal status in dental fluorosis subjects using community periodontal index of treatment needs. Ind J Dent Res 8:67–71

Wenzel WW, Blum WEH (1992) Fluoride speciation and mobility in fluoride concentration soil and minerals. Soil Sci 153:357–364

WHO (1994) Fluoride and oral health. WHO technical report series 846. World Health Organization, Geneva

World Health Organization (1993) Guidelines for drinking water quality, recommendations, 2nd edn. Geneva

Yearsley LE, Crowder RE (1990) State-of-the-art borehole geophysics applied to hydrology. In: Canadian/American conference on hydrology, 18–20 Sept 1990

Yur F, Belge F, Mert N, Yoruk I (2003) Change in erythrocyte parameters of fluoride sheep. Fluoride 36:152–156

Author Index

D. Saha et al. (eds.), *Clean and Sustainable Groundwater in India*,
Springer Hydrogeology, https://doi.org/10.1007/978-981-10-4552-3

MIX
Papier aus verantwortungsvollen Quellen
Paper from responsible sources
FSC® C105338

If you have any concerns about our products,
you can contact us on
ProductSafety@springernature.com

In case Publisher is established outside the EU,
the EU authorized representative is:
Springer Nature Customer Service Center GmbH
Europaplatz 3, 69115 Heidelberg, Germany

Printed by Libri Plureos GmbH
in Hamburg, Germany